TONY DAWSON obtained his degree at the University of Southampton where he also completed a post-graduate Certificate in Education. Since then he has taught Mathematics at all levels in various Surrey Schools and Colleges. He was Head of the Mathematics Department at Woking Grammar School for Boys and Senior Tutor at Woking Sixth Form College. He is at present the Deputy Principal at Woking Sixth Form College.

ROD PARSONS graduated from Bristol University and completed a post-graduate certificate in Education at Southampton University. Since then he has taught in schools and sixth form colleges in Surrey. He has taught both modern and traditional syllabuses and has written for S.M.P. and lectured to teachers at S.M.P. conferences. He was Head of the Mathematics Department at Woking Sixth Form College but has recently been appointed Vice Principal of Esher Sixth Form College.

GCE A-Level Passbooks

BIOLOGY, H. Rapson, B.Sc.

CHEMISTRY, J. E. Chandler, B.Sc. and
R. J. Wilkinson, Ph.D.

PHYSICS, J. Garrood, M.A., Ph.D.

PURE MATHEMATICS, R. Parsons, B.Sc.
and A. G. Dawson, B.Sc.

PURE AND APPLIED MATHEMATICS,
R. A. Parsons, B.Sc. and A. G. Dawson, B.Sc.

APPLIED MATHEMATICS, E. M. Peet, B.A.

ECONOMICS, Roger Maile, B.A.

GEOGRAPHY, R. Bryant, B.A., Ph.D.,
R. Knowles, M.A. and J. Wareing, B.A., M.Sc.

GCE A-Level Passbook

Pure Mathematics

R. A. Parsons, B.Sc.
and
A. G. Dawson, B.Sc.

Published by Charles Letts and Co Ltd
London, Edinburgh and New York

First published 1980 by Intercontinental Book Productions

Published 1982 by Charles Letts & Co Ltd
Diary House, Borough Road, London SE1 1DW

1st edition   7th impression
© Charles Letts & Co Ltd
Made and printed by Charles Letts (Scotland) Ltd
ISBN 0 85097 515 8

Contents

Introduction

This book is a notebook for students of Pure Mathematics at advanced level. The Authors have intended to cover most of the topics that are studied at advanced level both for modern and traditional syllabuses. The differences between modern and traditional syllabuses are now decreasing and the topics in this book include most of those that an advanced level student would encounter, no matter which syllabus he or she is following.

At advanced level a student usually covers a little of each topic at a time, and as the course progresses the topics become inter-dependent and more related to each other. For ease of reference and clarity this book is written as twelve different chapters, although there are cross-references between the chapters.

Each topic is explained thoroughly with formulae derived as though the student were doing the work himself. Frequent examples are included to enable the student to assimilate a process thoroughly and to understand its applications.

Various approaches to learning and teaching each topic have been given so that students are not handicapped by seeing the work presented in a completely different way. Indeed a new method of presentation can often lead to greater understanding.

The style of examinations for the different Examining Boards can vary and students are best advised to study past examina-tion papers which are obtainable from the Examination Boards. (A list of addresses is given on page 318). Some examinations at advanced level include multi-choice papers, with papers of short or long questions. Other papers have a compulsory section A of short questions with a choice of longer questions in section B.

Key facts and definitions are emphasized in bold print and these appear in the index at the back of the book. There is a short summary at the end of each chapter, but the student is advised by the authors to consult the full text whenever possible.

Chapter I
Algebra

If $A = 2x + 1$ and $B = x + 3$ then

$$A + B = 3x + 4$$

$$A - B = (2x + 1) - (x + 3) = x - 2$$

$$A \times B = (2x + 1)(x + 3) = 2x^2 + 7x + 3$$

$$A \div B = \frac{2x + 1}{x + 3} = \frac{2x + 6 - 5}{x + 3} = \frac{2x + 6}{x + 3} - \frac{5}{x + 3} = 2 - \frac{5}{x + 3}$$

A is an example of a linear function since the graph of $y = 2x + 1$ or $f(x) = 2x + 1$ is a straight line. A is also a **Polynomial** of degree 1 while $A \times B = 2x^2 + 7x + 3$ is a Polynomial of degree 2 since the highest power of x occurring in the expression is x^2. When drawing the graph of $y = 2x^2 + 7x + 3$ it may be helpful to know its *Factors* are $2x + 1$ and $x + 3$.

Example 1 Draw the graph of
$$g(x) = x^2 - 3x + 2 = (x - 1)(x - 2).$$

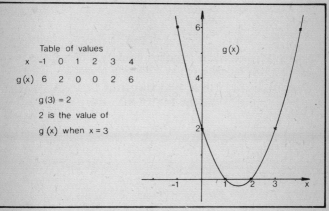

Table of values

x	-1	0	1	2	3	4
g(x)	6	2	0	0	2	6

$g(3) = 2$

2 is the value of

$g(x)$ when $x = 3$

Figure 1

$x - 1$ and $x - 2$ are the factors of $g(x) = x^2 - 3x + 2$
$x = 1$ and $x = 2$ are the zero values (or zeros) of $g(x)$ since $g(1) = 0$ and $g(2) = 0$
$x = 1$ and $x = 2$ are the solutions (roots) of the equation $g(x) = x^2 - 3x + 2 = 0$

9

Polynomials are added and subtracted by considering corresponding terms.

Example 2 If $g(x) = x^2 - 3x + 2$ and $h(x) = 2x + 1$.

$$g(x) + h(x) = x^2 - x + 3 \qquad g(x) - h(x) = x^2 - 3x + 2 - (2x + 1)$$
$$= x^2 - 5x + 1$$

Polynomials are multiplied by multiplying out the brackets.

$$g(x) \times h(x) = (x^2 - 3x + 2)(2x + 1)$$
$$= x^2(2x + 1) - 3x(2x + 1) + 2(2x + 1)$$
$$= 2x^3 + x^2 - 6x^2 - 3x + 4x + 2 = 2x^3 - 5x^2 + x + 2$$

Long multiplication process

$$
\begin{array}{r}
x^2 - 3x + 2 \\
2x + 1 \times \\
\hline
2x^3 - 6x^2 + 4x \\
x^2 - 3x + 2 \\
\hline
2x^3 - 5x^2 + x + 2
\end{array}
\qquad \text{OR} \qquad
\begin{array}{rrrr}
1 & -3 & 2 & \\
 & 2 & 1 & \times \\
\hline
2 & -6 & 4 & 0 \\
1 & -3 & 2 & \\
\hline
2 & -5 & 1 & 2
\end{array}
$$

Example 3 Multiply $x^3 + 2x^2 + 3x + 4$ by $x^2 - x + 2$.

$$
\begin{array}{rrrrrr}
1 & 2 & 3 & 4 & & \\
 & & 1 & -1 & 2 & \times \\
\hline
1 & 2 & 3 & 4 & 0 & 0 \\
-1 & -2 & -3 & -4 & 0 & \\
 & 2 & 4 & 6 & 8 & \\
\hline
1 & 1 & 3 & 5 & 2 & 8
\end{array}
\qquad
\begin{array}{l}
(x^3 + 2x^2 + 3x + 4)(x^2 - x + 2) \\
\\
= x^5 + x^4 + 3x^3 + 5x^2 + 2x + 8
\end{array}
$$

Dividing polynomials is more involved. It is best achieved by setting the polynomials out as a division process.

$$
\begin{array}{r}
21 \text{ rem } 1 \\
12 \overline{) 253} \\
\underline{24} \\
13 \\
\underline{12} \\
1
\end{array}
\qquad
\begin{array}{c}
\text{QUOTIENT \& REMAINDER} \\
\hline
\text{DIVISOR) DIVIDEND}
\end{array}
\qquad
\begin{array}{r}
2x + 1 \text{ rem } 1 \\
x + 2 \overline{) 2x^2 + 5x + 3} \\
\underline{2x^2 + 4x} \\
0 + x + 3 \\
\underline{x + 2} \\
1
\end{array}
$$

Example 4 Divide $2x^3 - 5x^2 + x + 2$ by $x + 2$.

$$
\begin{array}{r}
2x^2 - 9x + 19 \text{ rem} - 36 \\
x + 2 \overline{) 2x^3 - 5x^2 + \quad x + 2} \\
\text{subtract } \underline{2x^3 + 4x^2} \\
-9x^2 + \quad x \\
\text{subtract } \underline{-9x^2 - 18x} \\
19x + 2 \\
\underline{19x + 38} \\
-36 \quad \text{REMAINDER}
\end{array}
$$

So $2x^3 - 5x^2 + x + 2 = (x + 2)(2x^2 - 9x + 19) - 36$

Example 5 Divide $2x^3 - 5x^2 + x + 2$ by $x - 1$.

$$
\begin{array}{r}
2x^2 - 3x - 2 \\
x - 1 \overline{)\,2x^3 - 5x^2 + \ x + 2} \\
\underline{2x^3 - 2x^2} \\
-3x^2 + \ x \\
\underline{-3x^2 + 3x} \\
-2x + 2 \\
\underline{-2x + 2} \\
0
\end{array}
$$

There is no remainder, which shows $x - 1$ is a factor of $2x^3 - 5x^2 + x + 2$

There is a simple way to find the remainder when a polynomial is divided by a linear factor. This is known as the **Remainder theorem.**

If $f(x)$ is divided by $x - a$ then the remainder is $f(a)$.

Example 6 From example 4 above if $f(x) = 2x^3 - 5x^2 + x + 2$

$$f(-2) = 2(-2)^3 - 5(-2)^2 + -2 + 2 = -36$$

and in example 5, $f(1) = 2 - 5 + 1 + 2 = 0$ showing a remainder of 0.

Proof of Remainder theorem

$f(x)$ when divided by $x - a$ gives a quotient of $Q(x)$ and a remainder R, i.e. $f(x) = (x - a)Q(x) + R$.

So $f(a) = 0.\ Q(a) + R = R$ so Remainder $= f(a)$

This is most useful when factorizing a polynomial, for if $f(a) = 0$ then $x - a$ is a factor.

Example 7 Factorize $f(x) = 2x^3 - 5x^2 + x + 2$.

Possible factors are $x \pm 1$, $x \pm 2$, $2x \pm 1$ by examining the first and last terms.

$$
\begin{array}{llll}
f(1) &=& 2 - 5 + 1 + 2 &= 0 \\
f(-1) &=& -2 - 5 - 1 + 2 &= -6 \\
f(2) &=& 16 - 20 + 2 + 2 &= 0 \\
f(-2) &=& -16 - 20 - 2 + 2 &= -36 \\
f(-\tfrac{1}{2}) &=& -\tfrac{1}{4} - \tfrac{5}{4} - \tfrac{1}{2} + 2 &= 0
\end{array}
$$

There are three linear factors $x - 1$, $x - 2$, $2x + 1$ so

$$2x^3 - 5x^2 + x + 2 = (x - 1)(x - 2)(2x + 1)$$

There is no need to test for $2x + 1$. Once two linear factors have been found, the third can be deduced. If one linear factor, say

11

$x - 1$, can be found it is better to divide this into $f(x)$ to get

$$2x^3 - 5x^2 + x + 2 = (x - 1)(2x^2 - 3x - 2)$$

either by the division process or by inspection.
Then $2x^2 - 3x - 2 = (x - 2)(2x + 1)$

It is easy to see whether a quadratic expression will factorize by using the formula for solving quadratic equations.
NB. The factors are unique

$$2x^3 - 5x^2 + x + 2 = (x - 2)(2x^2 - x - 1)$$
$$= (x - 2)(x - 1)(2x + 1)$$

Factor theorem
$x - a$ is a factor of $f(x)$ if $f(a) = 0$.

Dividing a cubic polynomial by a quadratic will, in general, leave a remainder of a linear polynomial.

Example 8 Divide $2x^3 - 5x^2 + x + 2$ by $x^2 + x + 1$.

$$
\begin{array}{r}
2x - 7 \\
x^2 + x + 1 \overline{) 2x^3 - 5x^2 + \ x + 2} \\
2x^3 + 2x^2 + 2x \\
\hline
0 \ -7x^2 - \ x + 2 \\
-7x^2 - 7x - 7 \\
\hline
0 \ + 6x + 9
\end{array}
$$

$$2x^3 - 5x^2 + x + 2 = (2x - 7)(x^2 + x + 1) + 6x + 9$$

Special factors

$\mathbf{x^2 - a^2 = (x + a)(x - a)}$ $\quad f(x) = x^2 - a^2 \Rightarrow f(a) = a^2 - a^2 = 0$
$\Rightarrow x - a$ is a factor
$f(-a) = 0 \Rightarrow x + a$ is a factor

$\mathbf{x^2 + 2ax + a^2 = (x + a)^2}$ $\quad$ and $\quad$ $\mathbf{x^2 - 2ax + a^2 = (x - a)^2}$

$\mathbf{x^3 - a^3 = (x - a)(x^2 + ax + a^2)}$ $\quad f(x) = x^3 - a^3$
$\Rightarrow f(a) = a^3 - a^3 = 0$
$\Rightarrow x - a$ is a factor

$\mathbf{x^3 + a^3 = (x + a)(x^2 - ax + a^2)}$ $\quad f(x) = x^3 + a^3$
$\Rightarrow f(-a) = (-a)^3 + a^3 = 0$
$\Rightarrow x + a$ is a factor

$\mathbf{x^4 - a^4 = (x^2 - a^2)(x^2 + a^2) = (x + a)(x - a)(x^2 + a^2)}$

Roots of equations
Example 9 The quadratic equation $x^2 - 3x + 2 = 0$ has two solutions (or roots) $x = 1$ or $x = 2$.

$$x^2 - 3x + 2 = 0 \Rightarrow (x - 1)(x - 2) = 0$$
$$\Rightarrow x - 1 = 0 \text{ or } x - 2 = 0 \Rightarrow x = 1 \text{ or } x = 2$$

sum of roots = 3   product of roots = 2

If $x^2 + px + q = 0$ factorizes to $(x - \alpha)(x - \beta) = 0$ then the roots are α and β.

$$(x - \alpha)(x - \beta) = x^2 - (\alpha + \beta)x + \alpha\beta$$

Comparing with $x^2 + px + q = 0$

sum of roots = $\alpha + \beta = -p$ product of roots = $\alpha\beta = q$

In general, if $ax^2 + bx + c = 0$ has roots α and β then

$$\alpha + \beta = -\frac{b}{a} \quad \text{and} \quad \alpha\beta = \frac{c}{a}$$

Consider the **cubic equation**

$$ax^3 + bx^2 + cx + d = 0 \qquad (a \neq 0) \qquad (1)$$

If the roots are α, β and γ then $(x - \alpha)(x - \beta)(x - \gamma) = 0$

$$(x - \alpha)(x^2 - (\beta + \gamma)x + \beta\gamma) = 0$$
$$x^3 - (\alpha + \beta + \gamma)x^2 + (\alpha\beta + \alpha\gamma + \beta\gamma)x - \alpha\beta\gamma = 0$$

Comparing with (1)

$$\alpha + \beta + \gamma = -\frac{b}{a}; \qquad \alpha\beta + \beta\gamma + \alpha\gamma = +\frac{c}{a}; \qquad \alpha\beta\gamma = -\frac{d}{a}$$

Example 10 $2x^3 - 5x^2 + x + 2 = 0.$

$\Leftrightarrow (x - 1)(x - 2)(2x + 1) = 0$ Roots are $\alpha = 1$, $\beta = 2$, $\gamma = -\frac{1}{2}$

Sum of roots $\alpha + \beta + \gamma = 2\frac{1}{2}$ $\dfrac{-b}{a} = \dfrac{5}{2}$

Sum of roots in pairs $\alpha\beta + \beta\gamma + \alpha\gamma = 2 - 1 - \frac{1}{2} = \frac{1}{2}$ $\dfrac{c}{a} = \dfrac{1}{2}$

Product of roots $\alpha\beta\gamma = -1$ $\dfrac{-d}{a} = -1$

Partial fractions

$$\frac{1}{x + 1} + \frac{1}{x + 2} \equiv \frac{(x + 2) + (x + 1)}{(x + 1)(x + 2)} \equiv \frac{2x + 3}{(x + 1)(x + 2)}$$

$$\frac{1}{x - 1} - \frac{1}{x + 1} \equiv \frac{(x + 1) - (x - 1)}{(x + 1)(x - 1)} \equiv \frac{2}{(x + 1)(x - 1)} \equiv \frac{2}{x^2 - 1}$$

It is useful when studying graphs and integration to perform the reverse operation, that is to express $\dfrac{5}{(x+2)(x-3)}$ in simpler fractions.

Example 11 $\dfrac{5}{(x+2)(x-3)} \equiv \dfrac{a}{x+2} + \dfrac{b}{x-3}$

Multiplying through by $(x+2)(x-3)$ gives

$$5 \equiv a(x-3) + b(x+2)$$

$x = 3$ gives $\quad 5 = 0 + 5b \Rightarrow b = 1$

$x \equiv -2$ gives $\quad 5 = -5a + 0 \Rightarrow a = -1$

$$\dfrac{5}{(x+2)(x-3)} \equiv \dfrac{-1}{x+2} + \dfrac{1}{x-3}$$

Written in this form the expression is said to be in partial fractions.

If the degree of the numerator is greater than or equal to the degree of the denominator it is necessary to divide the denominator into the numerator first.

Example 12

$$\dfrac{(x+1)(x+2)}{(x-1)(x+3)} \equiv \dfrac{x^2+3x+2}{x^2+2x-3} \equiv \dfrac{x^2+2x-3}{x^2+2x-3} + \dfrac{x+5}{x^2+2x-3}$$

$$= 1 + \dfrac{x+5}{(x-1)(x+3)}$$

$$\dfrac{x+5}{(x-1)(x+3)} \equiv \dfrac{a}{x-1} + \dfrac{b}{x+3}$$

$$\Leftrightarrow x+5 \equiv a(x+3) + b(x-1)$$

$x = 1$ gives $6 = 4a \Leftrightarrow a = 1\frac{1}{2}$;
$x = -3$ gives $2 = -4b \Leftrightarrow b = -\frac{1}{2}$

$$\dfrac{(x+1)(x+2)}{(x-1)(x+3)} \equiv 1 + \dfrac{3}{2(x-1)} - \dfrac{1}{2(x+3)}$$

Example 13 Express $\dfrac{11x+12}{(2x+3)(x+2)(x-3)}$ in partial fractions.

$$\dfrac{11x+12}{(2x+3)(x+2)(x-3)} \equiv \dfrac{a}{2x+3} + \dfrac{b}{x+2} + \dfrac{c}{x-3}$$

14

$$\Leftrightarrow 11x + 12 \equiv a(x+2)(x-3) + b(2x+3)(x-3) + c(2x+3)(x+2)$$

$x = 3$ gives $45 = 45c \Leftrightarrow c = 1$;
$x = -2$ gives $-10 = +5b \Leftrightarrow b = -2$

$$x = -\frac{3}{2} \text{ gives } -\frac{9}{2} = -\frac{9a}{4} \Leftrightarrow a = 2$$

So $\dfrac{11x+12}{(2x+3)(x+2)(x-3)} \equiv \dfrac{2}{2x+3} - \dfrac{2}{x+2} + \dfrac{1}{x-3}$

If the degree of the numerator is less than the degree of the denominator the fraction is proper and its partial fractions will also be proper.

The above shows that $a = \dfrac{11x+12}{(x+2)(x-3)}$ when $x = -\dfrac{3}{2}$, $b = \dfrac{11x+12}{(2x+3)(x-3)}$ when $x = -2$ and $c = \dfrac{11x+12}{(2x+3)(x+2)}$ when $x = 3$. This indicates a simple process for determining a, b and c. To determine a, the numerator of $2x+3$ in the partial fractions expression of $F(x) = \dfrac{11x+12}{(2x+3)(x+2)(x-3)}$, work out $F\left(-\dfrac{3}{2}\right)$ ignoring the factor $(2x+3)$. To determine b work out $F(-2)$ ignoring the factor $(x+2)$ in the denominator, which can be covered up and $F(-2)$ evaluated without it. This covering up rule gives a quick easy way for finding linear partial fractions. Care must be taken when the denominator contains repeated factors like $(x+2)^3$, or quadratic factors like $x^2 + 1$ which have no linear factors.

Example 14 $\quad \dfrac{2x+3}{(x+1)(x^2+1)} \equiv \dfrac{a}{x+1} + \dfrac{f(x)}{x^2+1}$

Multiplying through by $(x+1)(x^2+1)$ gives

$$2x + 3 \equiv a(x^2+1) + (x+1)f(x) \qquad (1)$$

Substituting $x = -1$ gives $1 = 2a + 0$ so $a = \frac{1}{2}$ which is the value given by using the covering up rule.
Rearranging (1) gives $2x + 3 - \frac{1}{2}(x^2+1) = (x+1)f(x)$

so $\qquad 2(x+1)f(x) = -x^2 + 4x + 5 = -(x+1)(x-5)$

$$f(x) = -\tfrac{1}{2}(x-5)$$

$$\dfrac{2x+3}{(x+1)(x^2+1)} \equiv \dfrac{\frac{1}{2}}{x+1} - \dfrac{\frac{1}{2}(x-5)}{x^2+1} \equiv \dfrac{1}{2(x+1)} - \dfrac{x-5}{2(x^2+1)}$$

Now that we have established that $f(x)$ is a linear function it is more straightforward to replace $f(x)$ by $bx + c$.

$$\frac{2x + 3}{(x + 1)(x^2 + 1)} = \frac{a}{x + 1} + \frac{bx + c}{x^2 + 1}$$
$$2x + 3 = a(x^2 + 1) + (bx + c)(x + 1)$$

Substituting $x = -1$ gives $a = \frac{1}{2}$.

Equating coefficients of x^2 gives $0 = a + b = \frac{1}{2} + b$ so $b = -\frac{1}{2}$.
Equating coefficients of x gives $2 = b + c = -\frac{1}{2} + c$ so $c = 2\frac{1}{2}$.

$$\frac{2x + 3}{(x + 1)(x^2 + 1)} = \frac{\frac{1}{2}}{x + 1} + \frac{-\frac{1}{2}x + 2\frac{1}{2}}{x^2 + 1} \quad \text{as before.}$$

Repeated factors

Example 15

$$\frac{x + 1}{(x + 2)^2} \equiv \frac{x + 2 - 1}{(x + 2)^2} \equiv \frac{x + 2}{(x + 2)^2} - \frac{1}{(x + 2)^2} \equiv \frac{1}{x + 2} - \frac{1}{(x + 2)^2}$$

This can be achieved more easily by substituting $y = x + 2$.

$$\frac{x + 1}{(x + 2)^2} \equiv \frac{y - 1}{y^2} \equiv \frac{y}{y^2} - \frac{1}{y^2} \equiv \frac{1}{y} - \frac{1}{y^2} \equiv \frac{1}{x + 2} - \frac{1}{(x + 2)^2}$$

Example 16 $\quad \dfrac{4x + 5}{(x - 1)(x + 2)^2} \equiv \dfrac{a}{x - 1} + \dfrac{f(x)}{(x + 2)^2}$

$4x + 5 \equiv a(x + 2)^2 + (x - 1)f(x) \quad x = 1$ gives $a = 1$
$4x + 5 - (x + 2)^2 = (x - 1)f(x) \equiv -x^2 + 1 \equiv -(x + 1)(x - 1)$ so
$f(x) = -(x + 1)$

Hence $\dfrac{4x + 5}{(x - 1)(x + 2)^2} \equiv \dfrac{1}{x - 1} - \dfrac{x + 1}{(x + 2)^2} \equiv \dfrac{1}{x - 1} - \dfrac{1}{x + 2} + \dfrac{1}{(x + 2)^2}$

using Example 15 to separate the last fraction.
The result suggests that repeated factors are separated like this:

$$\frac{4x + 5}{(x - 1)(x + 2)^2} \equiv \frac{a}{x - 1} + \frac{b}{x + 2} + \frac{c}{(x + 2)^2}$$

Hence $4x + 5 \equiv a(x + 2)^2 + b(x - 1)(x + 2) + c(x - 1)$.
$x = 1$ gives $9 = 9a$, so $a = 1$ just like the covering up method.
$x = -2$ gives $-3 = -3c$, so $c = 1$ and this is the value given by covering $(x + 2)^2$.
Equating coefficients of x^2 gives $0 = a + b$, so $b = -a = -1$.

Alternatively, substituting $y = x + 2$ gives

$$\frac{4x+5}{(x-1)(x+2)^2} \equiv \frac{4y-3}{(y-3)(y^2)} \equiv \frac{a}{y-3} + \frac{by+c}{y^2}$$

$a = 1$ by covering $4y - 3 = y^2 + (by + c)(y - 3)$ gives $b = -1$ and $c = 1$ by comparing coefficients. Hence the same result is obtained on substituting back for x. In the following example there is only a repeated factor with no other factors. In this case it is easy to separate directly.

Example 17

$$\frac{x^2+3x+3}{(x+1)^3} \equiv \frac{x^2+2x+1}{(x+1)^3} + \frac{x+1}{(x+1)^3} + \frac{1}{(x+1)^3}$$

$$\equiv \frac{1}{x+1} + \frac{1}{(x+1)^2} + \frac{1}{(x+1)^3}$$

Inequalities

In fig. 2 three sets of points are shown.

Set 1. Points on the line $y = x + 2$

Set 2. Shaded area above the line for which $y > x + 2$

Set 3. Unshaded area below the line for which $y < x + 2$

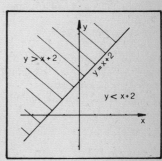

Figure 2

Solving inequalities

Example 18 $3x + 4 < 13$.

$3x < 9$ subtracting 4 from both sides

$x < 3$ dividing both sides by 3

Example 19 $3x + 4 > 2x - 3$.

$x + 4 > -3$ subtracting $2x$ from both sides

$x > -7$ subtracting 4 from both sides

Inequalities can be solved like equations with the following rules.

1. $p > q \Leftrightarrow p + r > q + r$ whether r is positive or negative.

For example $6 > 4 \Leftrightarrow 6 + 3 > 4 + 3$

$6 - 3 > 4 - 3$

2. $p > q \Leftrightarrow rp > rq$ if $r > 0$ e.g. $6 > 4 \Leftrightarrow 3 \times 6 > 3 \times 4$

3. $p > q \Leftrightarrow rp < rq$ if $r < 0$ e.g. $6 > 4 \Leftrightarrow -3 \times 6 < -3 \times 4$

Care must be taken when dealing with negative quantities because the inequality sign can change.

Example 20 $3x + 4 > 4x - 3$.

$-x + 4 > -3$ subtracting $4x$	OR	$3x + 4 > 4x - 3$
$-x > -7$ subtracting 4		$3x + 7 > 4x$ adding 3
$x < 7$ multiplying by -1		$7 > x$ subtracting $3x$

Inequalities and regions

$3x + 4y = 12$ represents a straight line (see fig. 3) passing through $(4, 0)$ and $(0, 3)$.

For the point $(1, 1)$
$3x + 4y = 3 + 4 = 7 < 12$
$(1, 1)$ lies in region $3x + 4y < 12$.

Region below line represents $3x + 4y < 12$ (shaded).

Region above line (unshaded) represents $3x + 4y > 12$.

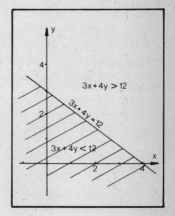

Figure 3

Simultaneous inequalities

Find the region satisfied by the inequalities $x \geqslant 0$, $y \geqslant 0$, $x + y \leqslant 4$, $y \geqslant 2x - 1$.

$x \geqslant 0$ and $y \geqslant 0$ restricts the solution set to the positive quadrant. It is most helpful to shade the region UNWANTED to focus attention on the WANTED region.

$x + y = 4$ is a straight line passing through $(0, 4)$ and $(4, 0)$.

$x + y < 4$ is the region below this line (including $(0, 0)$).

$y = 2x - 1$ passes through $(1, 1)$ and $(2, 3)$.

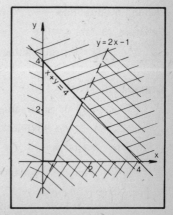

Figure 4

18

The point $(2, 1)$ satisfies $y < 2x - 1$ and lies in the unwanted region. The UNSHADED quadrilateral is the solution set of points (fig. 4). The boundaries are all included as the inequalities are greater than or equal to' or 'less than or equal to'.

Quadratic inequalities

Example 21 Find the values of x which satisfy $x^2 - 4x + 3 < 0$. Figure 5a shows the graph of $y = x^2 - 4x + 3 = (x - 1)(x - 3)$ which cuts the x axis at $(1, 0)$ and $(3, 0)$. When x lies between these values, i.e. $1 < x < 3$, $y = x^2 - 4x + 3 < 0$ and so the solution set is $1 < x < 3$, OR $x^2 - 4x + 3 = (x - 1)(x - 3) < 0$ which is true if one bracket is negative and the other positive. This happens when x is between 1 and 3 i.e. $1 < x < 3$.

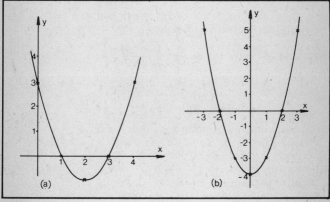

Figure 5

Example 22 Find the set of values for $x^2 > 4$.
Method 1 $x^2 > 4 \Leftrightarrow x^2 - 4 > 0$.   *Method 2* $x^2 > 4$
Draw graph of $y = x^2 - 4$ (fig. 5b) $x^2 = 4 \Leftrightarrow x = \pm 2$
$x^2 - 4 > 0$ (above x axis) when $x^2 > 4 \Leftrightarrow x > 2$ or $x < -2$
$x > 2$ or $x < -2$

Algebraic inequalities

Example 23 $(a - b)^2 > 0$ if a and b are real numbers and $a \neq b$.
$$\Leftrightarrow a^2 - 2ab + b^2 > 0 \Leftrightarrow a^2 + b^2 > 2ab$$

19

Writing $a^2 = x$ and $b^2 = y$ $\quad \dfrac{x+y}{2} > \sqrt{x}\sqrt{y} = \sqrt{xy}$

Thus the arithmetic mean of two numbers is greater than their geometric mean.

Example 24 Find the ranges of values of x and y for which there are no real points on the following locus $y^2 = x(1-x)$.

$y^2 = x - x^2 > 0$ for real values of y

$$\Leftrightarrow x > x^2 \Leftrightarrow 0 < x < 1$$
$$y^2 = x - x^2 \Leftrightarrow x^2 - x + y^2 = 0$$

Regarding this as a quadratic in x, real values of x occur when

$$1 > 4y^2 \quad (b^2 > 4ac)$$
$$\Leftrightarrow y^2 < \tfrac{1}{4} \Leftrightarrow -\tfrac{1}{2} < y < \tfrac{1}{2}$$

$y^2 = x - x^2 \Leftrightarrow y^2 + x^2 - x + \tfrac{1}{4} = \tfrac{1}{4} \Leftrightarrow y^2 + (x - \tfrac{1}{2})^2 = (\tfrac{1}{2})^2$ which represents a circle centre $(\tfrac{1}{2}, 0)$ radius $\tfrac{1}{2}$.

Example 25 Solve for x $\quad \dfrac{(x-1)(x-3)}{(x+1)(x-2)} > 0$.

$(x-1)(x-3) > 0$ if $(x+1)(x-2) > 0$ i.e. $x < -1$ or $x > 2$ (cf. fig. 5a)
$$(x-1)(x-3) > 0 \Leftrightarrow x < 1 \text{ or } x > 3.$$

These four conditions mean $x < -1$ or $x > 3$.
$(x-1)(x-3) < 0$ if $(x+1)(x-2) < 0$ i.e. $-1 < x < 2$.
$(x-1)(x-3) < 0 \Leftrightarrow 1 < x < 3$.
These conditions mean $1 < x < 2$.
Complete solution $x < -1$, $1 < x < 2$, $x > 3$.
Another method would be to consider the sign of $y = \dfrac{(x-1)(x-3)}{(x+1)(x-2)}$ in between the four critical values $x = -1, 1, 2$ and 3.

If $x < -1$, each bracket is negative so $y = \dfrac{(-) \times (-)}{(-) \times (-)} > 0$

If $-1 < x < 1$ $\quad y = \dfrac{(-) \times (-)}{(+) \times (-)} < 0$

If $1 < x < 2$ $\quad y = \dfrac{(+) \times (-)}{(+) \times (-)} > 0$

If $2 < x < 3$ $\quad y = \dfrac{(+) \times (-)}{(+) \times (+)} < 0$

If $3 < x \qquad y = \dfrac{(+) \times (+)}{(+) \times (+)} > 0$

Complete solution is: $x < -1$; $1 < x < 2$; $x > 3$.

Example 26 Show that $a^3 + b^3 + c^3 - 3abc = (a + b + c)$
$\times (a^2 + b^2 + c^2 - ab - bc - ac)$.

RHS $= a^3 + ab^2 + ac^2 - a^2 b - abc - a^2 c + a^2 b + b^3 + bc^2 - ab^2$
$\qquad - b^2 c - abc + a^2 c + b^2 c + c^3 - abc - bc^2 - ac^2$
$\qquad = a^3 + b^3 + c^3 - 3abc = $ LHS

Deduce that the arithmetic mean of three unequal positive numbers p, q, r is greater than their geometric mean.

Then $a^3 + b^3 + c^3 - 3abc$
$= (a + b + c)(a^2 + b^2 + c^2 - ab - bc - ac) > 0$ since
$(a^2 + b^2 + c^2 - ab - bc - ac) = \frac{1}{2}\{(a - b)^2 + (b - c)^2 + (c - a)^2\} > 0$
Writing $a^3 = p, b^3 = q, c^3 = r$

$$p + q + r - 3\sqrt[3]{p}\sqrt[3]{q}\sqrt[3]{r} > 0$$
$$\frac{p + q + r}{3} > \sqrt[3]{pqr}$$

Arithmetic mean > Geometric mean.

Indices and Logarithms

Rules of indices for positive integers
1. $2^3 \times 2^4 = 2^7$ in general $a^m \times a^n = a^{m+n}$ m, n positive integers
2. $2^5 \div 2^3 = 2^2$ in general $a^m \div a^n = a^{m-n}$ $m > n$
3. $(2^3)^4 = 2^{12}$ in general $(a^m)^n = a^{mn}$
Fractional and negative indices
4. $2^3 \div 2^3 = 2^0$ from rule 2 so $2^0 = 1$, in general $a^0 = 1$
5. $2^3 \div 2^5 = 2^{-2}$ from rule 2 so $2^{-2} = \dfrac{1}{2^2}$ in general $a^{-n} = \dfrac{1}{a^n}$
6. $2^{1/2} \times 2^{1/2} = 2^1$ from rule 1 so $2^{1/2} = \sqrt{2}$
   $2^{1/3} \times 2^{1/3} \times 2^{1/3} = 2^1$ from rule 1 so $2^{1/3} = \sqrt[3]{2}$
7. $8^{2/3} = (8^2)^{1/3} = \sqrt[3]{8^2} = 4$
   $8^{2/3} = (8^{1/3})^2 = (\sqrt[3]{8})^2 = 2^2 = 4$ in general $a^{m/n} = \sqrt[n]{a^m} = (\sqrt[n]{a})^m$

Logarithms

$16 = 2^4$, $32 = 2^5 \Leftrightarrow 16 \times 32 = 2^4 \times 2^5 = 2^{4+5} = 2^9 = 512$

$\left.\begin{array}{l} 16 = 2^4 \Leftrightarrow 4 = \log_2 16 \\ 32 = 2^5 \Leftrightarrow 5 = \log_2 32 \end{array}\right\} \log_2 16 + \log_2 32 = \log_2 16 \times 32 = \log_2 512$

$\log_{10} 16 + \log_{10} 32 = 1{\cdot}2041 + 1{\cdot}5051 = 2{\cdot}7092 = \log_{10} 512$

In general $a^x = p \Leftrightarrow x = \log_a p$

$$a^y = q \Leftrightarrow y = \log_a q$$

$$pq = a^x a^y = a^{x+y} \Leftrightarrow \log_a pq = x + y = \log_a p + \log_a q$$

$$\frac{p}{q} = \frac{a^x}{a^y} = a^{x-y} \Leftrightarrow \log_a \left(\frac{p}{q}\right) = x - y = \log_a p - \log_a q$$

$$p^n = (a^x)^n = a^{xn} \Leftrightarrow \log_a p^n = nx = n \log_a p$$

Example 27 Solve for x, $3^x = 8$.

Take logs of both sides $\log_{10} 3^x = \log_{10} 8$

$$x \log_{10} 3 = \log_{10} 8$$

$$x = \frac{\log_{10} 8}{\log_{10} 3} = \frac{0 \cdot 9031}{0 \cdot 4771} = \frac{1 \cdot 893 \text{ to three}}{\text{decimal places}}$$

Example 28 $\log_{10} 800 = \log_{10} 100 \times 8 = \log_{10} 100 + \log_{10} 8$
$$= 2 + 0 \cdot 9031$$
$$= 2 \cdot 9031$$

$$\log_{10} 0 \cdot 008 = \log_{10} 0 \cdot 001 \times 8 = \log_{10} 0 \cdot 001 + \log_{10} 8$$
$$= -3 + 0 \cdot 9031$$
$$= \bar{3} \cdot 9031 \text{ or } -2 \cdot 0969$$

Change of base of logarithms

From Example 27 $3^x = 8 \Leftrightarrow x = \log_3 8 = \dfrac{\log_{10} 8}{\log_{10} 3}$

In general, $a^x = b \Leftrightarrow x = \log_a b$

Taking logs to base c $\log_c a^x = \log_c b$

$$x \log_c a = \log_c b$$

$$x = \frac{\log_c b}{\log_c a} = \log_a b$$

This formula gives a method for changing from logs to base a into logs to base c.

Experimental laws

x 1 2 3 4 This table of values gives a straight line when
y 3 5 7 9 values of y are plotted against values of x. The equation of the straight line is $y = 2x + 1$ where 2 is the gradient and 1 the intercept on the y axis.

$$y = kx^n \Leftrightarrow \log y = \log k + \log x^n$$
$$= n \log x + \log k$$

If $\log x$ values are plotted against values of $\log y$, fig. 6a, a straight line results in which the gradient is n and the intercept on the y axis is $\log k$. The values of n and $\log k$ can be calculated from the graph and n and k determined.

$$y = kb^x \Leftrightarrow \log y = \log k + \log b^x$$
$$= x \log b + \log k$$

If values of x are plotted against values of $\log y$, a straight line results in which the gradient is $\log b$ and the intercept $\log k$. Hence k and b are determined.

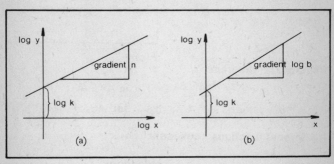

Figure 6

Key terms

Remainder theorem if $f(x)$ is divided by $x - a$ then the remainder is $f(a)$.
Factor theorem $f(a) = 0$ $x - a$ is a factor if $f(x)$.
Special factors $x^2 - a^2 = (x - a)(x + a)$;
$(x \pm a)^2 = x^2 \pm 2ax + a^2$
$x^3 - a^3 = (x - a)(x^2 + ax + a^2)$; $x^3 + a^3 = (x + a)(x^2 - ax + a^2)$
Sum and product of roots;
If α and β are the roots of the equation $ax^2 + bx + c = 0$ then

$$\alpha + \beta = \frac{-b}{a} \quad \text{and} \quad \alpha\beta = \frac{c}{a}$$

Indices and logarithms $a^x = b$ $x = \log_a b$

Change of base $\log_a b = \dfrac{\log_c b}{\log_c a}$

$y = kx^n$ $\log y = n \log x + \log k$. Plotting $\log x$ against $\log y$ gives gradient n and intercept $\log k$.

$y = kb^x$ $\log y = x \log b + \log k$. Plotting x against $\log y$ gives gradient $\log b$ and intercept $\log k$.

Chapter 2
Trigonometry

In the right-angled triangle in fig. 7 the trigonometrical ratios are

sine $\quad \sin A = \dfrac{a}{c}$

cosine $\quad \cos A = \dfrac{b}{c}$

tangent $\quad \tan A = \dfrac{a}{b}$

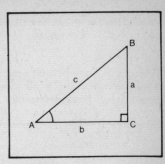

Figure 7

Pythagoras' theorem

$$a^2 + b^2 = c^2; \quad \frac{a^2}{c^2} + \frac{b^2}{c^2} = 1; \quad (\sin A)^2 + (\cos A)^2 = 1$$

$(\sin A)^2$ is written as $\sin^2 A$, so that $\quad \sin^2 A + \cos^2 A = 1 \quad (1)$

Reciprocal functions **cosecant** $\quad \operatorname{cosec} A = \dfrac{c}{a} = \dfrac{1}{\sin A}$

secant $\quad \sec A = \dfrac{c}{b} = \dfrac{1}{\cos A} \quad$ **cotangent** $\quad \cot A = \dfrac{b}{a} = \dfrac{1}{\tan A}$

Dividing (1) throughout by $\cos^2 A$ gives $\dfrac{\sin^2 A}{\cos^2 A} + \dfrac{\cos^2 A}{\cos^2 A} = \dfrac{1}{\cos^2 A}$ which simplifies to $\tan^2 A + 1 = \sec^2 A$.

Dividing (1) throughout by $\sin^2 A$ gives $\dfrac{\sin^2 A}{\sin^2 A} + \dfrac{\cos^2 A}{\sin^2 A} = \dfrac{1}{\sin^2 A}$ which simplifies to $1 + \cot^2 A = \operatorname{cosec}^2 A$.

Special triangles

In an isosceles right-angled triangle with $AC = 1 = BC$, fig. 8

$AB = \sqrt{2} \simeq 1 \cdot 414$ (Pythagoras)

$\sin 45° = \dfrac{1}{\sqrt{2}} = \cos 45°$

$\tan 45° = 1 = \cot 45°$

$\sec 45° = \sqrt{2} = \operatorname{cosec} 45°$

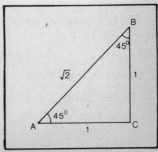

Figure 8

24

In fig. 9 triangle ACD is equilateral with $AD = 2$,
$AB = 1$, $BC = \sqrt{3} \approx 1.732$
(Pythagoras)

From triangle ABC

$$\sin 30° = \frac{1}{2} = \cos 60°$$

$$\cos 30° = \frac{\sqrt{3}}{2} = \sin 60°$$

$$\tan 30° = \frac{1}{\sqrt{3}} = \frac{1}{\tan 60°} = \cot 60°$$

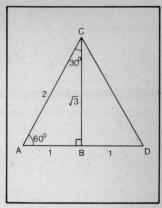

Figure 9

The general angle

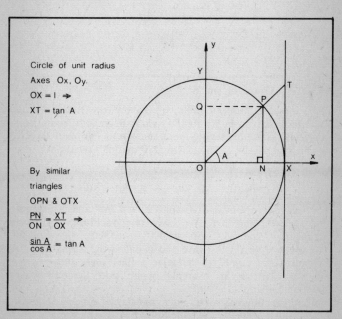

Circle of unit radius
Axes Ox, Oy.
$OX = 1 \Rightarrow$
$XT = \tan A$

By similar triangles
OPN & OTX
$$\frac{PN}{ON} = \frac{XT}{OX} \Rightarrow$$
$$\frac{\sin A}{\cos A} = \tan A$$

Figure 10

25

sin A is the length PN (y coordinate) as OP rotates anti-clockwise and is the projection of OP onto the y axis.

cos A is the length ON (x coordinate) as OP rotates anti-clockwise and is the projection of OP onto the x axis.

tan A is the length XT which becomes infinite when $A = 90°$ and is negative when $90° < A < 180°$.

Graphs of sin A, cos A, tan A

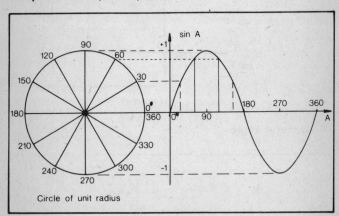

Figure 11. Graph of sin A

As A increases from $0°$ to $90°$, sin A increases from 0 to 1 and as A increases from $90°$ to $180°$, sin A decreases from 1 to 0. sin A is positive in the 2nd quadrant ($90°$ to $180°$) and negative in the 3rd ($180°$ to $270°$) and 4th ($270°$ to $360°$) quadrants.

From the symmetry of the graph sin $150° = $ sin $30° = +0.5$

$$\sin 210° = \sin 330° = -\sin 30° = -0.5$$

The sine curve repeats its values every $360°$ and is **periodic** the **period** being $360°$. We have one **oscillation** or **cycle** between $0°$ and $360°$. So sin $400° = $ sin $(360° + 40°) = $ sin $40°$. The greatest height is 1 unit and is called the **amplitude.** By extending the graph for negative angles sin $(-30°) = -0.5 = -\sin 30°$. This is an example of an **odd** function where sin $(-x) = -\sin x$.

An odd function has rotational symmetry of order 2 about the origin. Such a function is called odd because all the graphs of odd powers of x have this property.

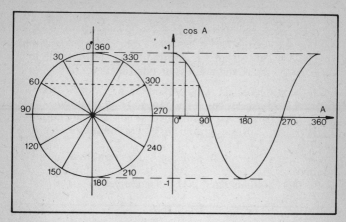

Figure 12. Graph of cos *A*

Since cos *A* is a projection onto a horizontal axis (*x* axis) to draw its curve we rotate the unit circle through $^+90°$ or start labelling our angles with 0° at the top.

The cosine curve is also periodic with period 360°, and is a translation of the sine curve through $^-90°$, so $\sin(A + 90°) = \cos A$. The cosine curve is symmetrical about the *y* axis so $\cos(-A) = \cos A$. cos *A* is an example of an **even** function where $\cos(-x) = \cos x$.

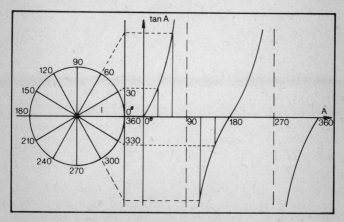

Figure 13. Graph of tan *A*

27

As A increases from $0°$ to $90°$, tan A increases from 0 to $+\infty$.
As A increases from $90°$ to $180°$, tan A increases from $-\infty$ to 0.
As A increases from $180°$ to $360°$, tan A repeats the values it took
for $0° \leqslant A \leqslant 180°$. So tan A is a **periodic** function of period $180°$.
tan $(-A) = -$ tan A and tan A is an odd function.

From the graphs of sin A, cos A and tan A, it can be seen that
if $0° < A < 90°$, sin A, cos A and tan A are all **positive**
if $90° < A < 180°$ sin A is **positive**, cos A and tan A are **negative**
if $180° < A < 270°$ sin A and cos A are **negative**, tan A is
positive
if $270° < A < 360°$ sin A and tan A are **negative**, cos A is
positive
These facts are illustrated in the diagram below.

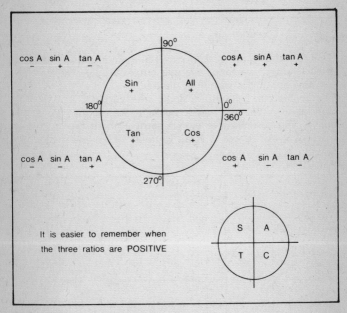

Figure 14

Radians
It is most useful to measure angles in radians rather than
degrees. When this is done the gradient of the graph of sin x
when $x = 0$ is equal to 1 and this has significant advantages
when differentiating.

Definition

A radian (1^c) is the angle subtended at the centre of a circle by an arc equal in length to the radius, fig. 15a.

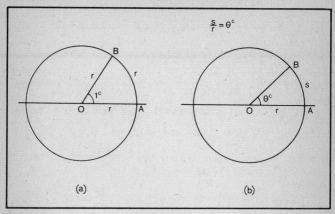

Figure 15

In general, fig. 15b, θ^c is the ratio of arc length s to radius r.

$$\frac{s}{r} = \theta^c \quad \text{or} \quad s = r\theta^c$$

If $s = 2\pi r$ (circumference) $\dfrac{s}{r} = 2\pi^c = 360°$

$$\pi^c = 180° \quad \text{i.e.} \quad 1^c = \frac{180°}{\pi} \simeq 57 \cdot 3°$$

Compound angles

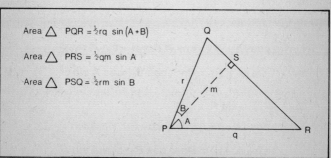

Area $\triangle$ PQR = $\frac{1}{2}rq$ sin (A + B)

Area $\triangle$ PRS = $\frac{1}{2}qm$ sin A

Area $\triangle$ PSQ = $\frac{1}{2}rm$ sin B

Figure 16

Area ΔPQR = area ΔPRS + area ΔPSQ

$\frac{1}{2}rq \sin(A+B) = \frac{1}{2}qm \sin A + \frac{1}{2}rm \sin B$

$rq \sin(A+B) = qr \cos B \sin A + rq \cos A \sin B$

$\sin(A+B) = \sin A \cos B + \cos A \sin B$

This is the simplest way of establishing $\sin(A+B)$ but it may be difficult if either A or B exceeds $90°$.
Replacing B by $-B$ gives

$\sin(A-B) = \sin A \cos(-B) + \cos A \sin(-B)$

$\sin(A-B) = \sin A \cos B - \cos A \sin B$

since $\cos(-B) = \cos B$ and $\sin(-B) = -\sin B$.

Another method uses matrices and vectors. Rotate the vector

$$\overrightarrow{OP} = \begin{pmatrix} \cos A \\ \sin A \end{pmatrix}$$

through an angle B to Q so

$$\overrightarrow{OQ} = \begin{pmatrix} \cos(A+B) \\ \sin(A+B) \end{pmatrix}$$

The matrix for a rotation of B about O is

$$\begin{pmatrix} \cos B & -\sin B \\ \sin B & \cos B \end{pmatrix}$$

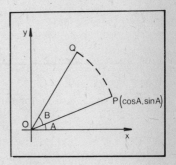

Figure 17

so

$$\begin{pmatrix} \cos(A+B) \\ \sin(A+B) \end{pmatrix} = \begin{pmatrix} \cos B & -\sin B \\ \sin B & \cos B \end{pmatrix} \begin{pmatrix} \cos A \\ \sin A \end{pmatrix}$$

$$= \begin{pmatrix} \cos B \cos A - \sin B \sin A \\ \sin B \cos A + \cos B \sin A \end{pmatrix}$$

$\cos(A+B) = \cos A \cos B - \sin A \sin B$ $\qquad$ (2)

$\sin(A+B) = \sin A \cos B + \cos A \sin B$ $\qquad$ (3)

Substituting $-B$ for B in (2) and (3) and using $\cos(-B) = \cos B$ and $\sin(-B) = -\sin B$ gives

$\cos(A-B) = \cos A \cos B + \sin A \sin B$ $\qquad$ (4)

$\sin(A-B) = \sin A \cos B - \cos A \sin B$ $\qquad$ (5)

Substituting $B = A$ in (3) gives

$$\sin(A + A) = \sin A \cos A + \cos A \sin A = 2 \sin A \cos A$$

$$\mathbf{\sin 2A = 2 \sin A \cos A} \tag{6}$$

Substituting $B = A$ in (2) gives

$$\cos(A + A) = \cos A \cos A - \sin A \sin A$$

$$\mathbf{\cos 2A = \cos^2 A - \sin^2 A} \tag{7}$$

Using equation (1), i.e. $\cos^2 A = 1 - \sin^2 A$

$$\cos 2A = 1 - \sin^2 A - \sin^2 A$$

$$\mathbf{\cos 2A = 1 - 2 \sin^2 A} \tag{8}$$

This is often useful in the form $\sin^2 A = \frac{1}{2}(1 - \cos 2A)$.
Using $\sin^2 A = 1 - \cos^2 A$ in (7) gives

$$\cos 2A = \cos^2 A - (1 - \cos^2 A) = \cos^2 A - 1 + \cos^2 A$$

$$\mathbf{\cos 2A = 2 \cos^2 A - 1} \quad \text{or} \quad \cos^2 A = \tfrac{1}{2}(1 + \cos 2A) \tag{9}$$

$$\tan(A + B) = \frac{\sin(A + B)}{\cos(A + B)} = \frac{\sin A \cos B + \cos A \sin B}{\cos A \cos B - \sin A \sin B}$$

Dividing each term top and bottom by $\cos A \cos B$ gives

$$\mathbf{\tan(A + B) = \frac{\tan A + \tan B}{1 - \tan A \tan B}}$$

If $B = A$

$$\mathbf{\tan 2A = \frac{2 \tan A}{1 - \tan^2 A}}$$

Example 1 Express $\sin 3A$ in terms of $\sin A$ only.

$$\begin{aligned}
\sin 3A = \sin(2A + A) &= \sin 2A \cos A + \cos 2A \sin A \\
&= 2 \sin A \cos A \cos A + (1 - 2 \sin^2 A) \sin A \\
&= 2 \sin A (1 - \sin^2 A) + \sin A - 2 \sin^3 A \\
&= 2 \sin A - 2 \sin^3 A + \sin A - 2 \sin^3 A
\end{aligned}$$

$$\mathbf{\sin 3A = 3 \sin A - 4 \sin^3 A}$$

A similar method gives $\mathbf{\cos 3A = 4 \cos^3 A - 3 \cos A}$, a similar result!

Example 2 Express $\sin 2A$ and $\cos 2A$ in terms of $\tan A$ only.

$$\sin 2A = 2 \sin A \cos A = \frac{2 \sin A \cos A}{\cos^2 A + \sin^2 A}$$

using $\cos^2 A + \sin^2 A = 1$ for the denominator.
Divide top and bottom by $\cos^2 A$

$$\sin 2A = \frac{2 \tan A}{1 + \tan^2 A} \qquad (10)$$

$$\cos 2A = \frac{\cos^2 A - \sin^2 A}{\cos^2 A + \sin^2 A} = \frac{1 - \tan^2 A}{1 + \tan^2 A} \qquad (11)$$

dividing top and bottom by $\cos^2 A$.

Also $\tan 2A = \dfrac{2 \tan A}{1 - \tan^2 A}$

Writing $t = \tan A$ we have

$$\sin 2A = \frac{2t}{1 + t^2}, \qquad \cos 2A = \frac{1 - t^2}{1 + t^2}, \qquad \tan 2A = \frac{2t}{1 - t^2}$$

Factor formulae

Adding equations (3) and (5) gives

$$\sin(A + B) + \sin(A - B) = 2 \sin A \cos B \qquad (12)$$

Subtracting (5) from (3) gives

$$\sin(A + B) - \sin(A - B) = 2 \cos A \sin B \qquad (13)$$

Adding (2) and (4),

$$\cos(A + B) + \cos(A - B) = 2 \cos A \cos B \qquad (14)$$

Subtracting (4) from (2),

$$\cos(A + B) - \cos(A - B) = -2 \sin A \sin B \qquad (15)$$

These formulae are most useful in integration to change products into sums and differences which are generally easier to integrate.

By writing $A + B = P$ and $A - B = Q$ i.e. $A = \dfrac{P + Q}{2}$,

$B = \dfrac{P - Q}{2}$ equations (12) to (14) become

$$\sin P + \sin Q = 2 \sin \frac{(P + Q)}{2} \cos \frac{(P - Q)}{2}$$

$$\sin P - \sin Q = 2 \cos \frac{(P + Q)}{2} \sin \frac{(P - Q)}{2}$$

$$\cos P + \cos Q = 2 \cos \frac{(P + Q)}{2} \cos \frac{(P - Q)}{2}$$

$$\cos P - \cos Q = -2 \sin \frac{(P + Q)}{2} \sin \frac{(P - Q)}{2}$$

Equations

In solving trigonometrical equations it is possible to obtain infinite numbers of solutions.

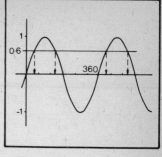

Figure 18

Example 3 Solve $\sin x = 0 \cdot 6$.
Four solutions are arrowed in fig. 18
$x = 36 \cdot 9°$ or $180° - 36 \cdot 9° = 143 \cdot 1°$
or $360° + 36 \cdot 9°$ or $360° + 143 \cdot 1°$

General solution is
$x = 360k + 36 \cdot 9°$ or $360k + 143 \cdot 1°$
$k = 0, \pm 1, \pm 2$, etc.

In the following examples solutions will be given in the range $0° \leqslant x \leqslant 360°$

Example 4 $\cos x = -0 \cdot 5$ $x = 120°$ or $360° - 120° = 240°$

General solution $x = 360k \pm 120°$

Example 5 $\tan x = 0 \cdot 4$ $x = 21 \cdot 8°$ or $180° + 21 \cdot 8° = 201 \cdot 8°$

General solution $x = 180k + 21 \cdot 8°$

Example 6 $4 \sin x + 3 \cos x = 1$

Method 1 Divide through by $\sqrt{4^2 + 3^2} = 5$.

$$\frac{4}{5} \sin x + \frac{3}{5} \cos x = \frac{1}{5} \qquad (1)$$

This will make LHS a compound angle. If $\frac{4}{5} = \cos A$, $\frac{3}{5} = \sin A$

$$\cos A \sin x + \sin A \cos x = 0 \cdot 2 \qquad (2)$$

Recognizing LHS as $\sin(x + A)$ gives $\sin(x + A) = 0 \cdot 2$.

$$x + A = 11 \cdot 5° \text{ or } 168 \cdot 5°$$

$\frac{4}{5} = \cos A \Rightarrow A = 36 \cdot 9°$ so $x + 36 \cdot 9° = 11 \cdot 5°$ or $168.5.°$

$$\begin{array}{ll} x = 11 \cdot 5° - 36 \cdot 9° & \text{or} \qquad x = 168 \cdot 5° - 36 \cdot 9° \\ = -25 \cdot 4° = 334 \cdot 6° & \qquad\qquad x = 131 \cdot 6° \end{array}$$

If $\frac{4}{5} = \sin B$ and $\frac{3}{5} = \cos B$ equation (1) becomes

$\sin B \sin x + \cos B \cos x = 0 \cdot 2$
$\cos(x - B) = 0 \cdot 2$ leading to the same result.

33

Method 2 From (10), $\sin x = \dfrac{2t}{1+t^2}$ where $t = \tan \frac{1}{2}x$

From (11), $\cos x = \dfrac{1-t^2}{1+t^2}$ where $t = \tan \frac{1}{2}x$

Substituting in $4 \sin x + 3 \cos x = 1$

$$4 \times \frac{2t}{1+t^2} + 3 \frac{(1-t^2)}{1+t^2} = 1$$

$\therefore \quad 8t + 3 - 3t^2 = 1 + t^2$

$4t^2 - 8t - 2 = 0$, i.e. $2t^2 - 4t - 1 = 0$

$\therefore \quad t = \dfrac{4 \pm \sqrt{16 + 4.2.1}}{4} = \dfrac{4 \pm \sqrt{24}}{4} = \dfrac{2 \pm \sqrt{6}}{2} = 2 \cdot 225 \text{ or } -0 \cdot 225$

$$t = \tan \frac{x}{2} = 2 \cdot 225 \quad \text{or} \quad -0 \cdot 225$$

$$\frac{x}{2} = 65 \cdot 8 \quad \text{or} \quad 180° - 12 \cdot 7° = 167 \cdot 3°$$

$$x = 131 \cdot 6° \quad \text{or} \quad 334 \cdot 6°$$

Example 7 $\sin 3x + 3 + 4 \sin x = 0$.

Use $\sin 3x = 3 \sin x - 4 \sin^3 x$. Then

$$3 \sin x - 4 \sin^3 x + 3 + 4 \sin x = 0$$

$$4 \sin^3 x - 7 \sin x - 3 = 0, \text{ a cubic in } \sin x.$$

It is not easy to solve a cubic equation. If one root can be found, the cubic expression can then be expressed as the product of a linear factor and a quadratic factor.
Consider $f(y) = 4y^3 - 7y - 3 = 0$ and use the **factor theorem.**

$$f(1) = 4 - 7 - 3 = -6$$

$$f(2) = 32 - 14 - 3 = 15$$

There is a root between 1 and 2.
This will not give us a solution since $\sin x$ would be > 1.

$$f(0) = -3$$

$$f(-1) = -4 + 7 - 3 = 0$$

Hence $y + 1$ is a factor of $f(y)$.

$$f(y) = (y + 1)(4y^2 - 4y - 3) = 0$$

34

By long division or inspection
$$(y + 1)(2y - 3)(2y + 1) = 0$$

$y + 1 = 0$	or	$2y - 3 = 0$	or	$2y + 1 = 0$
$y = -1$	or	$y = 1\cdot 5$	or	$y = -\frac{1}{2}$

$$\sin x = -1 \qquad \textit{NO SOLUTION} \qquad \sin x = -0\cdot 5$$
$$x = 270° \qquad\qquad\qquad\qquad\qquad x = 210° \text{ or } 330°$$

Example 8 $\sin x + \sin 2x - \sin 3x = 0$.

Using $\sin P - \sin Q = 2 \cos \dfrac{(P + Q)}{2} \sin \dfrac{(P - Q)}{2}$

$$\sin x - \sin 3x = 2 \cos \frac{(x + 3x)}{2} \sin \frac{(x - 3x)}{2}$$
$$= 2 \cos 2x \sin (-x)$$
$$= -2 \cos 2x \sin x$$

expression becomes

$$\sin 2x - 2 \cos 2x \sin x = 0$$
$$2 \sin x \cos x - 2 \cos 2x \sin x = 0$$
$$2 \sin x (\cos x - \cos 2x) = 0$$
$$\sin x = 0 \text{ or } \cos x - \cos 2x = 0$$
$$x = 0°, 180°, 360° \text{ or } \cos x - (2 \cos^2 x - 1) = 0$$
$$\Rightarrow 2 \cos^2 x - \cos x - 1 = 0$$
$$\Rightarrow (2 \cos x + 1)(\cos x - 1) = 0$$
$$\Rightarrow \cos x = -\tfrac{1}{2} \text{ or } \cos x = 1$$
$$\Rightarrow x = 120°, 240° \text{ or } 0°, 360°$$

Solutions are $x = 0°, 120°, 180°, 240°$ or $360°$.

Inverse trigonometric functions

$$f(x) = \sin x = 0\cdot 5 \Rightarrow x = \sin^{-1}(0\cdot 5) = 30° = \frac{\pi^c}{6}$$

the angle whose sine is $0\cdot 5$.

$$f(30°) = 0\cdot 5 \Rightarrow f^{-1}(0\cdot 5) = 30°$$

$\sin^{-1} x$ (pronounced sine minus one) is the **inverse function** for $\sin x$, i.e. the angle whose sine is x.

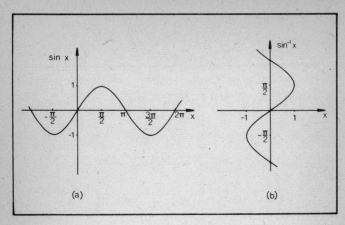

Figure 19

$\sin^{-1}(0.5) = \frac{\pi}{6}(=30°)$ or $\frac{5\pi}{6}(150°)$ and has two values.

To make sure that $\sin^{-1} x$ is a one-valued function we must restrict the range of values of $\sin^{-1} x$ to be between $-\frac{\pi}{2}$ and $+\frac{\pi}{2}$ and the values obtained are called the principle values of the function.

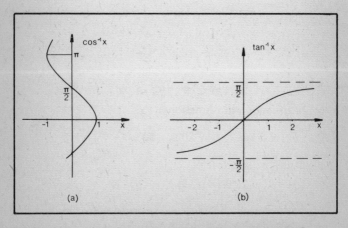

Figure 20

Figure 20a shows the graph of $\cos^{-1} x$.

$$\cos A = 0{\cdot}5 \Rightarrow A = \cos^{-1}(0{\cdot}5) = 60°$$

The principal values of $\cos^{-1} x$ lie between 0 and π^c. Figure 20b shows the graph of $\tan^{-1} x$.

$$\tan A = 1 \Rightarrow A = \tan^{-1}(1) = 45°.$$

The principal values of $\tan^{-1} x$ lie between $-\dfrac{\pi^c}{2}$ and $+\dfrac{\pi^c}{2}$.

Since $\sin^{-1} x$ is the inverse of $\sin x$, $\sin(\sin^{-1} x) = x$.

Do not confuse $\sin^{-1} x$ with $\dfrac{1}{\sin x} = \operatorname{cosec} x$.

The notation $\sin^{-1} x$ is reserved to stand for the inverse function of $\sin x$.

Key terms

$$\operatorname{cosec} A = \frac{1}{\sin A}; \qquad \sec A = \frac{1}{\cos A}; \qquad \cot A = \frac{1}{\tan A}$$

$$\sin^2 A + \cos^2 A = 1; \quad 1 + \tan^2 A = \sec^2 A; \quad 1 + \cot^2 A = \operatorname{cosec}^2 A$$

Special angles $\qquad \sin 45° = \cos 45° = \dfrac{1}{\sqrt{2}}; \qquad \tan 45° = 1$

$$\sin 30° = \cos 60° = 0{\cdot}5; \qquad \sin 60° = \cos 30° = \frac{\sqrt{3}}{2};$$

$$\tan 60° = \frac{1}{\tan 30°} = \sqrt{3}$$

All angles

$$\sin(180° - A) = \sin A \qquad \sin(180° + A) = \sin(360° - A) = -\sin A$$
$$\cos(180° \pm A) = -\cos A \qquad \cos(360° \pm A) = \cos A$$
$$\tan(180° + A) = \tan A \qquad \tan(180° - A) = \tan(360° - A) = -\tan A$$

Compound angles

$$\sin(A \pm B) = \sin A \cos B \pm \cos A \sin B$$
$$\cos(A \pm B) = \cos A \cos B \pm \sin A \sin B$$

$$\tan(A + B) = \frac{\tan A + \tan B}{1 - \tan A \tan B} \qquad \tan(A - B) = \frac{\tan A - \tan B}{1 + \tan A \tan B}$$

$$\sin 2A = 2 \sin A \cos A; \qquad \tan 2A = \frac{2 \tan A}{1 - \tan^2 A}$$

$$\cos 2A = \cos^2 A - \sin^2 A; \quad = 2 \cos^2 A - 1 \quad = 1 - 2 \sin^2 A$$

Chapter 3
Differentiation

Differentiation is the name given to a process which determines the rate of change of one variable with respect to another. It is a section of that branch of mathematics called the calculus. In it, we consider the rates of change rather than the amount of change and do so by investigating the effect of very small increases or decreases in the variables concerned.

Before proceeding further we need to examine the idea of a limit of a function since this is required for our discussion of the process of differentiation.

Limits

Consider the function given by $y = f(x) = \dfrac{2}{x}$. The graph of this function is shown in fig. 21. It can be clearly seen that

(a) as $x \to \infty$, $f(x) \to 0$
(b) as $x \to -\infty$, $f(x) \to 0$

but $f(x) \neq 0$ for any value of x and we say that $f(x)$ tends to a limit of zero.
Statements (a) and (b) above are written

$$\lim_{x \to \infty} f(x) = 0 \quad \text{and} \quad \lim_{x \to -\infty} f(x) = 0$$

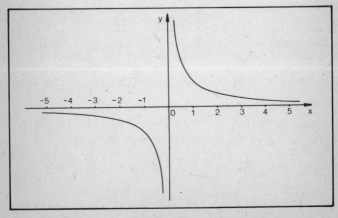

Figure 21

Note that as $x \to \pm\infty$, $f(x) \to 0$ but in one case the values of $f(x)$ are just positive and in the other they are just negative. It is usual to say that $f(x)$ approaches zero from above or below as the case may be.

The curve is discontinuous at $x = 0$ and we say that at $x = 0$ there is a point of discontinuity.

Example 1 Sketch the graph of the function $y = f(x) = (4x - 1)/(2x + 1)$. Using the given equation we see that when $x = 0$, $y = -1$ and when $x = \frac{1}{4}$, $y = 0$. When $x \to -\frac{1}{2}$, $y \to \infty$.

By dividing denominator into numerator by algebraic long division the equation can be written as $y = 2 - \dfrac{3}{2x + 1}$. In this form it becomes clear that as $x \to \infty$, $y \to 2$ from below and as $x \to -\infty$, $y \to 2$ from above. This means that the line $y = 2$ forms an **asymptote** to the curve.

The curve is sketched in fig. 22 and we can see that

$$\lim_{x \to \infty} f(x) = 2 \quad \text{and} \quad \lim_{x \to -\infty} f(x) = 2$$

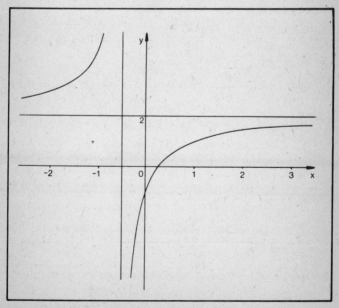

Figure 22

This example makes it clear that it is not always necessary to sketch the function to determine any limits. If it can be expressed in a convenient form the limit is easily recognizable.

Thus in the form $y = 2 - \dfrac{3}{2x+1}$ it is clear that as x gets larger the value of $\dfrac{3}{2x+1}$ decreases and hence the value of y becomes nearer to 2. Hence $\lim\limits_{x \to \infty} \left[2 - \dfrac{3}{2x+1} \right] = 2$.

Example 2 Evaluate the limits of the functions

$$\text{(i)} \ \lim_{x \to 3} \left(\frac{x-1}{x+3} \right), \quad \text{(ii)} \ \lim_{x \to -2} \left(\frac{x^3+8}{x+2} \right).$$

(i) Consider the function $\dfrac{x-1}{x+3}$ near to $x = 3$.

Let $x = 3 + h$ where h is small.

Clearly the value of the function when $x = 3 + h$ is

$$\frac{(3+h)-1}{(3+h)+3}$$

i.e. $\dfrac{2+h}{6+h}$ and as h becomes very small this approaches $2/6 = 1/3$.

Hence $\lim\limits_{h \to 0} \left(\dfrac{2+h}{6+h} \right) = \dfrac{1}{3}$ but as $h \to 0$, $x \to 3$ and we can write

$$\lim_{x \to 3} \left(\frac{x-1}{x+3} \right) = \frac{1}{3}$$

(ii) Now $\dfrac{x^3+8}{x+2} = \dfrac{(x+2)(x^2-2x+4)}{(x+2)} = x^2 - 2x + 4$ provided that $x + 2 \neq 0$.

Hence for all values of x except -2, $\dfrac{x^3+8}{x+2} = x^2 - 2x + 4$

$\therefore$ As $x \to -2$ the value of the function approaches 12.

$\therefore \ \lim\limits_{x \to -2} \left(\dfrac{x^3+8}{x+2} \right) = 12.$

In example 2(i) we could have obtained the result by direct substitution of $x = 3$, i.e. $\lim\limits_{x \to 3} \left(\dfrac{x-1}{x+3} \right) = \dfrac{3-1}{3+3} = \dfrac{2}{6} = \dfrac{1}{3}$ but in example 2(ii) this would not have given us a definite answer, for $\lim\limits_{x \to -2} \left(\dfrac{x^3 + 8}{x + 2} \right) = \dfrac{0}{0}$ which is indeterminate. When this situation arises another form has to be found as shown in the example.

We can illustrate the idea that, as x approaches a given value, a function $f(x)$ approaches a fixed limit by an arithmetical process.

Consider the function $f(x) = \dfrac{x^2 - 4}{x - 2}$ as $x \to 2$.

If we substitute $x = 2$ we find $f(2) = \dfrac{0}{0}$ which is indeterminate, but if we evaluate values of $f(x)$ when x approaches 2 we find $f(x)$ takes a sequence of values which approach a definite limit.

If $x = 2 \cdot 1$, $\quad f(x) = \dfrac{4 \cdot 41 - 4}{2 \cdot 1 - 2} \qquad = \dfrac{0 \cdot 41}{0 \cdot 1} \qquad = 4 \cdot 1$

If $x = 2 \cdot 01$, $f(x) = \dfrac{4 \cdot 0401 - 4}{2 \cdot 01 - 2} \qquad = \dfrac{0 \cdot 0401}{0 \cdot 01} = 4 \cdot 01$

If $x = 2 \cdot 001$, $\quad f(x) = \dfrac{4 \cdot 004001 - 4}{2 \cdot 001 - 2} = \dfrac{0 \cdot 004001}{0 \cdot 001} = 4 \cdot 001$

and hence as $x \to 2$ the value of $f(x) \to 4$ but will never actually equal 4.

We write $\lim\limits_{x \to 2} \left(\dfrac{x^2 - 4}{x - 2} \right) = 4$

We usually effect the above process by adapting to an algebraic method.

Hence if $f(x) = \dfrac{x^2 - 4}{x - 2}$ consider the value of $f(2 + h)$

We have $f(2 + h) = \dfrac{(2 + h)^2 - 4}{(2 + h) - 2} = \dfrac{h^2 + 4h}{h} = 4 + h$

Now as h tends to 0 the value of x tends to 2 and hence $f(2) = 4$.

To show that $\lim\limits_{\theta \to 0} \left(\dfrac{\sin \theta}{\theta} \right) = 1$

Sometimes we are required to find limits that involve the trigonometrical functions. One important result is $\sin \theta \simeq \theta$ when θ is small. Remember that θ must be measured in radians.

Consider fig. 23 in which two radii OA and OC enclose a sector of a circle centre O. Draw a tangent at A and produce OC to meet the tangent at B. Join AC.

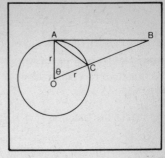

Let the radius of the circle be r and angle $AOC = \theta$ (radians). In $\triangle OAB$, $AB = r\tan\theta$ $\therefore$ area $\triangle OAB = \frac{1}{2}r^2\tan\theta$ and area sector $OAC = \frac{1}{2}r^2\theta$ and area $\triangle OAC = \frac{1}{2}r^2\sin\theta$.

Figure 23

Clearly area $\triangle OAB >$ area sector $OAC >$ area $\triangle OAC$
i.e. $\frac{1}{2}r^2\tan\theta > \frac{1}{2}r^2\theta > \frac{1}{2}r^2\sin\theta$.
Since $r^2 > 0$ we have $\tan\theta > \theta > \sin\theta$.
Dividing by $\sin\theta$, which is positive since θ is acute

$$\frac{1}{\cos\theta} > \frac{\theta}{\sin\theta} > 1$$

Hence as $\theta \to 0$ $\quad \dfrac{1}{\cos\theta} \to 1$

$\therefore$ $\dfrac{\theta}{\sin\theta} \to 1$ or $\theta \simeq \sin\theta$ and we write $\displaystyle\lim_{\theta \to 0}\left(\frac{\sin\theta}{\theta}\right) = 1$

This also gives $\cos\theta \to 1$ as $\theta \to 0$ but by using the identity $\cos\theta = 1 - 2\sin^2(\theta/2)$ we can obtain a better approximation, namely

$$\cos\theta \simeq 1 - 2\left(\frac{\theta}{2}\right)^2 = 1 - \tfrac{1}{2}\theta^2$$

Example 3 Find the $\displaystyle\lim_{\theta \to 0}\left[\frac{3\theta\sin\theta}{1 - \cos 2\theta}\right]$.

Now $\displaystyle\lim_{\theta \to 0}\left[\frac{3\theta\sin\theta}{1 - \cos 2\theta}\right] = \lim_{\theta \to 0}\left[\frac{3\theta\sin\theta}{2\sin^2(\theta/2)}\right]$

$\qquad = \displaystyle\lim_{\theta \to 0}\left[\frac{6\theta\sin(\theta/2)\cos(\theta/2)}{2\sin^2(\theta/2)}\right] = \lim_{\theta \to 0}\left[\frac{3\theta\cos(\theta/2)}{\sin(\theta/2)}\right]$

$\qquad = 6$ since $\cos(\theta/2) \to 1$ and $\dfrac{(\theta/2)}{\sin(\theta/2)} \to 1$

In this example we have assumed in the final stages that the limit of a product is the same as the product of the separate limits. There are a number of properties of limits which should be remembered. These are given below

(i) $\lim_{x \to a} [f(x) + g(x)] = \lim_{x \to a} f(x) + \lim_{x \to a} g(x)$

(ii) $\lim_{x \to a} [kf(x)] = k \lim_{x \to a} f(x)$ where k is a constant

(iii) $\lim_{x \to a} [f(x)g(x)] = [\lim_{x \to a} f(x)][\lim_{x \to a} g(x)]$

The derived function

The gradient of a linear function is constant but when the graph is a curve the **gradient is different at different points** on the curve. A general function which represents the gradient of the curve is called the **derived function** and the process of finding this function is called **differentiation.**

First consider a geometrical approach to a limit. If we take a point P on a curve and a second point Q at a different point then the gradient of the chord PQ can be found as this is a straight line (see page 138). We now reduce the increment from P to Q so that Q moves to a new position nearer P, i.e. Q'. The gradient of PQ' can now be found. Repeating this process we can find the gradient of PQ'' and so on until Q becomes a point adjacent to P. In this case the gradient of the chord between P and a neighbouring point Q is very close to the gradient of the tangent to the curve drawn at P (fig. 24).

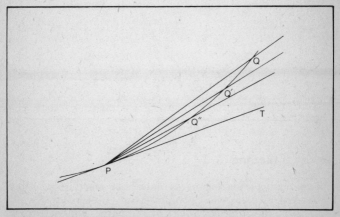

Figure 24

We say that the **gradient of a curve at a given point is the gradient of the tangent to the curve at that point.** It can be determined by finding the limit of the gradient of a chord PQ as $Q \to P$.

Example 4 Find the gradient of the tangent to the curve $y = 3x^2$ at the point $P(1, 3)$.

Consider fig. 25. Let Q be a point on the curve near to P whose x coordinate is given by $x = 1 + h$. Hence the y coordinate of Q will be $3(1 + h)^2$.

By considering $\triangle QPN$ the gradient of $PQ = \dfrac{3(1 + h)^2 - 3}{(1 + h) - 1}$

$$= \frac{6h + 3h^2}{h} = 6 + 3h$$

Thus as $h \to 0$, i.e. the point Q approaches the point P on the curve the gradient of $PQ \to 6$.

$\therefore$ The gradient of the tangent at $(1, 3)$ is 6.

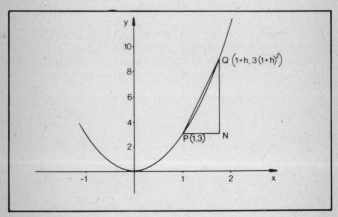

Figure 25

General Method

We can apply the above process a little more generally by letting P be a general point whose coordinates are (x, y). In this case we use a special notation instead of h by writing the coordinates of Q as $(x + \delta x, y + \delta y)$.

Note δx simply means an increment in the variable x and δy an increment in the variable y. It does not mean the product of δ and x.

Consider the curve $y = x^3$, where P is a fixed point (x, y) and Q is a neighbouring point whose coordinates are given by $(x + \delta x, y + \delta y)$.

Since Q lies on the curve, $y + \delta y = (x + \delta x)^3$.
Since P lies on the curve, $y = x^3$.
Subtracting, $\delta y = (x + \delta x)^3 - x^3$
dividing by δx we have

the gradient of the chord $= \dfrac{\delta y}{\delta x} = \dfrac{(x + \delta x)^3 - x^3}{\delta x}$

$$= \frac{x^3 + 3x^2(\delta x) + 3x(\delta x)^2 + (\delta x)^3 - x^3}{\delta x}$$

$$\therefore \quad \frac{\delta y}{\delta x} = \frac{3x^2(\delta x) + 3x(\delta x)^2 + (\delta x)^3}{\delta x}$$

$$\therefore \quad \frac{\delta y}{\delta x} = 3x^2 + 3x(\delta x) + (\delta x)^2$$

Now the gradient of the tangent is the limit as $\delta x \to 0$ and is denoted by $\dfrac{dy}{dx}$.

$$\therefore \quad \frac{dy}{dx} = \lim_{\delta x \to 0} \left(\frac{\delta y}{\delta x} \right) = \lim_{\delta x \to 0} [3x^2 + 3x(\delta x) + (\delta x)^2]$$

Hence the derived function is given by $\dfrac{dy}{dx} = 3x^2$.

Note that when $x = 2$ the gradient of the tangent is $3(2)^2 = 12$.
        when $x = 0$ the gradient of the tangent $= 0$
and is thus parallel to the x axis.
The above process is known as **differentiation from first principles.**

Example 5 Differentiate from first principles the function $y = ax^n$ where a and n are constants.

Consider a given point $P(x, y)$ of the curve and a neighbouring point $Q(x + \delta x, y + \delta y)$.
Since Q lies on the curve, $y + \delta y = a(x + \delta x)^n$.
Since P lies on the curve, $y = ax^n$.
Subtracting $\delta y = a(x + \delta x)^n - ax^n$.
Expanding $(x + \delta x)^n$ by the binomial theorem we have

$$\delta y = a\left[x^n + nx^{n-1}(\delta x) + \frac{n(n-1)}{2!}x^{n-2}(\delta x)^2 + \cdots + (\delta x)^n\right] - ax^n$$

$$\therefore \quad \delta y = nax^{n-1}(\delta x) + \frac{an(n-1)}{2!}x^{n-2}(\delta x)^2 + \cdots + a(\delta x)^n$$

Hence dividing by δx, the gradient of the chord PQ which is given by $\dfrac{\delta y}{\delta x} = nax^{n-1} + \dfrac{n(n-1)a}{2!}x^{n-2}(\delta x) + \cdots + a(\delta x)^{n-1}$

Since every term of the RHS except the first contains δx or a power of δx they will tend to zero as $\delta x \to 0$.

$$\therefore \quad \text{The gradient of the curve } \frac{dy}{dx} = \lim_{\delta x \to 0}\left(\frac{\delta y}{\delta x}\right) = nax^{n-1}.$$

This gives a general rule for differentiation of a function of the form ax^n.

If $y = ax^n$ $\qquad\qquad \dfrac{dy}{dx} = nax^{n-1}$.

We can apply this rule to functions of this form to obtain their gradients without working through from first principles. It applies even if n is negative or fractional.

Example 6 Find the gradients of the following functions at the points stated (i) $y = 6x^3$ at $(2, 48)$; (ii) $y = \dfrac{1}{x}$ at $(1, 1)$;

(iii) $y = \sqrt{x}$ at $(4, 2)$.

(i) Since $y = 6x^3$, $\dfrac{dy}{dx} = 18x^2 \Rightarrow$ gradient at $(2, 48) = 72$.

(ii) Since $y = \dfrac{1}{x} = x^{-1}$, $\dfrac{dy}{dx} = -1x^{-2} = \dfrac{-1}{x^2} \Rightarrow$ gradient at $(1, 1) = -1$.

(iii) Since $y = \sqrt{x} = x^{1/2}$, $\dfrac{dy}{dx} = \dfrac{1x^{-1/2}}{2} = \dfrac{1}{2\sqrt{x}} \Rightarrow$ gradient at $(4, 2) = \dfrac{1}{4}$.

By considering an alternative approach we can derive the gradient function in a different form though it ultimately produces the same solution.

Consider a mapping diagram for $x \to 2x$ for the interval $\{x : 0 \leqslant x \leqslant 4\}$ as shown in fig. 26a.

It is clear that in fig. 26a we are considering a doubling

function and the mapping is a uniform stretch of scale factor 2. Whereas in fig. 26b, which shows the mapping $x \rightarrow 3x^2$ over the same interval, it is clear that no constant stretching is taking place.

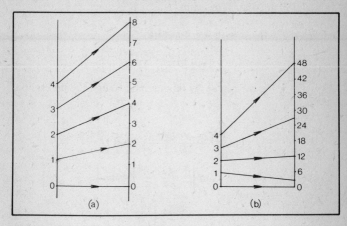

(a)   (b)

Figure 26

In this case we consider an average scale factor. For a function $f(x)$ over an interval $a \leqslant x \leqslant b$ the average scale factor is given by $\dfrac{f(b) - f(a)}{b - a}$

For $f(x) = 3x^2$ this factor would give $\dfrac{f(4) - f(0)}{4 - 0}$

over an interval $0 < x < 4$.

$$= \frac{48 - 0}{4 - 0} = 12$$

Now we can find the stretch experienced by a particular point by introducing a limiting process.

Consider again $x \rightarrow 3x^2$ as we approach $x = 1$ from above, i.e. the interval will be $1 \leqslant x \leqslant 1 + h$.

$\therefore$ Average scale factor is $\dfrac{3(1 + h)^2 - 3(1)^2}{(1 + h) - 1} = \dfrac{6h + 3h^2}{h} = 6 + 3h$

Now as $h \rightarrow 0$ the factor $\rightarrow 6$. Compare our method and result on page 44 and clearly we have formed the derivative of the function $3x^2$ at $(1, 3)$.

The derivative of $f(x)$ is written as $f'(x)$ and the derivative at a given point $x = a$ is given as $f'(a)$.

Hence over an interval $a \leqslant x \leqslant b$ the derivative of $f(x)$ at $x = a$ is defined as $f'(a) = \lim\limits_{b \to a} \left[\dfrac{f(b) - f(a)}{b - a} \right]$

or, by replacing b with $a + h$ we have

$$f'(a) = \lim_{h \to 0} \left[\frac{f(a+h) - f(a)}{h} \right]$$

For $f(x) = 1/x$ we can apply this method to find the differential at $(2, \frac{1}{2})$.

$$f'(2) = \lim_{h \to 0} \left[\frac{f(2+h) - f(2)}{h} \right] = \lim_{h \to 0} \left[\frac{\dfrac{1}{2+h} - \dfrac{1}{2}}{h} \right]$$

$$= \lim_{h \to 0} \left[\frac{2 - (2+h)}{2(2+h)h} \right] = \lim_{h \to 0} \left[\frac{-h}{2(2+h)h} \right]$$

$$= \lim_{h \to 0} \left[\frac{-1}{2(2+h)} \right] = \frac{-1}{4}$$

Polynomials

For a polynomial in x, say $y = 3x^2 + 7x - 6$ we can differentiate term by term using the general rule. Let y be a function which is the sum of two terms u and v, both of which are functions of x.

Then $y = u + v$.   (1)

Now an increment of δx in x would cause increments of δu, δv and δy in u, v and y respectively.

Hence $y + \delta y = (u + \delta u) + (v + \delta v)$.   (2)

Subtracting (1) from (2) we have

$$\delta y = \delta u + \delta v$$

Dividing by δx $\dfrac{\delta y}{\delta x} = \dfrac{\delta u}{\delta x} + \dfrac{\delta v}{\delta x}$

As $\delta x \to 0$, δu, δv and δy all approach 0.

$$\therefore \frac{dy}{dx} = \lim_{\delta x \to 0} \left(\frac{\delta y}{\delta x} \right) = \lim_{\delta x \to 0} \left[\frac{\delta u}{\delta x} + \frac{\delta v}{\delta x} \right]$$

$$\Rightarrow \frac{dy}{dx} = \frac{du}{dx} + \frac{dv}{dx}$$

This can be extended to any number of terms in the sum.

Hence for $y = 3x^2 + 7x - 6$, we have $\dfrac{dy}{dx} = 6x + 7$.

Products

Consider the product $y = uv$ where u and v are both functions of x. An increment of δx in x will cause increments of δu, δv and δy in u, v and y respectively.

Hence $y + \delta y = (u + \delta u)(v + \delta v)$

$\therefore \qquad y + \delta y = uv + u(\delta v) + v(\delta u) + (\delta u)(\delta v)$

But $y = uv$

$\therefore$ Subtracting, $\quad \delta y = u(\delta v) + v(\delta u) + (\delta u)(\delta v)$

Dividing by δx we have

$$\frac{\delta y}{\delta x} = u \frac{\delta v}{\delta x} + v \frac{\delta u}{\delta x} + \frac{\delta u}{\delta x}(\delta v)$$

In the limit as $\delta x \to 0$, δu, δv, δy all $\to 0$.

$$\therefore \quad \frac{dy}{dx} = u \frac{dv}{dx} + v \frac{du}{dx}$$

Example 7 Differentiate $y = (x^2 - 3x + 7)(2x^3 - 5x)$.

Now $u = x^2 - 3x + 7 \quad \therefore \quad \dfrac{du}{dx} = 2x - 3$

$v = 2x^3 - 5x \qquad \dfrac{dv}{dx} = 6x^2 - 5$

Using the result for a product

$$\frac{dy}{dx} = u \frac{dv}{dx} + v \frac{du}{dx}$$

$$= (x^2 - 3x + 7)(6x^2 - 5) + (2x^3 - 5x)(2x - 3)$$

Quotients

Consider the quotient $y = u/v$ where u and v are both functions of x. An increment of δx in x causes increments of δu, δv, δy in u, v and y respectively.

Hence $y + \delta y = \dfrac{u + \delta u}{v + \delta v}$

But since $y = \dfrac{u}{v}$, subtracting gives $\quad \delta y = \dfrac{u + \delta u}{v + \delta v} - \dfrac{u}{v}$

$$\therefore \quad \delta y = \frac{(u + \delta u)v - u(v + \delta v)}{v(v + \delta v)} = \frac{v\delta u - u\delta v}{v(v + \delta v)}$$

Dividing by δx we have, $\qquad \dfrac{\delta y}{\delta x} = \dfrac{v\dfrac{\delta u}{\delta x} - u\dfrac{\delta v}{\delta x}}{v(v + \delta v)}$

In the limit as $\delta x \to 0$, δu, δv, $\delta y \to 0$

$$\therefore \quad \frac{dy}{dx} = \frac{v\dfrac{du}{dx} - u\dfrac{dv}{dx}}{v^2}$$

Example 8 Differentiate $y = \dfrac{x}{x^2 + 2}$.

Now $u = x \qquad \dfrac{du}{dx} = 1$

$v = x^2 + 2 \quad \dfrac{dv}{dx} = 2x$

Using the quotient result $\qquad \dfrac{dy}{dx} = \dfrac{(x^2 + 2)1 - x(2x)}{(x^2 + 2)^2}$

$$\frac{dy}{dx} = \frac{2 - x^2}{(x^2 + 2)^2}$$

Function of a function

Consider $(3x^2 - 7)^4$. This is a function (i.e. the fourth power) of $3x^2 - 7$ which is itself a function of x. Hence $(3x^2 - 7)^4$ is called a function of a function of x.

If we let $y = (3x^2 - 7)^4$ and $u = 3x^2 - 7$ then $y = u^4$. Consider an increment of δx in x which will cause increments of δu and δy in u and y respectively.
Since δu, δy, δx are finite quantities (although small)

$$\frac{\delta y}{\delta x} = \frac{\delta y}{\delta u} \times \frac{\delta u}{\delta x}$$

Clearly in the limit as $\delta x \to 0$, δu, $\delta y \to 0$

$$\frac{dy}{dx} = \frac{dy}{du} \times \frac{du}{dx}$$

Hence in the above example $y = u^4 \quad \therefore \quad \dfrac{dy}{du} = 4u^3$

$$u = 3x^2 - 7 \quad \frac{du}{dx} = 6x$$

∴ Using the function of a function result

$$\frac{dy}{dx} = 4u^3 \times 6x = 24x(3x^2 - 7)^3$$

Example 9 Differentiate $\sqrt{\dfrac{1+x}{2+x}}$ with respect to x.

This problem can be solved in two ways, (i) by using the quotient rule and (ii) by using the product rule.

(i) Let $y = \dfrac{(1+x)^{1/2}}{(2+x)^{1/2}}$

In which case

$$\left. \begin{array}{ll} u = (1+x)^{1/2} & \therefore \quad \dfrac{du}{dx} = \tfrac{1}{2}(1+x)^{-1/2} \cdot 1 \\[2mm] v = (2+x)^{1/2} & \dfrac{dv}{dx} = \tfrac{1}{2}(2+x)^{-1/2} \cdot 1 \end{array} \right\} \quad \begin{array}{l} \text{using function} \\ \text{of a function} \end{array}$$

Using the quotient rule

$$\frac{dy}{dx} = \frac{(2+x)^{1/2} \cdot \tfrac{1}{2}(1+x)^{-1/2} - (1+x)^{1/2} \cdot \tfrac{1}{2}(2+x)^{-1/2}}{(2+x)}$$

Multiplying both numerator and denominator by $(1+x)^{1/2}(2+x)^{1/2}$

$$\frac{dy}{dx} = \frac{\tfrac{1}{2}(2+x) - \tfrac{1}{2}(1+x)}{(2+x)^{3/2}(1+x)^{1/2}} = \frac{1}{2(2+x)^{3/2}(1+x)^{1/2}}$$

or

$$\frac{dy}{dx} = \frac{1}{2\sqrt{(1+x)(2+x)^3}}$$

(ii) By writing the function in the form $y = (1+x)^{1/2}(2+x)^{-1/2}$ it is possible to use the product formula.

Now $u = (1+x)^{1/2}, \quad \dfrac{du}{dx} = \tfrac{1}{2}(1+x)^{-1/2} \times 1 = \tfrac{1}{2}(1+x)^{-1/2}$

$v = (2+x)^{-1/2}, \quad \dfrac{dv}{dx} = -\tfrac{1}{2}(2+x)^{-3/2} \times 1 = -\tfrac{1}{2}(2+x)^{-3/2}$

By the product rule

$$\therefore \quad \frac{dy}{dx} = (1+x)^{1/2}[-\tfrac{1}{2}(2+x)^{-3/2}] + (2+x)^{-1/2}[\tfrac{1}{2}(1+x)^{-1/2}]$$

$$= -\tfrac{1}{2}(1+x)^{1/2}(2+x)^{-3/2} + \tfrac{1}{2}(1+x)^{-1/2}(2+x)^{-1/2}$$

Extracting a factor of $(1+x)^{-1/2}(2+x)^{-3/2}$ we obtain

$$\frac{dy}{dx} = \frac{-(1+x)+(2+x)}{2(1+x)^{1/2}(2+x)^{3/2}} = \frac{1}{2\sqrt{(1+x)(2+x)^3}}$$

Implicit functions

An equation such as $3x^2 - xy + 3y = 7$ is such that y is not defined directly in terms of x. Such an equation gives y as an **implicit function** of x.

In this case the equation can be rearranged to give y as an explicit function of x, i.e.

$$y(3-x) = 7 - 3x^2 \quad \therefore \quad y = \frac{7-3x^2}{3-x}$$

However, some equations cannot be dealt with in this way for example, $x^2 + xy - y^2 = 8$.

In such examples we differentiate with respect to x, term by term.

i.e. $\dfrac{d}{dx}(x^2) + \dfrac{d}{dx}(xy) - \dfrac{d}{dx}(y^2) = \dfrac{d}{dx}(8)$

$\therefore \quad 2x + \left(x\dfrac{dy}{dx} + y\right) - 2y\dfrac{dy}{dx} = 0$

Remember that (i) $\dfrac{d}{dx}(xy) = x\dfrac{dy}{dx} + y$ since xy is a product.

$\qquad$ (ii) $\dfrac{d}{dx}(y^2) = 2y\dfrac{dy}{dx}$

since y^2 is a function of a function.

Hence $\qquad (x-2y)\dfrac{dy}{dx} = -y - 2x \Rightarrow \dfrac{dy}{dx} = -\dfrac{2x+y}{x-2y}$

If y is a function of x, an increment of δx in x will cause an increment of δy in y.

Since δy and δx are finite, small quantities

$$\frac{\delta y}{\delta x} = 1 \Big/ \left(\frac{\delta x}{\delta y}\right)$$

As $\delta x \to 0$, $\delta y \to 0$, so taking the limit

$$\frac{dy}{dx} = 1 \Big/ \left(\frac{dx}{dy}\right)$$

Example 10 Find the gradient of the parabola given by the equation $y^2 = 4ax$. We can tackle this in a number of ways producing an answer in different forms.

(i) Since $y^2 = 4ax$, $x = y^2/4a$.

Differentiating with respect to y,　　$\dfrac{dx}{dy} = \dfrac{2y}{4a} = \dfrac{y}{2a}$

But $\dfrac{dy}{dx} = 1 \Big/ \left(\dfrac{dx}{dy}\right) = \dfrac{2a}{y}$

(ii) Since $y^2 = 4ax$, $y = 2a^{1/2}x^{1/2}$

Differentiating with respect to x,　　$\dfrac{dy}{dx} = \dfrac{2a^{1/2}x^{-1/2}}{2} = \sqrt{\dfrac{a}{x}}$

(iii) By differentiating implicity with respect to x

$$\frac{d}{dx}(y^2) = \frac{d}{dx}(4ax)$$

$$2y\frac{dy}{dx} = 4a \qquad \therefore \quad \frac{dy}{dx} = \frac{2a}{y} = \frac{2a}{2a^{1/2}x^{1/2}} = \sqrt{\frac{a}{x}}$$

Parametric form

Sometimes, instead of writing an implicit equation relating two variables x and y, each is given in terms of a third variable, called a **parameter**. For example, in the equation $y^2 = 4ax$ above we could write two equations $x = at^2$, $y = 2at$ which give the variable x and y in terms of t, the parameter. Eliminating the parameter, if possible, will give the usual equation.

i.e. $x = a\left(\dfrac{y}{2a}\right)^2 \quad \therefore \quad x = \dfrac{y^2}{4a} \Rightarrow y^2 = 4ax$

To find the gradient of curves given in parametric form we use the function of a function rule.

Example 11 Find the gradient of the parabola whose equations are given parametrically as $x = at^2$, $y = 2at$.

Now $\dfrac{dx}{dt} = 2at$ and $\dfrac{dy}{dt} = 2a$. $\quad \therefore \quad \dfrac{dy}{dx} = \dfrac{dy}{dt} \times \dfrac{dt}{dx} = \dfrac{2a}{2at} = \dfrac{1}{t}$.

This gives yet another form for the gradient of the parabola $y^2 = 4ax$, this time in terms of a parameter t.

Trigonometrical functions

Consider the function $y = \sin x$. We can find the gradient from first principles by applying the process for algebraic functions.

Let $P(x, y)$ and $Q(x + \delta x, y + \delta y)$ be two neighbouring points on the curve where δx is an increment in x, causing an increment δy in y. Therefore $y + \delta y = \sin(x + \delta x)$

and $y = \sin x$

Subtracting $\delta y = \sin(x + \delta x) - \sin x$

$$= 2\cos(x + \tfrac{1}{2}\delta x)\sin(\tfrac{1}{2}\delta x) \text{ by factor formula.}$$

Hence, dividing by δx, the gradient of the chord PQ is given by

$$\frac{\delta y}{\delta x} = \frac{2\cos(x + \tfrac{1}{2}\delta x)\sin(\tfrac{1}{2}\delta x)}{\delta x} = \cos(x + \tfrac{1}{2}\delta x) \times \frac{\sin\tfrac{1}{2}\delta x}{\tfrac{1}{2}\delta x}$$

In the limit as $\delta x \to 0$, $\delta y \to 0$

$$\frac{dy}{dx} = \lim_{\delta x \to 0}\left(\frac{\delta y}{\delta x}\right) = \lim_{\delta x \to 0}\left[\cos(x + \tfrac{1}{2}\delta x) \times \frac{\sin\tfrac{1}{2}\delta x}{\tfrac{1}{2}\delta x}\right]$$

$$= \left[\lim_{\delta x \to 0}\cos(x + \tfrac{1}{2}\delta x)\right] \times \left[\lim_{\delta x \to 0}\frac{\sin\tfrac{1}{2}\delta x}{\tfrac{1}{2}\delta x}\right]$$

$$= \cos x, \quad \text{since } \lim_{\theta \to 0}\left(\frac{\sin\theta}{\theta}\right) = 1$$

Hence $\dfrac{d}{dx}(\sin x) = \cos x$.

That this result is correct can be seen if we draw the graph of $y = \sin x$ and measure the gradient of the curve at different points and then plot this gradient against x. See fig. 27.

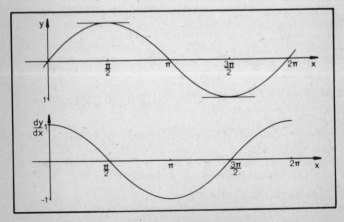

Figure 27

54

It can be seen that the gradient of $\sin x$ at $x = \pi/2$, $3\pi/2$ is zero and hence the second graph cuts the x axis at these points.

Also for $0 \leqslant x \leqslant \pi$ $\quad \dfrac{dy}{dx}$ decreases continuously from $+1$ to -1

for $\pi \leqslant x \leqslant 2\pi$ $\quad \dfrac{dy}{dx}$ increases continuously from -1 to $+1$

i.e. a pattern satisfied by the cosine curve.

Remember that in the derivation of the gradient above, x *must* be measured in radians in order that the limit of $\dfrac{\sin \theta}{\theta}$ as $\theta \to 0$ may be applied.

A similar method can be used to show that $\dfrac{d}{dx}(\cos x) = -\sin x$.

To find the differentials of **composite functions** we can use the methods discussed for products, quotients and the function of a function. The following examples illustrate the techniques.

Example 12 Find the differential coefficient of $\tan x$.

Let $y = \tan x = \dfrac{\sin x}{\cos x}$

Using the quotient rule $u = \sin x$, $\dfrac{du}{dx} = \cos x$

$$v = \cos x, \frac{dv}{dx} = -\sin x$$

$$\therefore \quad \frac{dy}{dx} = \frac{v\dfrac{du}{dx} - u\dfrac{dv}{dx}}{v^2} = \frac{\cos x(\cos x) - \sin x(-\sin x)}{\cos^2 x}$$

$$= \frac{\cos^2 x + \sin^2 x}{\cos^2 x} = \frac{1}{\cos^2 x} \Rightarrow \frac{d}{dx}(\tan x) = \sec^2 x.$$

Example 13 Differentiate with respect to x, $\sin 4x$.

Let $y = \sin 4x$ and $u = 4x$.

$\therefore \quad y = \sin u$

Since $\dfrac{dy}{du} = \cos u$ and $\dfrac{du}{dx} = 4$ we have

$$\frac{dy}{dx} = \frac{dy}{du} \cdot \frac{du}{dx} = 4 \cos u \Rightarrow \frac{dy}{dx} = 4 \cos 4x$$

Example 14 Differentiate $\sin^3 x$ with respect to x.

Let $y = \sin^3 x$.
If $u = \sin x$, $y = u^3$.

$$\therefore \quad \frac{du}{dx} = \cos x \qquad \frac{dy}{du} = 3u^2$$

Hence $\dfrac{dy}{dx} = \dfrac{dy}{du} \cdot \dfrac{du}{dx} = 3u^2 \cos x = 3 \sin^2 x \cos x$

$$\therefore \quad \frac{d}{dx}(\sin^3 x) = 3 \sin^2 x \cos x$$

Example 15 Differentiate with respect to x the following functions (i) $x^2 \sin x$, (ii) $x^2/\cos 2x$.

(i) Let $y = x^2 \sin x$. Since this is a product where $u = x^2$ and $v = \sin x$ we can use the product rule.

Now $\dfrac{du}{dx} = 2x$ and $\dfrac{dv}{dx} = \cos x$

$$\therefore \quad \frac{dy}{dx} = u \frac{dv}{dx} + v \frac{du}{dx} = x^2 \cos x + 2x \sin x$$

Hence $\dfrac{d}{dx}(x^2 \sin x) = x(x \cos x + 2 \sin x)$

(ii) Let $y = \dfrac{x^2}{\cos 2x}$

This is a quotient with $u = x^2$ and $v = \cos 2x$.
Now $\dfrac{du}{dx} = 2x$ and $\dfrac{dv}{dx} = -2 \sin 2x$.

$$\therefore \quad \frac{dy}{dx} = \frac{v \dfrac{du}{dx} - u \dfrac{dv}{dx}}{v^2} = \frac{\cos 2x(2x) - x^2(-2 \sin 2x)}{\cos^2 2x}$$

$$= \frac{2x \cos 2x + 2x^2 \sin 2x}{\cos^2 2x}$$

Hence $\dfrac{d}{dx}\left(\dfrac{x^2}{\cos 2x}\right) = \dfrac{2x(\cos 2x + x \sin 2x)}{\cos^2 2x}$

It is wise to remember a number of standard differentials so that they may be used in more complicated problems without difficulty. They are listed here for reference and the reader should check the derivation of each.

$$\frac{d}{dx}(\sin x) = \cos x \qquad \frac{d}{dx}(\cos x) = -\sin x$$

$$\frac{d}{dx}(\tan x) = \sec^2 x \qquad \frac{d}{dx}(\sec x) = \sec x \tan x$$

$$\frac{d}{dx}(\cot x) = -\operatorname{cosec}^2 x \qquad \frac{d}{dx}(\operatorname{cosec} x) = -\operatorname{cosec} x \cot x$$

Inverse circular functions

If $y = \sin^{-1} x$ then y is the angle whose sine is x. The function $\sin^{-1} x$ is called the inverse sine, and similar functions exist for $\cos^{-1} x$, $\tan^{-1} x$ and so on. These have been defined and discussed on page 35 and the graphs drawn. They can also be written as arc sin x, arc cos x, etc.

To differentiate $\sin^{-1} x$ we proceed as follows.
Let $y = \sin^{-1} x$, in which case $x = \sin y$.

Differentiating with respect to y we have $\dfrac{dx}{dy} = \cos y$

Now $\dfrac{dy}{dx} = \dfrac{1}{\dfrac{dx}{dy}} = \dfrac{1}{\cos y} = \dfrac{1}{\sqrt{1 - \sin^2 y}} = \dfrac{1}{\sqrt{1 - x^2}}$

Hence $\dfrac{d}{dx}(\sin^{-1} x) = \dfrac{1}{\sqrt{1 - x^2}}$

Note that the sign of the differential coefficient may be positive or negative for a given value of x. This is clear from the graph on page 36 as $\sin^{-1} x$ is a many valued function, although if we restrict the range by taking only the **principal values** the gradient will always be positive.

The gradient of $y = \cos^{-1} x$ can be derived in a similar way by writing $x = \cos y$ and differentiating with respect to y. This produces the result

$$\frac{d}{dx}(\cos^{-1} x) = \frac{-1}{\sqrt{1 - x^2}}$$

By reference to the graph of this function on page 36 it can be seen that if the principal values are chosen the gradient will always be negative.

The gradient of $y = \tan^{-1} x$ can also be found in this way and will give the result

$$\frac{d}{dx}(\tan^{-1} x) = \frac{1}{1+x^2}$$

Clearly the gradient of $\tan^{-1} x$ is always positive and the graph on page 36 will confirm this.

For $y = \mathbf{sec^{-1}\, x}$ we apply the same technique.

Since $y = \sec^{-1} x$, $x = \sec y$.

Differentiating with respect to y, $\dfrac{dx}{dy} = \sec y \tan y$

$$\therefore \quad \frac{dy}{dx} = \frac{1}{\sec y \tan y} = \frac{1}{\sec y \sqrt{\sec^2 y - 1}}$$

$$\therefore \quad \frac{dy}{dx} = \frac{1}{x\sqrt{x^2 - 1}} \quad \text{and we write} \quad \frac{d}{dx}(\sec^{-1} x) = \frac{1}{x\sqrt{x^2 - 1}}$$

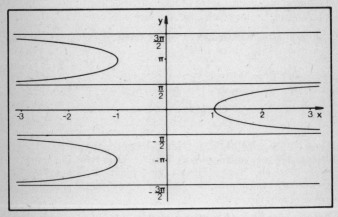

Figure 28

The graph of this function which is shown in fig. 28 shows a discontinuous curve with no part lying between $x = \pm 1$. If we restrict the range by taking the principal values, i.e. $0 \leqslant \sec^{-1} x \leqslant \pi$, then the gradient is always positive.

A list of the differential coefficients of the standard inverse circular functions is listed on the following page for reference. They should be learnt thoroughly so that they can be applied accurately.

$$\frac{d}{dx}(\sin^{-1} x) = \frac{1}{\sqrt{1-x^2}} \qquad \frac{d}{dx}\left(\sin^{-1} \frac{x}{a}\right) = \frac{1}{\sqrt{a^2-x^2}}$$

$$\frac{d}{dx}(\cos^{-1} x) = \frac{-1}{\sqrt{1-x^2}} \qquad \frac{d}{dx}\left(\cos^{-1} \frac{x}{a}\right) = \frac{-1}{\sqrt{a^2-x^2}}$$

$$\frac{d}{dx}(\tan^{-1} x) = \frac{1}{1+x^2} \qquad \frac{d}{dx}\left(\tan^{-1} \frac{x}{a}\right) = \frac{a}{a^2+x^2}$$

$$\frac{d}{dx}(\cot^{-1} x) = \frac{-1}{1+x^2} \qquad \frac{d}{dx}\left(\cot^{-1} \frac{x}{a}\right) = \frac{-a}{a^2+x^2}$$

$$\frac{d}{dx}(\sec^{-1} x) = \frac{1}{x\sqrt{x^2-1}} \qquad \frac{d}{dx}\left(\sec^{-1} \frac{x}{a}\right) = \frac{a}{x\sqrt{x^2-a^2}}$$

$$\frac{d}{dx}(\operatorname{cosec}^{-1} x) = \frac{-1}{x\sqrt{x^2-1}} \qquad \frac{d}{dx}\left(\operatorname{cosec}^{-1} \frac{x}{a}\right) = \frac{-a}{x\sqrt{x^2-a^2}}$$

Successive differentiation

Given y as a function of x we have seen how to form the gradient or differential coefficient. Unless y is a linear function the gradient function will also be a function of x and therefore we could differentiate this as well. We would form the gradient of the gradient function which would give the rate of change of the gradient.

This is called the second differential of y and is written

$$\frac{d}{dx}\left(\frac{dy}{dx}\right) \quad \text{or, more usually,} \quad \frac{d^2 y}{dx^2}$$

If we use the notation of $f(x)$ for the function of x the first differential is $f'(x)$ and the second $f''(x)$.

This notation can be extended to cover the third differential of the original function by writing $\frac{d^3 y}{dx^3}$ or $f'''(x)$.

Note, for higher derivatives, i.e. $\frac{d^6 y}{dx^6}$ or $\frac{d^n y}{dx^n}$ we would write $f^{(6)}(x)$ and $f^{(n)}(x)$ or sometimes y_6 and y_n respectively. The last of these forms does not usually cause any ambiguity (with series, for example) for the context is usually clear.

Example 16 Find $\frac{dy}{dx}, \frac{d^2 y}{dx^2}$ for the functions given by

(i) $y = x^3 + 3x^2 + 4x$, (ii) $y = x^2 \tan x$.

59

(i) Since $y = x^3 + 3x^2 + 4x$

$$\frac{dy}{dx} = 3x^2 + 6x + 4 \quad \text{and} \quad \frac{d^2y}{dx^2} = 6x + 6$$

(ii) Since $y = x^2 \tan x$

$$\frac{dy}{dx} = x^2 \sec^2 x + 2x \tan x \quad \text{(using product rule)}$$

$$\frac{d^2y}{dx^2} = x^2(2 \sec x \sec x \tan x) + 2x \sec^2 x + 2x \sec^2 x + 2 \tan x$$

$$= 2x^2 \sec^2 x \tan x + 4x \sec^2 x + 2 \tan x$$

Care must be taken if the given equations are expressed in terms of a parameter. If $x = f(t)$ and $y = g(t)$ then $\frac{dy}{dx}$ will be found in terms of t.

Hence $\dfrac{d^2y}{dx^2} = \dfrac{d}{dx}\left(\dfrac{dy}{dx}\right) = \dfrac{d}{dt}\left(\dfrac{dy}{dx}\right) \times \dfrac{dt}{dx}$.

Example 17 Find $\dfrac{dy}{dx}$ and $\dfrac{d^2y}{dx^2}$ if $x = (t^2 - 1)^2$ and $y = t^3$.

$$\therefore \quad \frac{dx}{dt} = 4t(t^2 - 1) \text{ and } \frac{dy}{dt} = 3t^2 \text{ giving } \frac{dy}{dx} = \frac{3t}{4(t^2 - 1)}$$

Hence $\dfrac{d^2y}{dx^2} = \dfrac{d}{dx}\left(\dfrac{dy}{dx}\right) = \dfrac{d}{dt}\left[\dfrac{3t}{4(t^2 - 1)}\right] \times \dfrac{dt}{dx}$

$$= \left[\frac{4(t^2 - 1)3 - 3t \cdot 8t}{16(t^2 - 1)^2}\right] \times \frac{1}{4t(t^2 - 1)}$$

$$\therefore \quad \frac{d^2y}{dx^2} = \frac{-12(1 + t^2)}{64t(t^2 - 1)^3} = \frac{-3(t^2 + 1)}{16t(t^2 - 1)^3}$$

It is sometimes possible to find a differential of high order by noticing a pattern in the successive differentials.

Example 18 Find the first two differential coefficients and the nth derivatives of the following functions.

1. $y = e^{ax}$

Clearly $\dfrac{dy}{dx} = ae^{ax}$ and $\dfrac{d^2y}{dx^2} = a^2e^{ax}$ (see page 120).

In this example only the power of a is changing and hence $\dfrac{d^ny}{dx^n} = a^n e^{ax}$.

2. $y = (3x + 2)^{-1}$

Denoting $\frac{dy}{dx} = y_1$ and in general $\frac{d^n y}{dx^n}$ by y_n we have

$$y_1 = -1(3)(3x + 2)^{-2}$$

$$\text{and } y_2 = (-1)(-2)(3)^2(3x + 2)^{-3}$$

Clearly $y_3 = (-1)(-2)(-3)(3)^3(3x + 2)^{-4}$

Hence we conclude that

$$y_n = (-1)(-2)(-3) \cdots (-n)(3)^n(3x + 2)^{-(n+1)}$$
$$= (-1)^n n! 3^n (3x + 2)^{-(n+1)}$$

3. $y = \sin x$

With this function we need to modify the differential at each stage to produce a sine function.

Now $y_1 = \cos x = \sin\left(x + \frac{\pi}{2}\right)$.

$$\therefore \quad y_2 = \cos\left(x + \frac{\pi}{2}\right) = \sin\left[\left(x + \frac{\pi}{2}\right) + \frac{\pi}{2}\right] = \sin\left(x + \frac{2\pi}{2}\right).$$

$$\therefore \quad y_3 = \cos\left(x + \frac{2\pi}{2}\right) = \sin\left(x + \frac{3\pi}{2}\right).$$

In general $y_n = \sin\left(x + \frac{n\pi}{2}\right)$.

We can extend this technique to cover a product.

4. $y = e^{3x} \cos 4x$

$\therefore \quad y_1 = -4e^{3x} \sin 4x + 3e^{3x} \cos 4x$

$\quad = e^{3x}(3 \cos 4x - 4 \sin 4x) = 5e^{3x}(\frac{3}{5} \cos 4x - \frac{4}{5} \sin 4x)$

But this can be written as $y_1 = 5e^{3x} \cos(4x + \alpha)$ where $\tan \alpha = \frac{4}{3}$.

$$\therefore \quad y_2 = 5[3e^{3x} \cos(4x + \alpha) - 4e^{3x} \sin(4x + \alpha)]$$
$$= 5^2 e^{3x}[\tfrac{3}{5} \cos(4x + \alpha) - \tfrac{4}{5} \sin(4x + \alpha)]$$
$$= 5^2 e^{3x} \cos(4x + 2\alpha).$$

Hence $y_n = 5^n e^{3x} \cos(4x + n\alpha)$ where $\tan \alpha = 4/3$.

The Theorem of Leibnitz

This is used to find the nth derivative of a product.

Let $y = uv$ where u and v are functions of x and let u_r and v_r denote $\frac{d^r u}{dx^r}$ and $\frac{d^r v}{dx^r}$ respectively.

Then if $y = uv$, $\quad y_1 = u_1v + uv_1$

$$y_2 = u_2v + 2u_1v_1 + uv_2$$
$$y_3 = u_3v + 3u_2v_1 + 3u_1v_2 + uv_3$$

We note that the coefficients of each term are beginning to correspond to those in the binomial expansion. This suggests that

$$y_n = u_nv + {}^nC_1u_{n-1}v_1 + {}^nC_2u_{n-2}v_2 + \cdots + {}^nC_ru_{n-r}v_r + \cdots + uv_n \quad (1)$$

It is usual to prove this by induction.
Assume the result is true for $n = k$

$$y_k = u_kv + {}^kC_1u_{k-1}v_1 + {}^kC_2u_{k-2}v_2 + \cdots + {}^kC_ru_{k-r}v_r + \cdots + uv_k$$

By differentiating with respect to x, we obtain

$$y_{k+1} = u_{k+1}v + (1 + {}^kC_1)u_kv_1 + ({}^kC_1 + {}^kC_2)u_{k-1}v_2 + \cdots$$
$$\cdots + ({}^kC_{r-1} + {}^kC_r)u_{k-r+1}v_r + \cdots + ({}^kC_{k-1} + 1)u_1v_k + uv_{k+1}$$

But ${}^kC_{r-1} + {}^kC_r = {}^{k+1}C_r$ see page 208.

$$\therefore \quad y_{k+1} = u_{k+1}v + {}^{k+1}C_1u_kv_1 + {}^{k+1}C_2u_{k-1}v_2 + \cdots$$
$$\cdots + {}^{k+1}C_ru_{k-r+1}v_r + \cdots + {}^{k+1}C_ku_1v_k + uv_{k+1}$$

Now this result is the form expected by substituting $n = k + 1$ in (1). So if the result is true for $n = k$ it is also true for $n = k + 1$.
But the theorem is clearly satisfied by $n = 1, 2, 3$ as shown by straightforward differentiation and therefore it is true for all positive integral values of n.

Example 19 Find the nth differential of x^3e^{2x} with respect to x.
Using Leibnitz's theorem, where $u = e^{2x}$ and $v = x^3$

$$u_1 = 2e^{2x}, \; u_2 = 4e^{2x}, \; u_n = 2^ne^{2x}$$
and $v_1 = 3x^2, \; v_2 = 6x, \; v_3 = 6, \; v_n = 0$ for $n > 3$

$$\therefore \quad \frac{d}{dx}(x^3e^{2x}) = 2^ne^{2x}x^3 + {}^nC_12^{n-1}e^{2x}3x^2$$
$$+ {}^nC_22^{n-2}e^{2x}.6x + {}^nC_32^{n-3}e^{2x}.6$$

$$\frac{d}{dx}(x^3e^{2x}) = e^{2x}\left(2^nx^3 + n2^{n-1}.3x^2 + \frac{n(n-1)}{2!}6x.2^{n-2}\right.$$
$$\left. + \frac{6n(n-1)(n-2)}{3!}2^{n-3}\right)$$

$$= e^{2x}2^{n-3}[8x^3 + 12nx^2 + 6n(n-1)x + n(n-1)(n-2)]$$

Example 20 If $y = \dfrac{\sin^{-1}x}{\sqrt{1-x^2}}$ show that $(1-x^2)\dfrac{dy}{dx} = xy + 1$.

Hence show that $(1-x^2)y_{n+1} - (2n+1)xy_n - n^2y_{n-1} = 0$ where $y_r = \dfrac{d^r y}{dx^r}$.

If $y = \dfrac{\sin^{-1} x}{\sqrt{1-x^2}}$ it follows that $\sqrt{1-x^2}\,y = \sin^{-1} x$.

Differentiating with respect to x

$$\sqrt{1-x^2}\,\frac{dy}{dx} - y\left(\frac{x}{\sqrt{1-x^2}}\right) = \frac{1}{\sqrt{1-x^2}}$$

Multiplying by $\sqrt{1-x^2}$ we have

$$(1-x^2)\frac{dy}{dx} = xy + 1 \qquad (1)$$

Applying Leibnitz's theorem, the nth derivative of the LHS will equal the nth derivative of the RHS.

$\therefore$ from equation (1) $(1-x^2)y_1 = xy + 1$

Applying Leibnitz's theorem, using $u = y_1$ and $v = 1 - x^2$ on the LHS and $u = y$ and $v = x$ on the RHS

$$(1-x^2)y_{n+1} + {}^nC_1 y_n(-2x) + {}^nC_2 y_{n-1}(-2) = xy_n + {}^nC_1 y_{n-1}(1)$$

$\therefore$ $(1-x^2)y_{n+1} - 2nxy_n - \dfrac{n(n-1)}{2!} \cdot 2y_{n-1} = xy_n + ny_{n-1}$

$\therefore$ $(1-x^2)y_{n+1} - (2n+1)xy_n - n^2y_{n-1} = 0$

Velocity and acceleration

By using the calculus we can extend the idea of travel graphs which are familiar in ordinary level mathematics. Consider a space–time graph which shows displacement against time.

Now the average velocity between two points P and Q is given by $\dfrac{\text{total displacement}}{\text{total time}}$ i.e. $\dfrac{RQ}{PR}$ which is the gradient of the chord PQ.

Now if we decrease the time interval, Q moves closer to P and we have a process similar to the one developed for differentiation from first principles, and the gradient of

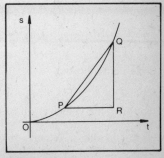

Figure 29

63

the chord PQ approaches the gradient of the curve at P. Hence the velocity at an instant is the limit of the average velocity over a small interval following that instant as the interval tends to zero.

Clearly, in the calculus notation, the velocity at a given instant is given by $\dfrac{ds}{dt}$.

An alternative approach is to use vectors. Let a particle move from P (position vector $\mathbf{r}_1$) to Q (position vector $\mathbf{r}_2$) in a time interval $t_2 - t_1$.
The average velocity over this interval is $\dfrac{\overrightarrow{PQ}}{t_2 - t_1}$.

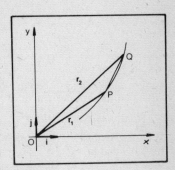

Figure 30

$\therefore$ Actual velocity at P is given by the limit as $(t_2 - t_1) \to 0$

i.e. velocity at P is $\lim\limits_{\delta t \to 0} \dfrac{\overrightarrow{PQ}}{\delta t}$

where $\delta t = t_2 - t_1$.

If $\mathbf{i}$ and $\mathbf{j}$ are unit vectors along Ox and Oy respectively, the position vectors $\mathbf{r}_1$ and $\mathbf{r}_2$ can be expressed in terms of $\mathbf{i}$ and $\mathbf{j}$. Let
$$\mathbf{r} = f(t)\mathbf{i} + g(t)\mathbf{j}$$

Hence if the particle is at P at time $t = a$ and at Q at time $t = b$ then the average velocity over this interval is given by

$$\frac{\mathbf{r}_2 - \mathbf{r}_1}{b - a} = \frac{f(b)\mathbf{i} + g(b)\mathbf{j} - f(a)\mathbf{i} - g(a)\mathbf{j}}{b - a}$$

$$= \frac{f(b) - f(a)}{b - a}\mathbf{i} + \frac{g(b) - g(a)}{b - a}\mathbf{j}$$

In the limit as $b \to a$.

Velocity at P is $f'(a)\mathbf{i} + g'(a)\mathbf{j}$

In general the velocity will be given by $\mathbf{v} = f'(t)\mathbf{i} + g'(t)\mathbf{j}$.

By considering a velocity–time graph we can deduce similar results for the average acceleration and acceleration at an instant.

64

The average acceleration over the interval PQ is

change in velocity
time interval

Actual acceleration at P is the limit of the average acceleration over a small interval after P as the interval tends to zero. In calculus notation this is $\dfrac{d\mathbf{v}}{dt}$ or $\dfrac{d^2\mathbf{r}}{dt^2}$.

In vector form the acceleration is given by $f''(t)\mathbf{i} + g''(t)\mathbf{j}$.

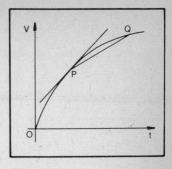

Figure 31

Example 21 If the position vector of a particle is given by $\mathbf{r} = t^2\mathbf{i} + 2t^3\mathbf{j}$ where $\mathbf{i}$ and $\mathbf{j}$ are perpendicular unit vectors find the velocity and acceleration of the particle after 3 seconds.

Since $\mathbf{r} = t^2\mathbf{i} + 2t^3\mathbf{j}$ the velocity is given by

$$\mathbf{v} = \frac{d\mathbf{r}}{dt} = 2t\mathbf{i} + 6t^2\mathbf{j}$$

$\therefore$ when $t = 3$, velocity $\mathbf{v} = 6\mathbf{i} + 54\mathbf{j}$.

The acceleration is given by $\mathbf{a} = \dfrac{d\mathbf{v}}{dt} = 2\mathbf{i} + 12t\mathbf{j}$

$\therefore$ when $t = 3$, acceleration $\mathbf{a} = 2\mathbf{i} + 36\mathbf{j}$.

Example 22 A particle travels in a straight line so that its distance, x metres, from a fixed point O after time t is given by $x = 3\cos t + 4\sin t$. Find (i) the distance from O when it first comes to rest instantaneously, (ii) its acceleration at this moment, (iii) its maximum velocity.

(i) Since $x = 3\cos t + 4\sin t$

Velocity is $\dfrac{dx}{dt} = -3\sin t + 4\cos t$ (1)

When the particle is at rest $\dfrac{dx}{dt} = 0$

i.e. $-3\sin t + 4\cos t = 0 \Rightarrow \tan t = \dfrac{4}{3}$

$\therefore$ the particle is first at rest when $t = \tan^{-1}\dfrac{4}{3} = 0 \cdot 927$ sec

$\therefore$ $\cos t = 3/5$ and $\sin t = 4/5$.

Hence distance of particle from O is $3(3/5) + 4(4/5) = 5$ m.

(ii) Acceleration is $\dfrac{dv}{dt} = -3\cos t - 4\sin t$

$\therefore$ when $\cos t = 3/5$ and $\sin t = 4/5$ the acceleration is given by $-3(3/5) - 4(4/5) = -5$ m s^{-2}.

(iii) The maximum velocity occurs when $\dfrac{dv}{dt} = 0$,

i.e. the acceleration is zero.

$\therefore$ $-3\cos t - 4\sin t = 0 \Rightarrow \tan t = -3/4$

Maximum velocity is found by substituting in equation (1). The maximum velocity is $-3(-3/5) + 4(+4/5) = +5$ m s^{-1}.

Maximum and minimum points

Points on a graph where the gradient is zero are called **stationary points**. They are either a maximum or a minimum point considered in a localized sense. Consider the curve shown in fig. 32 on which have been drawn the tangents at various points.

By considering the gradients of these tangents we observe that at each of the points A, B, C and D the gradient is zero since the tangent is parallel to the x axis. Also, we note that the gradient changes sign as x increases through each of the points.

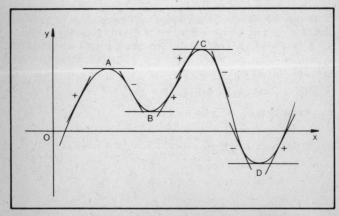

Figure 32

For A and C the gradient changes from positive to negative. For B and D the gradient changes from negative to positive.

Also for A and C, the curve has its greatest value in that region so that if points are chosen on the curve close to and on either side of A or C the value of the ordinate will be less than the ordinate of A or C.

Similarly B and D provide the lowest points of the curve in that region and any points chosen close to and on either side of B or D will have ordinates greater than the ordinates of B or D.

We say that A and C are **maximum points** of the curve and that B and D are **minimum points**.

Now at a **maximum point**, say A, in fig. 32 we can see that:

(i) The function is increasing in value before A and decreasing after A.

(ii) The gradient $\dfrac{dy}{dx}$ changes from positive to negative.

(iii) Thus from (ii) the gradient is decreasing and hence the rate of change of the gradient is negative, i.e. $\dfrac{d^2y}{dx^2} < 0$.

For a **minimum point**, say B, in fig. 32 we note that:

(i) The function is decreasing in value before B and increasing after B.

(ii) The gradient $\dfrac{dy}{dx}$ changes from negative to positive.

(iii) Hence the gradient is increasing and the rate of change of the gradient is positive, i.e. $\dfrac{d^2y}{dx^2} > 0$.

In application of the above we normally use either (ii) or (iii) to distinguish between maximum and minimum points. It is important to remember that Method (ii) can be useful if the second differential is difficult.

Example 23 Find the maximum and minimum values of the function $2x^3 - 9x^2 + 12x$.

Let $y = 2x^3 - 9x^2 + 12x$ $\quad \therefore \quad \dfrac{dy}{dx} = 6x^2 - 18x + 12$ $\qquad$ (1)

The curve will have a stationary point when $\dfrac{dy}{dx} = 0$

i.e.
$$6x^2 - 18x + 12 = 0$$
$$x^2 - 3x + 2 = 0$$
$$(x - 1)(x - 2) = 0$$
$$\therefore \quad x = 1 \text{ or } x = 2$$

When $x = 1$, $\quad y = 2(1)^3 - 9(1)^2 + 12(1) = 5$

When $x = 2$, $\quad y = 2(2)^3 - 9(2)^2 + 12(2) = 4$

The curve has stationary points at $(1, 5)$ and $(2, 4)$. To distinguish between them consider the second differential from

(1) $\dfrac{d^2y}{dx^2} = 12x - 18$

When $x = 1$, $\qquad \dfrac{d^2y}{dx^2} < 0 \qquad \therefore \quad$ a maximum point

When $x = 2$, $\qquad \dfrac{d^2y}{dx^2} > 0 \qquad \therefore \quad$ a minimum point

$\therefore \quad$ The curve has a maximum at $(1, 5)$ and a minimum at $(2, 4)$. If we use the method of considering the gradient on each side of $x = 1$ and $x = 2$ it is best to draw up a table. From equation

(1) $\dfrac{dy}{dx} = 6(x^2 - 3x + 2) = 6(x - 1)(x - 2)$

x:	$1 - h$	1	$1 + h$
$\dfrac{dy}{dx}$:	$+$ve	0	$-$ve

x:	$2 - h$	2	$2 + h$
$\dfrac{dy}{dx}$:	$-$ve	0	$+$ve

Hence at $x = 1$ the gradient changes from positive to negative and gives a maximum. At $x = 2$ the gradient changes from negative to positive and gives a minimum.

Points of inflexion

Consider the graph of $y = 2x^3 - 9x^2 + 12x$. We have already found a maximum point at $(1, 5)$ and a minimum at $(2, 4)$ so, knowing that it passes through the origin its sketch can be drawn. This is shown in fig. 33a.

We notice that at point A in fig. 33a the curvature of the curve is changing from concave downwards to concave upwards.

Now when the curve is concave downwards $\dfrac{dy}{dx}$ is decreasing and hence $\dfrac{d^2y}{dx^2} < 0$.

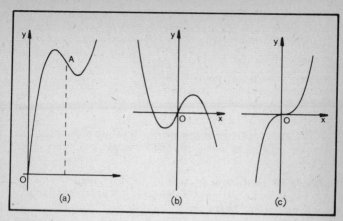

Figure 33

When the curve is concave upwards $\frac{dy}{dx}$ is increasing and hence $\frac{d^2y}{dx^2} > 0$.

∴ At the point of change, called the **point of inflexion**, $\frac{d^2y}{dx^2} = 0$ and $\frac{d^2y}{dx^2}$ changes sign at this point.

Hence at a point of inflexion

(i) The curvature changes from concave downwards to concave upwards (fig. 33a) or vice versa (fig. 33b).

(ii) $\frac{dy}{dx}$ is increasing before and decreasing after, or vice versa.

(iii) This means that $\frac{d^2y}{dx^2}$ changes sign on either side of the point.

(iv) $\frac{dy}{dx}$ will be a maximum or a minimum ∴ $\frac{d^2y}{dx^2} = 0$

In general it is not necessary for $\frac{dy}{dx} = 0$ although this is often the case as in fig. 33c which shows the curve $y = x^3$.

∴ If $y = 2x^3 - 9x^2 + 12$

$$\frac{dy}{dx} = 6x^2 - 18x + 12 \qquad \frac{d^2y}{dx^2} = 12x - 18$$

Now $\dfrac{d^2y}{dx^2} = 0$ when $12x - 18 = 0$, i.e. $x = 1.5$.

Since on either side of $x = 1.5$, $\dfrac{d^2y}{dx^2}$ changes sign, the point where $x = 1.5$ defines a point of inflexion.

Remember this can be checked by noting that the gradient of $\dfrac{d^2y}{dx^2}$ is nonzero!

The following example will illustrate how we can apply the methods for finding maximum and minimum points to other problems.

Example 24 Find the least area of metal required to make a cylindrical container from thin sheet metal in order that it might have a capacity of 2000π cc.

Let the radius be r cm, the height h cm and the total surface area S cm^2.

$$\text{Area of two ends} = 2\pi r^2; \text{ area of curved surface} = 2\pi rh$$

$$\text{Total surface area } S = 2\pi r^2 + 2\pi rh$$

$$\text{But volume} = \pi r^2 h = 2000\pi$$

$$\therefore \quad h = \frac{2000}{r^2} \tag{1}$$

Hence Total surface area $S = 2\pi r^2 + 2\pi r \cdot \dfrac{2000}{r^2}$

$$\therefore \quad S = 2\pi r^2 + \frac{4000\pi}{r} \tag{2}$$

To find the minimum area differentiate and equate to zero

$$\frac{dS}{dr} = 4\pi r - \frac{4000\pi}{r^2}$$

when $\dfrac{dS}{dr} = 0$, i.e. $4\pi r - \dfrac{4000\pi}{r^2} = 0$ then $r^3 = 1000 \Rightarrow r = 10$.

Now $\dfrac{d^2S}{dr^2} = 4\pi + \dfrac{8000\pi}{r^3}$ which is positive for $r = 10$

$\therefore$ When $r = 10$, the surface area is a minimum. In this case $h = 20$, from equation (1). Substituting in equation (2).

$$\text{Total surface area} = 200\pi + 400\pi = 600\pi \text{ cm}^2$$

Small changes

Since we defined $\dfrac{dy}{dx}$ as the limit of $\dfrac{\delta y}{\delta x}$ as $\delta x \to 0$

We can say that $\dfrac{\delta y}{\delta x} \simeq \dfrac{dy}{dx}$. Hence $\delta y \simeq \dfrac{dy}{dx} \cdot \delta x$.

We can use this approximation to find the effect on one variable of a small change in the other.

Example 25 An error of 2% is made in measuring the radius of a circle. What is the error in the area?

$$\text{Now } \frac{dA}{dr} \simeq \frac{\delta A}{\delta r} \quad \text{or} \quad \delta A \simeq \frac{dA}{dr} \cdot \delta r \tag{1}$$

Now $\delta r = \dfrac{2r}{100}$ Also $A = \pi r^2$ $\therefore \dfrac{dA}{dr} = 2\pi r$

$\therefore$ in equation (1) we have $\delta A \simeq 2\pi r \cdot \dfrac{2r}{100} = \dfrac{4\pi r^2}{100}$

$\therefore$ % error in the area $= \dfrac{\delta A}{A} \times 100 = \dfrac{4\pi r^2}{100 \cdot \pi r^2} \times 100\% = 4\%$.

Key terms

$$y = ax^n \Rightarrow \frac{dy}{dx} = nax^{n-1}$$

$$y = f(x) \Rightarrow \frac{dy}{dx} = f'(x) = \lim_{b \to a}\left[\frac{f(b) - f(a)}{b - a}\right] = \lim_{h \to 0}\left[\frac{f(a + h) - f(a)}{h}\right]$$

Product rule $y = u(x) \times v(x) \Rightarrow \dfrac{dy}{dx} = u\dfrac{dv}{dx} + v\dfrac{du}{dx}$

Quotient rule $y = \dfrac{u(x)}{v(x)} \Rightarrow \dfrac{dy}{dx} = \dfrac{v\dfrac{du}{dx} - u\dfrac{dv}{dx}}{v^2}$

Composite functions $y = y(u)$ and $u = u(x) \Rightarrow \dfrac{dy}{dx} = \dfrac{dy}{du} \times \dfrac{du}{dx}$

Maximum point $\dfrac{dy}{dx} = 0$ and $\dfrac{d^2y}{dx^2} < 0$

Minimum point $\dfrac{dy}{dx} = 0$ and $\dfrac{d^2y}{dx^2} > 0$

Point of inflexion $\dfrac{d^2y}{dx^2} = 0$ AND $\dfrac{d^2y}{dx^2}$ changes sign on either side of the point.

Chapter 4
Integration

This branch of the calculus is concerned, in one of its aspects, with the converse of differentiation. That is, if the gradient of a function is known, is it possible to find the function from which that gradient was derived?

Now we know from Chapter 3 that $\frac{d}{dx}(x^2+3) = 2x$ and $\frac{d}{dx}(x^2-7) = 2x$.

Indeed any function of the form $x^2 \pm k$ will have a derived function of $2x$.

Clearly then if we are given the gradient of a function as $2x$ the integral function must be of the form $x^2 + c$ where c is some positive or negative constant. In general it is not possible to determine the value of this constant and the resulting function is called an **indefinite integral**.

Example 1 If $f'(x) = x^4$ find $f(x)$.

If $f'(x) = x^4$ then $f(x) = \frac{1}{5}x^5 + c$ where c is an arbitrary constant.

Figure 34 gives a graphical interpretation of the above process.

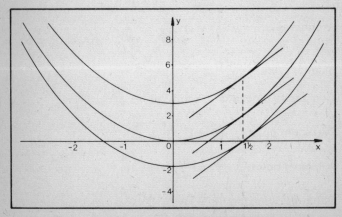

Figure 34

Each of the curves $y = x^2$, $y = x^2 + 3$, $y = x^2 - 2$ is shown with its gradient at the point where $x = 1\frac{1}{2}$. All of these curves have the same gradient function $2x$ represented by parallel tangents.

To determine the value of the arbitrary constant a set of conditions must be given. This may be the coordinates of one point on the curve.

Example 2 A curve has a gradient function of $3x^2$ and passes through the point (2, 10). Find the equation of the curve.

If $\dfrac{dy}{dx} = 3x^2$ then $y = x^3 + c$.

As the curve passes through the point (2, 10) these values satisfy the equation, i.e. $10 = (2)^3 + c$ and $c = 2$.
∴   the equation is $y = x^3 + 2$.
A special notation is used to denote the operation of integration. This is ∫ which is derived from the old English form of the letter s, being the first letter of the word sum. The idea of a sum is an aspect of integration which will be discussed later.

We write $\displaystyle\int 3x^2\, dx = x^3 + c$.

The meaning of the dx will also be made clear from the summation process but it does indicate with respect to which variable we are integrating.

i.e. $\displaystyle\int f(x)\, dx$ indicates that the function $f(x)$ is to be integrated with respect to x.

In general, using the result $\dfrac{d}{dx}(ax^n) = nax^{n-1}$ we can deduce

that
$$\int ax^n\, dx = \frac{ax^{n+1}}{(n+1)} + c.$$

This will hold for all positive, negative, integral or fractional values of n except $n = -1$.
The integral of the sum of any number of terms is equal to the sum of the separate integrals.

Example 3 Find $\displaystyle\int (x^3 - 3x^2 + 8x + 7)\, dx$.

Integrating with respect to x we obtain

$$\int (x^3 - 3x^2 + 8x + 7)\, dx = \frac{x^4}{4} - x^3 + 4x^2 + 7x + c$$

When $n = -1$ the general rule breaks down. We have shown on page 117 that $\frac{d}{dx}(\ln x) = \frac{1}{x}$

Hence $\int \frac{1}{x} dx = \ln(x) + c = \ln(kx)$ by writing $c = \ln k$.

Area under a curve

Another view of integration can be obtained by considering the area under a given curve. The technique involves dividing the area into a number of elementary strips and summing over the complete range. This process will involve some error but by increasing the number of elements a better approximation to the actual area can be found. Indeed, if the number of elements is allowed to tend to infinity the calculated area will, in the limit, be the actual area.

Consider $f(x) = 2x + 3$. If we wish to find the area bounded by the graph and the x axis between $x = 0$ and $x = 10$ we could divide the area into n strips of equal width, as shown in fig. 35.

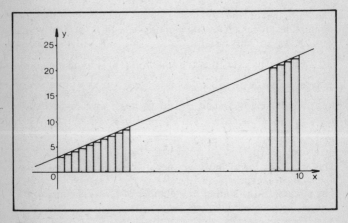

Figure 35

If the area is divided into n strips the width of each is $\frac{10}{n}$.

We calculate the ordinate for each value of x so that the area of each rectangle can be found.

when $x = 0$, $f(x) = 3$ i.e. $f(0) = 3$

when $x = \dfrac{10}{n}$, $f\left(\dfrac{10}{n}\right) = 2 \cdot \dfrac{10}{n} + 3$

when $x = 2 \cdot \dfrac{10}{n}$, $f\left(2 \cdot \dfrac{10}{n}\right) = 2 \cdot \dfrac{20}{n} + 3$

when $x = \left(10 - \dfrac{10}{n}\right)$, $f\left(10 - \dfrac{10}{n}\right) = 2\left(10 - \dfrac{10}{n}\right) + 3$

Hence the sum of all the rectangles is

$$3 \cdot \dfrac{10}{n} + \left(\dfrac{20}{n} + 3\right)\dfrac{10}{n} + \left(\dfrac{40}{n} + 3\right)\dfrac{10}{n} + \left(\dfrac{60}{n} + 3\right)\dfrac{10}{n}$$

$$+ \cdots + \left(20 - \dfrac{20}{n} + 3\right)\dfrac{10}{n}$$

$$= \dfrac{10}{n}\left[3 + \left(\dfrac{20}{n} + 3\right) + \left(\dfrac{40}{n} + 3\right) + \left(\dfrac{60}{n} + 3\right) + \cdots + \left(\dfrac{20(n-1)}{n} + 3\right)\right]$$

The terms inside the square bracket form an A.P. with a common difference of $\dfrac{20}{n}$ and with a first term of 3. Since there are n terms the sum is given by $\dfrac{n}{2}\left[6 + (n-1)\dfrac{20}{n}\right]$

$\therefore$ The sum of areas of the rectangles is

$$\dfrac{10}{n} \cdot \dfrac{n}{2}\left[6 + (n-1)\dfrac{20}{n}\right] = 5\left(26 - \dfrac{20}{n}\right) = 130 - \dfrac{100}{n}$$

From this form of our answer we can conclude that as $n \to \infty$ the area $\to 130$.

Hence the area under this straight line = 130.

This could, of course, be found from the area of the trapezium $\frac{1}{2}(3 + 23) \times 10 = 26 \times 5 = 130$.

We can also apply this process to a nonlinear function. Consider $f(x) = x^2$ as shown in fig. 36. Again divide the required area into n elementary rectangles of width h. A typical element $PQRS$ will have an area given by $(rh)^2 \times h = r^2 h^3$.

Hence the total area is $\displaystyle\sum_{r=1}^{n-1} r^2 h^3 = h^3 \sum_{r=1}^{n-1} r^2$.

$\therefore$ Area $= h^3 \times \dfrac{(n-1)}{6}(n)(2n-1)$. See the result for $\sum r^2$ on page 202 with n replaced by $n - 1$.

$\therefore$ Area $= h^3 n(n-1)(2n-1)/6$

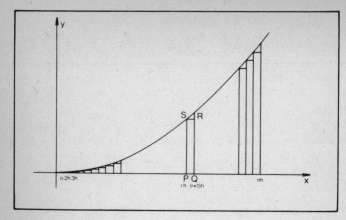

Figure 36

Now if the maximum value of $x = b \Rightarrow h = b/n$

$$\text{Area} = \frac{b^3 n}{6n^3}(n-1)(2n-1) = \frac{b^3}{6}\left(1-\frac{1}{n}\right)\left(2-\frac{1}{n}\right)$$

As $n \to \infty$, area $\to \frac{1}{3}b^3$. We say that the area is given by the limit as $n \to \infty$ and is equal to $\frac{1}{3}b^3$.

To generalize this method we consider one typical element obtained by dividing an area under a given curve into a large number of rectangles of width δx. Let the curve be given by the function $y = f(x)$ and let one typical element be formed between the ordinates x and $x + \delta x$. This is shown in fig. 37. Clearly, the chosen element *PRST* is formed by selecting two general points $P(x, y)$ and $Q(x + \delta x, y + \delta y)$ on the curve and drawing *PR* parallel to the x axis. Thus the area of *PRTS* is $y\delta x$. Hence the total area between $x = a$ and $x = b$ will be the sum of all such elements

i.e. $$\sum_{x=a}^{x=b} y\delta x$$

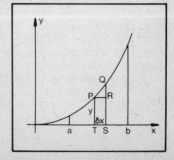

Figure 37

We now increase the number of rectangles by letting $\delta x \to 0$. Hence the area is given by

$\lim\limits_{\delta x \to 0} \sum\limits_{x=a}^{x=b} y \delta x$. For this we use the integral notation and the final form is given as follows.

The area under a curve $y = f(x)$ over the range $x = a$ to $x = b$

is given by $\int_a^b f(x)\,dx$

Since the evaluation of this result will give a definite numerical answer the integral is called a **definite integral** and no arbitrary constant is necessary.[1]

If $F(x)$ is the integral function of $f(x)$ then

$\int_a^b f(x)\,dx = F(b) - F(a)$

This can be seen from fig. 38 where $x = c$ is a lower boundary. Then, if the definite integral of $f(x)$ is known over the interval c to b, as b varies the integral function takes different values.

i.e. an integral function $F(x)$ is formed which maps b onto the integral

$\int_c^b f(x)\,dx.$

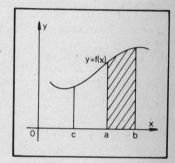

Figure 38

∴ We can write $F(b) = \int_c^b f(x)\,dx$ Hence, it follows that

$F(a) = \int_c^a f(x)\,dx$ and $\int_a^b f(x)\,dx = F(b) - F(a)$

A special notation is used to represent the value of

$F(b) - F(a)$. We write $F(b) - F(a) = \left[F(x) \right]_a^b$

Example 4 $\int_1^3 (x^2 + 3x)\,dx = \left[\dfrac{x^3}{3} + \dfrac{3x^2}{2} \right]_1^3$

$$= \left(\frac{27}{3} + \frac{27}{2} \right) - \left(\frac{1}{3} + \frac{3}{2} \right) = 20\frac{2}{3}$$

In the derivation of the above results we have considered the sum of the inner rectangles when finding the total area. We could have taken, instead, outer rectangles as shown by fig. 39. This would have given an area of $(y + \delta y)\,\delta x$ for the element $TQRS$ and hence by the same process as before the total area would be

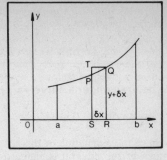

Figure 39

$$\lim_{\delta x \to 0} \sum_{x=a}^{x=b} (y + \delta y)\,\delta x$$

which will tend to the same limit. In this case we approach the limit from above instead of from below.

Example 5 Evaluate (i) $\displaystyle\int_0^2 f(x)\,dx$, (ii) $\displaystyle\int_2^3 f(x)\,dx$ and

(iii) $\displaystyle\int_0^3 f(x)\,dx$ where $f(x) = x^2 - 2x$.

Now (i) $\displaystyle\int_0^2 f(x)\,dx = \int_0^2 (x^2 - 2x)\,dx = \left[\frac{x^3}{3} - x^2\right]_0^2$

$$= \left(\frac{8}{3} - 4\right) - (0 - 0) = -1\tfrac{1}{3}$$

(ii) $\displaystyle\int_2^3 f(x)\,dx = \int_2^3 (x^2 - 2x)\,dx = \left[\frac{x^3}{3} - x^2\right]_2^3$

$$= \left(\frac{27}{3} - 9\right) - \left(\frac{8}{3} - 4\right) = 1\tfrac{1}{3}$$

(iii) $\displaystyle\int_0^3 f(x)\,dx = \int_0^3 (x^2 - 2x)\,dx = \left[\frac{x^3}{3} - x^2\right]_0^3$

$$= \left(\frac{27}{3} - 9\right) - (0 - 0) = 0$$

The solution to part (i) is negative which indicates that the area is below the axis, *not* that the area is negative. Since the results to parts (i) and (ii) are numerically equal, the areas are equal. Notice that when the integral over the whole range from 0 to 3 is evaluated the result is the algebraic sum of the first two answers, i.e. not the total actual area of $2\tfrac{2}{3}$.

This becomes clear if we sketch the graph of $y = x^2 - 2x$. The curve and the relevant areas are shown in fig. 40.

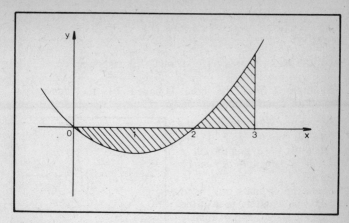

Figure 40

In general, if $y = f(x)$ is a function taking only negative values over an interval $x = a$ to $x = b$ then $y = -f(x)$ is a reflection of the curve in the x axis possessing the same area, i.e. an area given by

$$\int_a^b -f(x)\,dx = \left[-F(x)\right]_a^b = -\left[F(x)\right]_a^b = -\int_a^b f(x)\,dx$$

Thus the integral evaluates the negative of the area bounded by $y = f(x)$, $x = a$, $x = b$ and the x axis.

The three examples which follow illustrate the methods for evaluating an area not bounded by the x axis.

Example 6 Find the area enclosed by the curve $x = 2y^2$, the y axis and the line $y = 3$.

Refer to fig. 41. Choose an elementary strip parallel to the x axis $PQRS$ where $P(x, y)$ and $Q(x + \delta x, y + \delta y)$ are two neighbouring points on the curve. Hence the area of the element is approximately given by the area of the inner rectangle $RSPT$.

Area $\simeq x\delta y$

By summing and taking the limit as $\delta y \to 0$ we obtain a value for the total area.

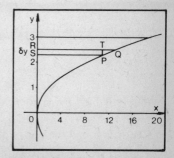

Figure 41

79

$$\lim_{\delta y \to 0} \sum_{y=0}^{y=3} x\delta y = \int_0^3 x\,dy = \int_0^3 2y^2\,dy$$

$\therefore$ the required area $= \left[\dfrac{2y^3}{3}\right]_0^3 = \left(2 \times \dfrac{27}{3} - 0\right) = \dfrac{54}{3}$

Example 7 Find the area bounded by the curve $y = x^2 - 2x + 2$ and the line $y = 5$.

By considering two neighbouring points $P(x, y)$ and $Q(x + \delta x, y + \delta y)$ on the curve a typical element $PQRS$ can be formed.

Since the ordinate of P is y, $SP = 5 - y$ and the area of the element is approximately $(5 - y)\,\delta x$.

To find limits for the summation we must find the points of intersection of the line and the curve.

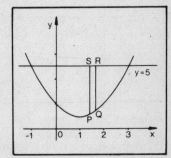

Figure 42

This occurs when

$$x^2 - 2x + 2 = 5 \Leftrightarrow x^2 - 2x - 3 = 0$$

$$\Leftrightarrow (x - 3)(x + 1) = 0 \Rightarrow x = -1 \text{ or } x = 3$$

Now by summing and taking the limit the required area is given by

$$\lim_{\delta x \to 0} \sum_{x=-1}^{3} (5 - y)\,\delta x = \int_{-1}^{3} (5 - y)\,dx$$

$\therefore$ Area $= \displaystyle\int_{-1}^{3} [5 - (x^2 - 2x + 2)]\,dx = \int_{-1}^{3} (3 + 2x - x^2)\,dx$

$$= \left[3x + x^2 - \frac{x^3}{3}\right]_{-1}^{3} = \left(9 + 9 - \frac{27}{3}\right) - \left(3(-1) + (-1)^2 - \frac{(-1)^3}{3}\right)$$

$$= 9 - \left(\frac{-5}{3}\right) = \frac{32}{3} = 10\tfrac{2}{3}$$

Example 8 Find the area enclosed by the curves $y = x^2 - 3x$, and $y = 9 - x^2$.

We must first find the points of intersection of the curves so that we know the range for the summation. The curves intersect when $x^2 - 3x = 9 - x^2$

$$\Leftrightarrow 2x^2 - 3x - 9 = 0 \Leftrightarrow (2x + 3)(x - 3) = 0$$

$$\Rightarrow x = -1\tfrac{1}{2} \text{ or } x = 3$$

$\therefore$ The curves intersect at $(-1\tfrac{1}{2}, 6\tfrac{3}{4})$ and $(3, 0)$.

By noting where the curves cut the axes a sketch can be drawn.

When $y = 0$, $x^2 - 3x = 0 \Rightarrow x = 0$ or 3

When $y = 0$, $9 - x^2 = 0 \Rightarrow x = \pm 3$

By letting $x = 0$ it is clear that the curves cut the y axis at the origin and $y = 9$ respectively. The completed sketch is shown in fig. 43.

In considering an element of area we must ensure that it is typical and hence a strip parallel to the y axis must be chosen.

e.g. *PQRS*

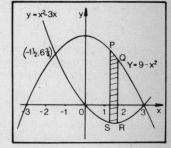

Figure 43

Now the height of *PS* is $Y - y$ where Y and y are the ordinates of P and S respectively. Note that it is the difference $Y - y$, since in the position shown y will be negative and will effectively produce the sum of Y and y as required. Since the width of the element is δx the area of $PQRS \simeq (Y - y)\,\delta x$.

By summing and taking the limit as $\delta x \to 0$, the total area is

$$\lim_{\delta x \to 0} \sum_{x=-1 1/2}^{3} (Y - y)\,\delta x = \int_{-3/2}^{3} (Y - y)\,dx$$

Now $Y = 9 - x^2$ and $y = x^2 - 3x$

$\therefore$ Area required $= \displaystyle\int_{-3/2}^{3} [(9 - x^2) - (x^2 - 3x)]\,dx$

$$= \int_{-3/2}^{3} (9 + 3x - 2x^2)\,dx = \left[9x + \frac{3x^2}{2} - \frac{2x^3}{3}\right]_{-3/2}^{3}$$

$$= \left(27 + \frac{27}{2} - 18\right) - \left(-\frac{27}{2} + \frac{27}{8} + \frac{9}{4}\right)$$

$$= \frac{45}{2} + \frac{63}{8} = 30\tfrac{3}{8}$$

Remember that in the calculation of area the choice of element is not necessarily unique but it must be typical of the whole region and must be such that it will produce a function which can be integrated, and preferably, easily.

Consider the curve $y = x^2$. If we require the area bounded by the curve, the y axis and the line $y = 4$, we can choose the element in two ways as shown in fig. 44a and 44b. Using the element in fig. 44a the area is $\simeq x\delta y$

$$\therefore \quad \text{total area} = \int_0^4 x\,dy = \int_0^4 y^{1/2}\,dy = \left[\frac{2}{3}y^{3/2}\right]_0^4 = \frac{16}{3} - 0 = \frac{16}{3}$$

Using the element in fig. 44b the area is $\simeq (4 - y)\,\delta x$.

$$\therefore \quad \text{total area} = \int_0^2 (4 - y)\,dx = \int_0^2 (4 - x^2)\,dx = \left[4x - \frac{x^3}{3}\right]_0^2 = \frac{16}{3}$$

However if we consider the curve $y = x^2 - x - 2$ as shown in fig. 44c and evaluate the area between the curve, the y axis and the line $y = 3$, we can only take the element parallel to the y axis.

In this case the total area is given by $\int_0^a (3 - y)\,dx$ where $a = \frac{1}{2}(1 + \sqrt{21})$. By substituting for y this can be evaluated. But if we take an element parallel to the x axis the area above the x axis would be given by $\int_0^3 x\,dy$ and it is not possible to find x as an explicit function of y in order to evaluate this integral.

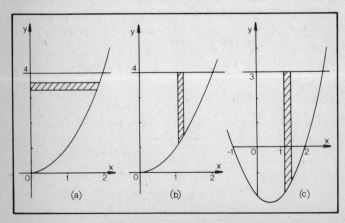

Figure 44

Standard integrals

We have considered the meaning of integration and now turn our attention to the integrals of specific functions. In Chapter 3 we derived the differentials of a number of different functions. By considering integration as the inverse operation to differentiation we can list some standard integrals which should be learnt thoroughly.

$$\frac{d}{dx}(ax^n) = nax^{n-1} \qquad\qquad \int ax^n \, dx = \frac{ax^{n+1}}{n+1} + c$$

$$(n \neq -1)$$

$$\frac{d}{dx}(\sin x) = \cos x \qquad\qquad \int \cos x \, dx = \sin x + c$$

$$\frac{d}{dx}(\cos x) = -\sin x \qquad\qquad \int \sin x \, dx = -\cos x + c$$

$$\frac{d}{dx}(\tan x) = \sec^2 x \qquad\qquad \int \sec^2 x \, dx = \tan x + c$$

$$\frac{d}{dx}(\cot x) = -\operatorname{cosec}^2 x \qquad\qquad \int \operatorname{cosec}^2 x \, dx = -\cot x + c$$

$$\frac{d}{dx}(\sec x) = \sec x \tan x \qquad\qquad \int \sec x \tan x \, dx = \sec x + c$$

$$\frac{d}{dx}(\operatorname{cosec} x) = \qquad\qquad \int \operatorname{cosec} x \cot x \, dx =$$
$$-\operatorname{cosec} x \cot x \qquad\qquad\qquad -\operatorname{cosec} x + c$$

The **inverse trignometrical functions** were also discussed in Chapter 3 and produced the following results.

$$\frac{d}{dx}\sin^{-1}\left(\frac{x}{a}\right) = \frac{1}{\sqrt{a^2 - x^2}} \qquad \int \frac{dx}{\sqrt{a^2 - x^2}} = \sin^{-1}\left(\frac{x}{a}\right) + c$$

$$\frac{d}{dx}\cos^{-1}\left(\frac{x}{a}\right) = \frac{-1}{\sqrt{a^2 - x^2}} \qquad \int \frac{dx}{\sqrt{a^2 - x^2}} = -\cos^{-1}\left(\frac{x}{a}\right) + c$$

$$\frac{d}{dx}\tan^{-1}\left(\frac{x}{a}\right) = \frac{a}{a^2 + x^2} \qquad \int \frac{a}{a^2 + x^2} \, dx = \tan^{-1}\left(\frac{x}{a}\right) + c$$

$$\frac{d}{dx}\cot^{-1}\left(\frac{x}{a}\right) = \frac{-a}{a^2 + x^2} \qquad \int \frac{a}{a^2 + x^2} \, dx = -\cot^{-1}\left(\frac{x}{a}\right) + c$$

$$\frac{d}{dx}\sec^{-1}\left(\frac{x}{a}\right) = \frac{a}{x\sqrt{x^2 - a^2}} \qquad \int \frac{a}{x\sqrt{x^2 - a^2}} \, dx = \sec^{-1}\left(\frac{x}{a}\right) + c$$

$$\frac{d}{dx}\operatorname{cosec}^{-1}\left(\frac{x}{a}\right) = \frac{-a}{x\sqrt{x^2-a^2}} \quad \int \frac{a}{x\sqrt{x^2-a^2}}\,dx$$

$$= -\operatorname{cosec}^{-1}\left(\frac{x}{a}\right) + c$$

The **exponential and logarithmic functions** are discussed fully in Chapter 5. The differentials and integrals of these functions will be quoted here and used later in the chapter.

$$\frac{d}{dx}(e^{ax}) = ae^{ax} \qquad \int e^{ax}\,dx = \frac{1}{a}e^{ax} + c$$

$$\frac{d}{dx}(\ln x) = \frac{1}{x} \qquad \int \frac{dx}{x} = \ln x + c = \ln(kx)$$

$$\frac{d}{dx}(a^x) = a^x \ln a \qquad \int a^x\,dx = \frac{a^x}{\ln a} + c$$

In Chapter 5 the **hyperbolic functions** and their derivatives are discussed. Again, the results are tabulated here for reference.

$$\frac{d}{dx}(\cosh x) = \sinh x \qquad\qquad \int \sinh x\,dx = \cosh x + c$$

$$\frac{d}{dx}(\sinh x) = \cosh x \qquad\qquad \int \cosh x\,dx = \sinh x + c$$

$$\frac{d}{dx}(\tanh x) = \operatorname{sech}^2 x \qquad\qquad \int \operatorname{sech}^2 x\,dx = \tanh x + c$$

$$\frac{d}{dx}(\operatorname{sech} x) = \qquad\qquad \int \operatorname{sech} x \tan x\,dx =$$
$$-\operatorname{sech} x \tanh x \qquad\qquad\qquad -\operatorname{sech} x + c$$

$$\frac{d}{dx}(\operatorname{cosech} x) = \qquad\qquad \int \operatorname{cosech} x \coth x\,dx =$$
$$-\operatorname{cosech} x \coth x \qquad\qquad\qquad -\operatorname{cosech} x + c$$

$$\frac{d}{dx}(\coth x) = -\operatorname{cosech}^2 x \qquad \int \operatorname{cosech}^2 x\,dx = \coth x + c$$

The **inverse hyperbolic functions** produce very similar results to those for the inverse circular functions.

$$\frac{d}{dx}\sinh^{-1}\left(\frac{x}{a}\right) = \frac{1}{\sqrt{x^2+a^2}} \quad \int \frac{dx}{\sqrt{x^2+a^2}} = \sinh^{-1}\left(\frac{x}{a}\right) + c$$

$$= \ln\left(x + \sqrt{x^2+a^2}\right) + c$$

$$\frac{d}{dx}\cosh^{-1}\left(\frac{x}{a}\right) = \frac{1}{\sqrt{x^2-a^2}} \qquad \int \frac{dx}{\sqrt{x^2-a^2}} = \cosh^{-1}\left(\frac{x}{a}\right) + c$$

$$= \ln\left(x + \sqrt{x^2-a^2}\right) + c$$

$$\frac{d}{dx}\tanh^{-1}\left(\frac{x}{a}\right) = \frac{a}{a^2-x^2} \qquad \int \frac{dx}{a^2-x^2} = \frac{1}{a}\tanh^{-1}\left(\frac{x}{a}\right) + c$$

$$= \frac{1}{2a}\ln\left(\frac{a+x}{a-x}\right) + c$$

$$\frac{d}{dx}\coth^{-1}\left(\frac{x}{a}\right) = \frac{-a}{x^2-a^2} \qquad \int \frac{dx}{x^2-a^2} = -\frac{1}{a}\coth^{-1}\left(\frac{x}{a}\right) + c$$

$$= \frac{1}{2a}\ln\left(\frac{x-a}{x+a}\right) + c$$

$$\frac{d}{dx}\operatorname{sech}^{-1}\left(\frac{x}{a}\right) = \frac{-a}{x\sqrt{a^2-x^2}} \qquad \int \frac{dx}{x\sqrt{a^2-x^2}} = \frac{-1}{a}\operatorname{sech}^{-1}\left(\frac{x}{a}\right) + c$$

$$= \frac{-1}{a}\ln\left[\frac{x+\sqrt{a^2-x^2}}{x}\right] + c$$

$$\frac{d}{dx}\operatorname{cosech}^{-1}\left(\frac{x}{a}\right) = \frac{-a}{x\sqrt{a^2+x^2}} \qquad \int \frac{dx}{x\sqrt{a^2+x^2}}$$

$$= \frac{-1}{a}\operatorname{cosech}^{-1}\left(\frac{x}{a}\right) + c = \frac{-1}{a}\ln\left[\frac{a+\sqrt{a^2+x^2}}{x}\right] + c$$

Methods of integration

The above list of integrals enables the student to integrate any function given in one of these forms. However, many functions which we are required to integrate are more complex and may well be made up of a combination of these functions. We now consider various special methods for dealing with specific types of functions.

Inspection

This method demands a sound knowledge of the function of a function technique used in differentiation (see page 50). To use this method the student must be able to recognize the presence of a function and its differential. The following examples will illustrate the method.

Example 9 Find $\int x(3x^2+7)^4\,dx$.

Consider $\frac{d}{dx}(3x^2+7)^5 = 5(3x^2+7)^4 \times 6x = 30x(3x^2+7)^4$

Since our result differs from the function we require to integrate only by a constant it follows that

$$\int x(3x^2 + 7)^4 \, dx = \frac{1}{30}(3x^2 + 7)^5 + c$$

Example 10 Find $\int 8 \sin^4 x \cos x \, dx$.

Since $\frac{d}{dx}(\sin^5 x) = 5 \sin^4 x \cos x$, clearly it follows that

$$\int 8 \sin^4 x \cos x \, dx = \frac{8}{5} \sin^5 x + c.$$

Example 11 Find $\int \sec^4 x \tan x \, dx$.

Now $\sec^4 x \tan x = \sec^3 x(\sec x \tan x)$ and

$$\frac{d}{dx}(\sec x) = \sec x \tan x.$$

This suggests an integral of the form $\sec^4 x$.

Consider $\frac{d}{dx}(\sec^4 x) = 4 \sec^3 x(\sec x \tan x) = 4 \sec^4 x \tan x$

Hence $\int \sec^4 x \tan x \, dx = \frac{1}{4} \sec^4 x + c$

Powers of sin x and cos x
There are two separate techniques. One for odd powers and one for even powers.

Odd powers In these integrals we split the function into an even power multiplied by a single sine or cosine. By using Pythagoras' theorem, $\cos^2 x + \sin^2 x = 1$ or the equivalent forms we can rearrange the integral to give functions that can be integrated by inspection.

Example 12 Find $\int \cos^5 x \, dx$.

$$\int \cos^5 x \, dx = \int \cos^4 x \cos x \, dx = \int (1 - \sin^2 x)^2 \cos x \, dx$$

$$= \int (1 - 2 \sin^2 x + \sin^4 x) \cos x \, dx$$

$$= \int (\cos x - 2 \sin^2 x \cos x + \sin^4 x \cos x) \, dx$$

$$\therefore \int \cos^5 x \, dx = \sin x - \frac{2}{3} \sin^3 x + \frac{1}{5} \sin^5 x + c$$

Even powers Functions of this type can be integrated by rearranging the integrand by using the double angle formulae, i.e. $\cos^2 x = \frac{1}{2}(1 + \cos 2x)$ and $\sin^2 x = \frac{1}{2}(1 - \cos 2x)$ (see page 31).

Example 13 Find $\int \cos^4 x \, dx$.

$$\int \cos^4 x \, dx = \int (\cos^2 x)^2 \, dx = \frac{1}{4} \int (1 + \cos 2x)^2 \, dx$$

$$= \frac{1}{4} \int (1 + 2\cos 2x + \cos^2 2x) \, dx$$

$$= \frac{1}{4} \int \left[1 + 2\cos 2x + \frac{1}{2}(1 + \cos 4x) \right] dx$$

$$= \frac{1}{4} \int \left(\frac{3}{2} + 2\cos 2x + \frac{1}{2}\cos 4x \right) dx$$

$$= \frac{1}{4} \left(\frac{3}{2} x + \sin 2x + \frac{1}{8} \sin 4x \right) + c$$

$$= \frac{3}{8} x + \frac{1}{4} \sin 2x + \frac{1}{32} \sin 4x + c$$

Functions of the type $f'(x)/f(x)$
Consider a function of the form $y = \ln f(x)$. In this case y is a function of a function of x and hence

$$\frac{dy}{dx} = \frac{1}{f(x)} \times f'(x) = \frac{f'(x)}{f(x)}$$

Hence $\int \frac{f'(x)}{f(x)} \, dx = \ln [f(x)] + c = \ln [kf(x)]$ where $\ln k = c$

A useful integral which can be evaluated by this method is

$$\int \sec x \, dx = \int \frac{\sec x(\sec x + \tan x)}{\sec x + \tan x} \, dx = \int \frac{\sec^2 x + \sec x \tan x}{\sec x + \tan x} \, dx$$

The numerator is now the differential of the denominator and hence

$$\int \sec x \, dx = \ln (\sec x + \tan x) + c$$

This method is again encountered in chapter 5.

Substitution

Suppose we wish to evaluate $\int f(x)\,dx$. We can sometimes effect this by making a substitution and evaluating a different integral. There are two basic ways in which we can make the substitution.

(i) by putting $u = h(x)$
(ii) by putting $x = g(u)$

(i) If $f(x)$ is a function of x and $y = \int f(x)\,dx$ then $\dfrac{dy}{dx} = f(x)$.

$$\therefore \quad \frac{dy}{du} = \frac{dy}{dx} \cdot \frac{dx}{du} = f(x)\frac{dx}{du} \quad \text{as} \quad \frac{dy}{dx} = f(x)$$

$$y = \int f(x)\left(\frac{dx}{du}\right) du$$

Example 14 Consider again the integral in example 9, i.e. $\int x(3x^2 + 7)^4\,dx$. Now if we make a substitution $u = 3x^2 + 7$

$$\frac{du}{dx} = 6x \quad \text{which gives} \quad \frac{dx}{du} = \frac{1}{6x}$$

Hence $\displaystyle\int x(3x^2 + 7)^4\,dx = \int x(3x^2 + 7)^4 \frac{dx}{du}\,du$

$$= \int \frac{1}{6} u^4\,du = \frac{u^5}{30} + c$$

$$\therefore \quad \int x(3x^2 + 7)^4\,dx = \frac{1}{30}(3x^2 + 7)^5 + c \quad \text{as before}$$

Example 15 Find $\displaystyle\int \frac{x}{\sqrt{x-2}}\,dx$.

Let $u = \sqrt{x-2} = (x-2)^{1/2}$ $\quad \therefore \quad u^2 = x - 2 \Leftrightarrow x = u^2 + 2$

Now $\dfrac{du}{dx} = \frac{1}{2}(x-2)^{-1/2} \Rightarrow \dfrac{dx}{du} = 2(x-2)^{1/2}$

Hence $\displaystyle\int \frac{x}{\sqrt{x-2}}\,dx = \int \frac{x}{\sqrt{x-2}} \frac{dx}{du}\,du = \int (u^2 + 2)2\,du$

$$= \tfrac{2}{3}u^3 + 4u + c = \tfrac{2}{3}u(u^2 + 6) + c$$

Replacing u by $(x-2)^{1/2}$

$$\int \frac{x}{\sqrt{x-2}}\,dx = \frac{2}{3}(x-2)^{1/2}(x-2+6) + c = \frac{2}{3}(x+4)\sqrt{x-2} + c$$

Example 16 Find $\int \sin^2 x \cos^3 x \, dx$.

Let $u = \sin x \Rightarrow \dfrac{du}{dx} = \cos x \quad$ i.e. $\dfrac{dx}{du} = \dfrac{1}{\cos x}$

$$\int \sin^2 x \cos^3 x \, dx = \int \sin^2 x \cos^3 x \, \frac{dx}{du} \, du$$

$$= \int \sin^2 x (1 - \sin^2 x) \cos x \, \frac{dx}{du} \, du$$

$$= \int u^2 (1 - u^2) \, du = \int (u^2 - u^4) \, du$$

$$= \frac{u^3}{3} - \frac{u^5}{5} + c = \frac{u^3}{15}(5 - 3u^2) + c.$$

Replacing u by $\sin x$ we obtain

$$\int \sin^2 x \cos^3 x \, dx = \frac{1}{15} \sin^3 x (5 - 3 \sin^2 x) + c.$$

This method is also applicable to definite integrals, in which case it is often more convenient to change the limits to those of the new variable.

Example 17 Evaluate $\int_1^2 \dfrac{8x}{(2x+1)^3} \, dx$.

Let $u = 2x + 1 \Rightarrow x = \frac{1}{2}(u - 1)$

$$\therefore \quad \frac{du}{dx} = 2 \Rightarrow \frac{dx}{du} = \frac{1}{2}$$

Under this transformation the interval $1 \leqslant x \leqslant 2$ maps onto $3 \leqslant u \leqslant 5$. Hence

$$\int_1^2 \frac{8x}{(2x+1)^3} \, dx = \int_{x=1}^{x=2} \frac{8x}{(2x+1)^3} \, \frac{dx}{du} \, du$$

$$= \int_3^5 \frac{4 \times \frac{1}{2}(u-1)}{u^3} \, du = \int_3^5 2(u^{-2} - u^{-3}) \, du$$

$$= 2 \left[-u^{-1} + \frac{1}{2} u^{-2} \right]_3^5$$

$$= 2 \left\{ \left(-\frac{1}{5} + \frac{1}{50} \right) - \left(-\frac{1}{3} + \frac{1}{18} \right) \right\} = 2 \left(-\frac{9}{50} + \frac{5}{18} \right)$$

$$= 2 \left(\frac{-81 + 125}{450} \right) = \frac{88}{450} = \frac{44}{225}$$

Example 18 Evaluate $\displaystyle\int_{\pi/6}^{\pi/2} \frac{\cot x}{\sqrt{\operatorname{cosec}^3 x}}\, dx$.

Let $u = \operatorname{cosec} x$. Thus

$$\frac{du}{dx} = -\operatorname{cosec} x \cot x \Rightarrow \frac{dx}{du} = \frac{-1}{\operatorname{cosec} x \cot x}$$

For transformation the interval $\dfrac{\pi}{6} \leqslant x \leqslant \dfrac{\pi}{2}$ maps onto $2 \leqslant u \leqslant 1$.

Hence $\displaystyle\int_{\pi/6}^{\pi/2} \frac{\cot x}{\sqrt{\operatorname{cosec}^3 x}}\, dx = \int_{x=\pi/6}^{x=\pi/2} \frac{\cot x}{\sqrt{\operatorname{cosec}^3 x}} \frac{dx}{du}\, du$

$$= \int_2^1 \frac{1}{u^{3/2}} \left(\frac{-1}{u}\right) du = -\int_1^2 - u^{-5/2}\, du$$

$$= \left[\frac{-2}{3} u^{-3/2}\right]_1^2 = \frac{-2}{3}(2^{-3/2} - 1)$$

$$= \frac{-2}{3}\left(\frac{1}{2\sqrt{2}} - 1\right) = \frac{-2}{3}\left(\frac{\sqrt{2}}{4} - 1\right)$$

$$= \frac{1}{6}(4 - \sqrt{2})$$

(ii) When we use the **substitution** $x = f(u)$ a very similar process evolves.

Consider $y = \displaystyle\int g(x)\, dx$ and let $x = f(u)$

Now $\dfrac{dy}{dx} = g(x) = g[f(u)]$

Hence $\dfrac{dy}{du} = \dfrac{dy}{dx} \cdot \dfrac{dx}{du} = g[f(u)] \cdot \dfrac{dx}{du}$ $\therefore$ $y = \displaystyle\int g[f(u)] \dfrac{dx}{du}\, du$

Example 19 Find $\displaystyle\int_0^3 \frac{3}{\sqrt{9 - x^2}}\, dx$.

Let $x = 3 \sin u \Rightarrow \dfrac{dx}{du} = 3 \cos u$.

Hence $\displaystyle\int_0^3 \frac{3}{\sqrt{9 - x^2}} \frac{dx}{du}\, du = \int_0^{\pi/2} \frac{3}{\sqrt{9 - 9\sin^2 u}} 3 \cos u\, du$

$$= \int_0^{\pi/2} 3\, du = \Big[3u\Big]_0^{\pi/2} = \frac{3\pi}{2}.$$

We can also use a substitution involving one of the hyperbolic functions.

Example 20 Find $\int \dfrac{dx}{a^2 - x^2}$.

Let $x = a \tanh \theta$, and thus $\dfrac{dx}{d\theta} = a \operatorname{sech}^2 \theta$.

$$\therefore \int \frac{dx}{a^2 - x^2} = \int \frac{1}{a^2 - x^2} \frac{dx}{d\theta} \, d\theta = \int \frac{1}{a^2 - a^2 \tanh^2 \theta} \cdot a \operatorname{sech}^2 \theta \, d\theta$$

$$= \int \frac{1}{a} \, d\theta = \frac{\theta}{a} + c$$

But $\theta = \tanh^{-1}\left(\dfrac{x}{a}\right) \qquad \therefore \int \dfrac{dx}{a^2 - x^2} = \dfrac{1}{a} \tanh^{-1}\left(\dfrac{x}{a}\right) + c.$

The use of partial fractions
We can often rearrange a function by using partial fractions before attempting the integration. The methods of partial fractions were discussed in Chapter 1 and we shall use the results in this section.

Example 21 Find $\int \dfrac{x - 2}{x^2 - 4x - 5} \, dx$.

Now $x^2 - 4x - 5 \equiv (x + 1)(x - 5)$

$$\therefore \quad \frac{x - 2}{(x + 1)(x - 5)} \equiv \frac{A}{(x + 1)} + \frac{B}{(x - 5)} = \frac{A(x - 5) + B(x + 1)}{(x + 1)(x - 5)}$$

Equating coefficients gives $A = \frac{1}{2}$, $B = \frac{1}{2}$.

$$\therefore \quad \int \frac{x - 2}{(x + 1)(x - 5)} \, dx = \int \frac{1}{2(x + 1)} \, dx + \int \frac{1}{2(x - 5)} \, dx$$

$$= \tfrac{1}{2} \ln (x + 1) + \tfrac{1}{2} \ln (x - 5) + c$$

$$= \tfrac{1}{2} \ln (x + 1)(x - 5) + c = \ln k \sqrt{x^2 - 4x - 5}.$$

Notice that this result could have been obtained by writing the integral in the form $f'(x)/f(x)$

i.e. $\int \dfrac{x - 2}{x^2 - 4x - 5} \, dx = \dfrac{1}{2} \int \dfrac{2x - 4}{x^2 - 4x - 5} \, dx = \dfrac{1}{2} \ln (x^2 - 4x - 5) + c.$

This method gives an alternative way of doing example 20.

$$\int \frac{dx}{a^2 - x^2} = \frac{1}{a} \int \left(\frac{1}{2(a - x)} + \frac{1}{2(a + x)} \right) dx$$

$$= -\frac{1}{2a} \ln(a-x) + \frac{1}{2a} \ln(a+x) + c = \frac{1}{2a} \ln\left(\frac{a+x}{a-x}\right) + c.$$

Combining these results we have

$$\int \frac{dx}{a^2 - x^2} = \frac{1}{2a} \ln\left(\frac{a+x}{a-x}\right) + c = \frac{1}{a} \tanh^{-1}\left(\frac{x}{a}\right) + c.$$

Remember that if the integrand is a rational function in which the numerator has equal or higher degree than the denominator it is best to divide out the fraction first.

Example 22 Find $\int \frac{x^2}{x+1}\, dx$.

Now $\frac{x^2}{x+1} \equiv x - 1 + \frac{1}{x+1}$ by division (see page 10).

$$\int \frac{x^2}{x+1}\, dx = \int \left(x - 1 + \frac{1}{x+1}\right) dx = \frac{1}{2}x^2 - x + \ln(x+1) + c.$$

Integration by parts

The product rule for differentiation was given on page 49.

i.e. if u and v are functions of x, $\dfrac{d}{dx}(uv) = u\dfrac{dv}{dx} + v\dfrac{du}{dx}$.

Integrating both sides with respect to x,

$$uv = \int u\frac{dv}{dx}\, dx + \int v\frac{du}{dx}\, dx \quad \text{or}$$

$$\int u\frac{dv}{dx}\, dx = uv - \int v\frac{du}{dx}\, dx \qquad (1)$$

This result provides an equivalent of the product rule for differentiation, in integration.

For a definite integral between values of a and b the result becomes.

$$\int_a^b u\frac{dv}{dx}\, dx = \left[uv\right]_a^b - \int_a^b v\frac{du}{dx}\, dx \qquad (2)$$

Example 23 Find $\int x\sin x\, dx$.

Let $u = x \Rightarrow \dfrac{du}{dx} = 1$ and $\dfrac{dv}{dx} = \sin x \Rightarrow v = -\cos x$

At this stage we are choosing a particular integral and thus we let the constant of integration be zero.

Using (1) $\int x \sin x \, dx = -x \cos x - \int -\cos x \, dx$

$$= -x \cos x + \sin x + c$$

Example 24 Evaluate $\int_0^1 x^2 e^x \, dx$.

Let $u = x^2 \Rightarrow \dfrac{du}{dx} = 2x$ and $\dfrac{dv}{dx} = e^x \Rightarrow v = e^x$

Integrating by parts

$$\int_0^1 x^2 e^x \, dx = \left[x^2 e^x \right]_0^1 - \int_0^1 2x e^x \, dx.$$

The second integral can be evaluated using parts again

$$u = 2x \Rightarrow \frac{du}{dx} = 2 \quad \frac{dv}{dx} = e^x \Rightarrow v = e^x$$

$$\therefore \int_0^1 x^2 e^x \, dx = \left[x^2 e^x \right]_0^1 - \left\{ \left[2x e^x \right]_0^1 - \int_0^1 2e^x \, dx \right\}$$

$$= \left[x^2 e^x - 2x e^x \right]_0^1 + \left[2e^x \right]_0^1$$

$$= e^1 - 2e^1 + 2e^1 - 2$$

$$= e - 2 \simeq 0 \cdot 718.$$

Some functions do not appear to be products but can be integrated by parts using 1 as one of the terms.

Example 25 $\int \tan^{-1} x \, dx$.

Let $u = \tan^{-1} x \Rightarrow \dfrac{du}{dx} = \dfrac{1}{1+x^2}$ and $\dfrac{dv}{dx} = 1 \Rightarrow v = x$

Integrating by parts

$$\int \tan^{-1} x \, dx = x \tan^{-1} x - \int \frac{x}{1+x^2} \, dx$$

$$= x \tan^{-1} x - \tfrac{1}{2} \ln (1 + x^2) + c.$$

One useful technique in integration by parts is to produce the original integral again thus enabling it to be combined with the integral we are trying to evaluate. This method is particularly applicable to the product of exponentials and trigonometrical functions.

Example 26 Evaluate $I = \int e^{2x} \cos 3x \, dx$.

Let $u = \cos 3x \Rightarrow \dfrac{du}{dx} = -3 \sin 3x \qquad \dfrac{dv}{dx} = e^{2x} \Rightarrow v = \dfrac{1}{2} e^{2x}$

Integrating by parts

$$I = \frac{1}{2} e^{2x} \cos 3x - \int -\frac{3}{2} e^{2x} \sin 3x \, dx$$

$$= \frac{1}{2} e^{2x} \cos 3x + \frac{3}{2} \int e^{2x} \sin 3x \, dx$$

Repeating this process on the second integral

$$u = \sin 3x \Rightarrow \frac{du}{dx} = 3 \cos 3x \text{ and } \frac{dv}{dx} = e^{2x} \Rightarrow v = \frac{1}{2} e^{2x}$$

$$\therefore \quad I = \frac{1}{2} e^{2x} \cos 3x + \frac{3}{2} \left\{ \frac{1}{2} e^{2x} \sin 3x - \int \frac{3}{2} e^{2x} \cos 3x \, dx \right\}$$

$$= \frac{1}{2} e^{2x} \cos 3x + \frac{3}{4} e^{2x} \sin 3x - \frac{9}{4} I$$

$$\therefore \quad \frac{13}{4} I = \frac{1}{2} e^{2x} \cos 3x + \frac{3}{4} e^{2x} \sin 3x + k$$

Hence $\int e^{2x} \cos 3x \, dx = \dfrac{2}{13} e^{2x} \left(\cos 3x + \dfrac{3}{2} \sin 3x \right) + k$

Reduction formulae

The process developed in example 26 can be extended to cover a more general situation. Integration by parts can be used to find a recurrence relation giving one integral in terms of another. This usually enables the original integral to be expressed in terms of a simpler integral. The expression derived for this purpose is called a **reduction** formula.

Example 27 Find a reduction formula for $I_n = \displaystyle\int_0^{\pi/2} \cos^n x \, dx$

and hence evaluate $\displaystyle\int_0^{\pi/2} \cos^8 x \, dx$.

$\displaystyle\int_0^{\pi/2} \cos^n x \, dx$ can be written as $\displaystyle\int_0^{\pi/2} \cos^{n-1} x \cos x \, dx$.

Integrate by parts

$$u = \cos^{n-1} x \Rightarrow \frac{du}{dx} = -\sin x \, (\cos^{n-2} x)(n-1)$$

$$\frac{dv}{dx} = \cos x \Rightarrow v = \sin x$$

$$\therefore \quad I_n = \left[\sin x \cos^{n-1} x \right]_0^{\pi/2} - \int_0^{\pi/2} -(n-1) \sin^2 x \cos^{n-2} x \, dx$$

$$= (n-1) \int_0^{\pi/2} (1 - \cos^2 x) \cos^{n-2} x \, dx$$

$$= (n-1) \int_0^{\pi/2} \cos^{n-2} x \, dx - (n-1) \int_0^{\pi/2} \cos^n x \, dx$$

Since the two integrals on the right hand side are of the same format as the original they can be written as I_{n-2} and I_n respectively.

$$\therefore \quad I_n = (n-1)I_{n-2} - (n-1)I_n$$

$$nI_n = (n-1)I_{n-2}$$

$$I_n = \left(\frac{n-1}{n} \right) I_{n-2} \quad \text{for } n \geqslant 2$$

This is the reduction formula which gives I_n in terms of I_{n-2}. From this result it follows that $I_{n-2} = \dfrac{n-3}{n-2} I_{n-4}$

$$\text{and} \quad I_{n-4} = \frac{n-5}{n-4} I_{n-6} \text{ etc.}$$

Thus, numerically $I_8 = \dfrac{7 \cdot 5 \cdot 3 \cdot 1}{8 \cdot 6 \cdot 4 \cdot 2} I_0$

Now $I_0 = \displaystyle\int_0^{\pi/2} 1 \, dx = \left[x \right]_0^{\pi/2} = \dfrac{\pi}{2}$

$$\therefore \quad \int_0^{\pi/2} \cos^8 x \, dx = \frac{7 \cdot 5 \cdot 3 \cdot 1}{8 \cdot 6 \cdot 4 \cdot 2} (\pi/2) = \frac{35\pi}{256}$$

This technique can be extended to cover a product of two terms.

Example 28 Find a reduction formula for $\displaystyle\int x^n \sin x \, dx$.

Let $I_n = \displaystyle\int x^n \sin x \, dx$ and integrate by parts.

Let $u = x^n \Rightarrow \dfrac{du}{dx} = nx^{n-1}$ $\quad \dfrac{dv}{dx} = \sin x \Rightarrow v = -\cos x$

$$\therefore \quad I_n = -x^n \cos x - \int -n \cos x \cdot x^{n-1} \, dx$$

Repeating on the second integral with $u = x^{n-1}$ and $\dfrac{dv}{dx} = \cos x$

we have $I_n = -x^n \cos x + n\left\{x^{n-1} \sin x - (n-1) \displaystyle\int x^{n-2} \sin x \, dx\right\}$

$$I_n = -x^n \cos x + nx^{n-1} \sin x - n(n-1)I_{n-2}$$

Example 29 Show that if $I_{m,n} = \displaystyle\int \cos^m x \sin nx \, dx$ then a reduction formula can be written in the form
$$(m+n)I_{m,n} = -\cos^m x \cos nx + mI_{m-1,n-1}.$$

Let $u = \cos^m x \Rightarrow \dfrac{du}{dx} = m \cos^{m-1} x(-\sin x)$

$$\frac{dv}{dx} = \sin nx \Rightarrow v = -\frac{\cos nx}{n}$$

Integrating by parts

$$I_{m,n} = -\frac{\cos^m x \cos nx}{n} - \int -\frac{m}{n} \cos^{m-1} x \cos nx(-\sin x) \, dx$$

$$I_{m,n} = -\frac{\cos^m x \cos nx}{n} - \frac{m}{n} \int (\cos nx \sin x) \cos^{m-1} x \, dx$$

Now $\sin(n-1)x = \sin nx \cos x - \cos nx \sin x$
i.e. $\cos nx \sin x = \sin nx \cos x - \sin(n-1)x$

$$\therefore \quad I_{m,n} = -\frac{\cos^m x \cos nx}{n}$$
$$- \frac{m}{n} \int [\sin nx \cos x - \sin(n-1)x] \cos^{m-1} x \, dx$$

$$= -\frac{\cos^m x \cos nx}{n}$$
$$- \frac{m}{n} \int [\sin nx \cos^m x - \sin(n-1)x \cos^{m-1} x] \, dx$$

$$\therefore \quad nI_{m,n} = -\cos^m x \cos nx - mI_{m,n} + mI_{m-1,n-1}$$

$$\therefore \quad (m+n)I_{m,n} = -\cos^m x \cos nx + mI_{m-1,n-1}$$

This can be used to evaluate specific integrals.

$$I_{2,4} = \int_0^{\pi/4} \cos^2 x \sin 4x \, dx = \frac{1}{6}\left\{\left[-\cos^2 x \cos 4x\right]_0^{\pi/4}\right.$$
$$\left. + 2\int_0^{\pi/4} \cos x \sin 3x \, dx\right\}$$

$$= \frac{1}{6}\left\{\frac{3}{2} + 2I_{1,3}\right\} = \frac{1}{4} + \frac{1}{3}I_{1,3}$$

$$= \frac{1}{4} + \frac{1}{3} \cdot \frac{1}{4}\left\{\left[-\cos x \cos 3x\right]_0^{\pi/4} + I_{0,2}\right\}$$

$$= \frac{1}{4} + \frac{1}{12}\left\{\frac{3}{2} + I_{0,2}\right\}$$

Since $I_{0,2} = \int_0^{\pi/4} \sin 2x \, dx = \left[-\frac{1}{2}\cos 2x\right]_0^{\pi/4} = \frac{1}{2}$

$$\int_0^{\pi/4} \cos^2 x \sin 4x \, dx = \frac{5}{12}.$$

Special Methods

In integration there are many special techniques which help in the evaluation of given integrals. The following examples illustrate a few of the more common ones.

1. Integrals of the type $\int \dfrac{dx}{ax^2 + bx + c}$ can be evaluated by completing the square in the denominator.

Example 30 Find $\int \dfrac{dx}{x^2 + 6x + 17}$.

Now $x^2 + 6x + 17 \equiv (x + 3)^2 + 8$

$$\int \frac{dx}{x^2 + 6x + 17} = \int \frac{dx}{(x + 3)^2 + (\sqrt{8})^2}$$

Since $x + 3$ is linear this integral is of the form $\tan^{-1}\left(\dfrac{x}{a}\right)$.

$$\int \frac{dx}{x^2 + 6x + 17} = \frac{1}{\sqrt{8}}\tan^{-1}\left(\frac{x + 3}{\sqrt{8}}\right) + c.$$

2. Integrals of the type $\int \dfrac{Ax + B}{ax^2 + bx + c}\, dx$ can often be solved by rearranging the numerator to include the differential of the denominator.

Example 31 Find $\int \dfrac{2x + 5}{x^2 + 4x + 5}\, dx$.

Now $\dfrac{d}{dx}(x^2 + 4x + 5) = 2x + 4$.

Hence $\int \dfrac{2x+5}{x^2+4x+5}\, dx = \int \dfrac{2x+4}{x^2+4x+5}\, dx + \int \dfrac{1}{x^2+4x+5}\, dx$

$$= \ln(x^2+4x+5) + \int \dfrac{1}{(x+2)^2+1^2}\, dx$$

$$= \ln(x^2+4x+5) + \tan^{-1}(x+2) + c.$$

3. Integrals of the form $\int \dfrac{dx}{a+b\cos x}$ can usually be solved by using the substitution $t = \tan \frac{1}{2}x$.

Now if $t = \tan\left(\dfrac{x}{2}\right)$, $\tan x = \dfrac{2t}{1-t^2}$ (from the double angle formula for $\tan 2A$).

$\sin x = \dfrac{2t}{1+t^2}$, $\cos x = \dfrac{1-t^2}{1+t^2}$ and $\dfrac{dt}{dx} = \dfrac{1}{2}\sec^2\dfrac{x}{2} \Rightarrow \dfrac{dx}{dt} = \dfrac{2}{1+t^2}$.

Example 32 Evaluate $\displaystyle\int_0^{\pi/2} \dfrac{dx}{3+5\cos x}$.

Using the substitution $t = \tan \frac{1}{2}x$

$$\int_0^{\pi/2} \dfrac{dx}{3+5\cos x} = \int_{x=0}^{x=\pi/2} \dfrac{1}{3+5\cos x}\dfrac{dx}{dt}\, dt$$

$$= \int_0^1 \dfrac{1}{3+5\left(\dfrac{1-t^2}{1+t^2}\right)}\left(\dfrac{2}{1+t^2}\right) dt$$

$$= \int_0^1 \dfrac{2\, dt}{3(1+t^2)+5(1-t^2)} = \int_0^1 \dfrac{2\, dt}{8-2t^2} = \int_0^1 \dfrac{dt}{4-t^2}$$

$$= \dfrac{1}{4}\int_0^1 \left(\dfrac{1}{2-t}+\dfrac{1}{2+t}\right) dt = \dfrac{1}{4}\Big[\ln(2+t)-\ln(2-t)\Big]_0^1$$

$$= \left[\dfrac{1}{4}\ln\left|\dfrac{2+t}{2-t}\right|\right]_0^1 = \dfrac{1}{4}\ln 3.$$

4. Integrals similar to type 3 but involving $\cos 2x$, $\sin 2x$, $\tan 2x$, $\cos^2 x$, $\sin^2 x$ or $\tan^2 x$ can be evaluated using the substitution $t = \tan x$.

If $t = \tan x$, $\sin x = \dfrac{t}{\sqrt{1+t^2}}$, $\cos x = \dfrac{1}{\sqrt{1+t^2}}$,

$$\dfrac{dt}{dx} = \sec^2 x \Rightarrow \dfrac{dx}{dt} = \dfrac{1}{1+t^2}$$

Example 33 Evaluate $\int_0^{\pi/4} \dfrac{\sin^2 x}{1 + \cos^2 x}\, dx$.

Let $t = \tan x$ be the substitution

$$\int_0^{\pi/4} \frac{\sin^2 x}{1 + \cos^2 x}\, dx$$

$$= \int_0^1 \left(\frac{t^2}{1+t^2}\right) \div \left(1 + \frac{1}{1+t^2}\right) \frac{dx}{dt}\, dt$$

$$= \int_0^1 \left(\frac{t^2}{1+t^2}\right)\left(\frac{1+t^2}{2+t^2}\right) \frac{1}{(1+t^2)}\, dt = \int_0^1 \frac{t^2\, dt}{(2+t^2)(1+t^2)}$$

$$= \int_0^1 \left(\frac{2}{2+t^2} - \frac{1}{1+t^2}\right) dt = \left[\sqrt{2}\tan^{-1}\left(\frac{t}{\sqrt{2}}\right) - \tan^{-1} t\right]_0^1$$

$$= \sqrt{2}\tan^{-1}\left(\frac{1}{\sqrt{2}}\right) - \tan^{-1}(1) = \sqrt{2}\tan^{-1}\left(\frac{1}{\sqrt{2}}\right) - \frac{\pi}{4}$$

Infinite integrals

There are two cases to be considered:

 (i) definite integrals where the range is infinite.
(ii) definite integrals in which the integrand becomes infinite within the range.

(i) We define $\displaystyle\int_a^\infty f(x)\, dx = \lim_{N\to\infty} \int_a^N f(x)\, dx$

Example 34 Evaluate $\displaystyle\int_0^\infty e^{-x}\cos x\, dx$.

Integrating by parts with $u = e^{-x}$, $\dfrac{dv}{dx} = \cos x$ we have

$$\int_0^N e^{-x}\cos x\, dx$$

$$= \left[e^{-x}\sin x\right]_0^N - \int_0^N -\sin x\, e^{-x}\, dx$$

$$= \left[e^{-x}\sin x\right]_0^N + \left[-\cos x\, e^{-x}\right]_0^N - \int_0^N -e^{-x}(-\cos x)\, dx$$

$$\therefore \int_0^N e^{-x}\cos x\, dx = \frac{1}{2}\left[e^{-x}\sin x - e^{-x}\cos x\right]_0^N$$

$$= \frac{1}{2} e^{-N}(\sin N - \cos N) - \frac{1}{2}(-1)$$

As $\displaystyle\lim_{N\to\infty} e^{-N} = 0$ and $\sin N$, $\cos N$ are finite

$$\int_0^\infty e^{-x} \cos x \, dx = \lim_{N \to \infty} \int_0^N e^{-x} \cos x \, dx = \frac{1}{2}$$

(ii) When the integrand is infinite, say at $x = c$, $a \leqslant c \leqslant b$, then we define the integral

$$\int_a^b f(x) \, dx = \lim_{\alpha \to 0} \int_a^{c-\alpha} f(x) \, dx + \lim_{\beta \to 0} \int_{c+\beta}^b f(x) \, dx$$

Example 35 Evaluate $\displaystyle\int_0^1 \frac{dx}{\sqrt{1-x^2}}$.

Since the function $1/(1-x^2)^{1/2}$ becomes infinite at $x = 1$

$$\int_0^{1-\alpha} \frac{dx}{\sqrt{1-x^2}} = \left[\sin^{-1} x \right]_0^{1-\alpha} = \sin^{-1}(1-\alpha) - 0$$

as $\alpha \to 0$, $\sin^{-1}(1-\alpha) \to \frac{1}{2}\pi$

$$\int_0^1 \frac{dx}{\sqrt{1-x^2}} = \lim_{\alpha \to 0} \int_0^{1-\alpha} \frac{dx}{\sqrt{1-x^2}} = \frac{1}{2}\pi$$

Volumes of revolution

We can use integration to calculate the volumes of solids by a technique similar to that developed for finding areas.

Consider a function $y = f(x)$. If the curve is rotated about the x axis it forms a surface and encloses a volume. For the shape shown in fig. 45 the rotation would form a bowl on a circular base which has a diameter AB. Clearly then, if we choose an element of area, shown shaded, and rotate it about the x axis, an element of volume is formed. If the points $P(x, y)$ and $Q(x + \delta x, y + \delta y)$ lie on the curve then the volume of this

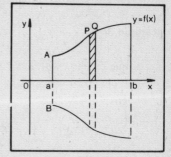

Figure 45

approximate cylinder is $\pi y^2 \delta x$. By summing all such elements in the range $x = a$ to $x = b$ and taking the limit as $\delta x \to 0$ we obtain an expression for the volume:

$$\text{Volume} = \lim_{\delta x \to 0} \left\{ \sum_{x=a}^{x=b} \pi y^2 \delta x \right\} = \int_a^b \pi y^2 \, dx$$

100

Example 36 Find the volume of a cone of height h and base radius r.

Let $f(x)$ in this case be given by $y = \dfrac{r}{h} x$. This will give a cone on rotation of radius r and height h. See fig. 46. Consider an element of volume formed by rotating the shaded area about the x axis. Volume of the element $= \pi y^2 \delta x$.

Summing all such elements between $x = 0$ and $x = h$ gives the total volume as $\displaystyle\sum_{x=0}^{x=h} \pi y^2 \delta x$.

In the limit as $\delta x \to 0$

$$\text{Volume} = \pi \int_0^h y^2 \, dx$$

$$= \pi \int_0^h \frac{r^2 x^2}{h^2} \, dx$$

$$= \pi \left[\frac{r^2 x^3}{3h^2} \right]_0^h$$

$$= \frac{1}{3} \pi r^2 h$$

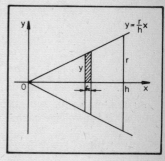

Figure 46

This method can be used even if the notation is not about the x axis.

Example 37 Find the volume generated when the curve $y = x^2 - 2x + 4$ is rotated about the line $y = 7$.

To find the points of intersection solve simultaneously.

$$\therefore \quad x^2 - 2x + 4 = 7$$

$$\Leftrightarrow x^2 - 2x - 3 = 0$$

$$\Leftrightarrow (x - 3)(x + 1) = 0$$

$$\Rightarrow x = -1 \text{ or } 3.$$

An element of volume is now formed by rotating the shaded element of area about $y = 7$.

Volume of element

$$= \pi (7 - y)^2 \, \delta x$$

$$= \pi (3 + 2x - x^2)^2 \, \delta x$$

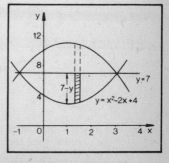

Figure 47

Total volume $= \pi \int_{-1}^{3} (3 + 2x - x^2)^2 \, dx$

$= \pi \int_{-1}^{3} (9 + 12x - 2x^2 - 4x^3 + x^4) \, dx.$

$= \pi \left[9x + 6x^2 - \frac{2}{3}x^3 - x^4 + \frac{1}{5}x^5 \right]_{-1}^{3}$

$= \pi \left\{ \left(27 + 54 - 18 - 81 + \frac{243}{5} \right) \right.$

$\left. - \left(-9 + 6 + \frac{2}{3} - 1 - \frac{1}{5} \right) \right\}$

$= \pi \left\{ \left(30\frac{3}{5} \right) - \left(-3\frac{8}{15} \right) \right\} = \pi \left(34\frac{2}{15} \right) = \frac{512}{15} \pi$

Note that, unlike the equivalent problem for finding areas, the calculation for volume cannot result in a negative answer.

Mean values

If $y = f(x)$ then the **mean or average value** of $f(x)$ over a given range $a \leqslant x \leqslant b$ is defined by

$$\frac{1}{(b-a)} \int_a^b f(x) \, dx$$

Consider $y = f(x)$ is shown in fig. 48 and let $y_1, y_2, y_3, y_4 \ldots y_n$ be the ordinates corresponding to $x = a$, $x = a + \delta x$, $x = a + 2\delta x$, $x = a + (n-1)\,\delta x$ where $n\delta x = b - a$. The mean of these values is

$\frac{1}{n}(y_1 + y_2 + \cdots + y_n)$

$= \frac{y_1\delta x + y_2\delta x + \cdots + y_n\delta x}{n\delta x}$

$\therefore$ Mean value $=$

$$\frac{1}{b-a} \sum_{x=a}^{x=b} y\delta x$$

If this result tends to a limit as $n \to \infty$ (i.e. $\delta x \to 0$) then the limit is $\dfrac{1}{b-a} \displaystyle\int_a^b y \, dx$.

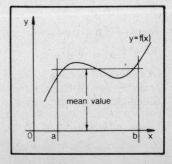

Figure 48

Geometrically, this can be seen as the height of the rectangle having the same area as the area under the graph over the interval $a \leq x \leq b$. See fig. 48.

Example 38 Find the mean value of the function $f(x) = \sin x$ over the interval $0 \leq x \leq \pi$.

The required mean $= \dfrac{1}{\pi - 0} \displaystyle\int_0^\pi \sin x \, dx = \dfrac{1}{\pi} \Big[-\cos x \Big]_0^\pi$

$$= \frac{1}{\pi}\{1 - (-1)\} = \frac{2}{\pi} \approx 0.637.$$

Velocity and acceleration

We can now complete the work on velocity and acceleration started in Chapter 3, page 63.

We derived two basic results.

(i) Velocity can be defined as $\dfrac{ds}{dt}$.

(ii) Acceleration is given by $\dfrac{dv}{dt} = \dfrac{d^2s}{dt^2} = v\dfrac{dv}{ds}$.

We can rewrite these in an integral format as

$$s = \int v \, dt \quad \text{and} \quad v = \int a \, dt$$

Example 39 A particle starts from rest at O and moves along a straight line OA so that its acceleration after t secs is $6t - 3t^2$ m s^{-2}. Determine (i) when it returns to O, and (ii) its maximum displacement from O in this interval.

Since $\dfrac{dv}{dt} = 6t - 3t^2$, $v = 3t^2 - t^3 + c$.

When $t = 0$, $v = 0$ (since it starts from rest) $\therefore c = 0$.

$$\therefore v = 3t^2 - t^3 \Rightarrow s = t^3 - \frac{t^4}{4} + B$$

When $t = 0$, $s = 0$, $\therefore B = 0$.

$\therefore$ displacement is given by $s = t^3 - \dfrac{t^4}{4}$ m

(i) When displacement is zero $t^3 - \dfrac{t^4}{4} = 0 \Rightarrow 4t^3 - t^4 = 0$

$$\Rightarrow t^3(4 - t) = 0 \Rightarrow t = 0 \text{ or } t = 4$$

Thus the particle returns to 0 after 4 secs.
(ii) The maximum displacement occurs when $v = 0$
i.e. $3t^2 - t^3 = 0 \Leftrightarrow t^2(3 - t) = 0 \Rightarrow t = 0$ or $t = 3$
Hence the maximum displacement $s = 3^3 - \dfrac{3^4}{4} = 27 - 20\frac{1}{4} = 6\frac{3}{4}$ m

Centre of mass

We can use a method similar to that developed for areas and volumes to find the position of the centre of mass of a lamina or solid body.

We know from mechanics that the centre of mass of a system of n discrete particles $m_1, m_2, \ldots m_n$ whose position vectors are $r_1, r_2, \ldots r_n$ is given by

$$r_g \sum_{i=1}^{i=n} m_i = \sum_{i=1}^{i=n} m_i r_i$$

We can extend this to cover the case of a lamina or solid body which can be considered as being made up of a large number of elementary areas or volumes. In this case we need to use limiting sums (i.e. integrals) rather than summations. Three examples will illustrate the method.

Example 40 Find the position of the centre of mass of a uniform rod of length $2a$.
If m is the mass per unit length, the total mass is $2am$. Consider an element of length δx distant x from one end, A. The mass of the element is $m\delta x$ and its moment about A is $mx\delta x$.

Hence over the whole rod $2am\bar{x} = \displaystyle\int_0^{2a} mx\,dx$

$$\therefore \quad 2a\bar{x} = \left[\frac{x^2}{2}\right]_0^{2a} \Rightarrow 2a\bar{x} = 2a^2 \Rightarrow \bar{x} = a$$

Thus the centre of mass of a uniform rod is at its centre.

Example 41 Find the position of the centre of mass of a uniform lamina bounded by $y^2 = 4x$ and the line $x = 4$.

Let m be the mass per unit area. The typical element has an area of $2y\delta x$ and a mass of $2my\delta x$. Its moment about the y axis is $2my\delta x.x$. The total moment is given by $\displaystyle\int_0^4 2mxy\,dx$

The total mass is given by $\displaystyle\int_0^4 2my\,dx$

Hence if $\bar{x}$ is the distance of the centre of mass from the y axis we have $\bar{x} \int_0^4 2my \, dx = \int_0^4 2mxy \, dx$

Now $y = 2x^{1/2}$, so after dividing by $2m$ we have

$$\bar{x} \int_0^4 2x^{1/2} \, dx = \int_0^4 2x^{3/2} \, dx$$

$$\bar{x} \left[\frac{2x^{3/2}}{3}\right]_0^4 = \left[\frac{2x^{5/2}}{5}\right]_0^4$$

$$\bar{x} \cdot \frac{16}{3} = \frac{64}{5} \qquad \bar{x} = \frac{12}{5}$$

By symmetry the centre of mass lies on the x axis and hence its position is given by

$$\left(\tfrac{12}{5}, 0\right).$$

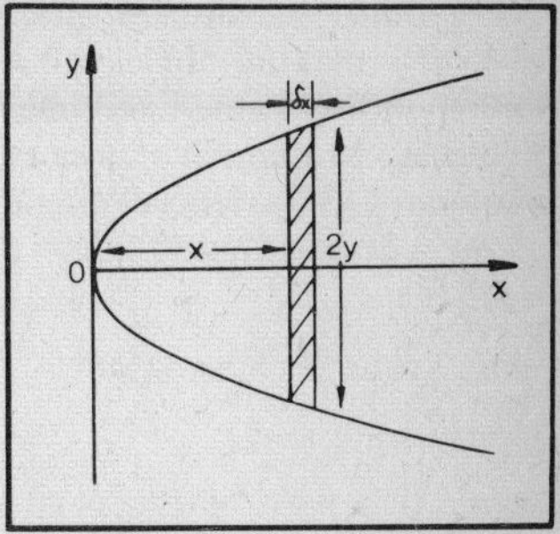

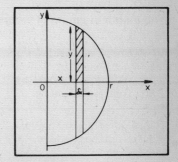

Figure 49

Example 42 Find the position of the centre of mass of a solid hemisphere obtained by rotating the part of the curve $x^2 + y^2 = r^2$ in the positive quadrant about the x axis.

Let m be the mass per unit volume. An element in the form of a disc is taken whose approximate volume is $\pi y^2 \delta x$.

Hence the total mass is $\int_0^r m\pi y^2 \, dx$ and the total moment about

the y axis is $\int_0^r \pi mxy^2 \, dx$. Now if $\bar{x}$ is the distance of the centre of

mass from the plane face

$$\bar{x} \int_0^r y^2 \, dx = \int_0^r xy^2 \, dx$$

Now $y^2 = r^2 - x^2$

$$\therefore \quad \bar{x} \int_0^r (r^2 - x^2) \, dx$$

$$= \int_0^r (r^2 x - x^3) \, dx$$

$$\bar{x} \left[r^2 x - \frac{x^3}{3}\right]_0^r = \left[\frac{r^2 x^2}{2} - \frac{x^4}{4}\right]_0^r$$

Figure 50

$$\bar{x}\left(r^3 - \frac{r^3}{3}\right) = \left(\frac{r^4}{2} - \frac{r^4}{4}\right) \text{ and } \frac{2r^3}{3}\bar{x} = \frac{r^4}{4} \Rightarrow \bar{x} = \frac{3r}{8}$$

By symmetry the centre of mass lies on the x axis and hence its position is $\frac{3}{8}r$ along a radius of symmetry.

Area of a sector

Areas can also be found using polar coordinates.

We require the area of the sector AOB between two radii OA, OB making angles of α and β with OX, the fixed line. Selecting an element of area as shown, OPQ, where angle $POQ = \delta\theta$ and angle $POX = \theta$ we have

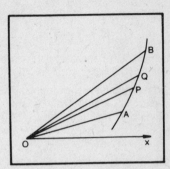

Area sector $POQ \simeq \frac{1}{2}r^2\delta\theta$

Hence total area

$$AOB = \int_\alpha^\beta \frac{1}{2}r^2 \, d\theta$$

obtained after summing be-
tween $\theta = \alpha$ and $\theta = \beta$ and   *Figure 51*
taking the limit as $\delta\theta \to 0$

Example 43 Find the area of the loop of the curve $r = 3\sin 2\theta$ between $\theta = 0$ and $\theta = \pi/2$.

Now area $= \displaystyle\int_0^{\pi/2} \frac{1}{2}r^2 \, d\theta = \int_0^{\pi/2} \frac{9}{2}\sin^2 2\theta \, d\theta$

$$= \int_0^{\pi/2} \frac{9}{2} \cdot \frac{1}{2}(1 - \cos 4\theta) \, d\theta = \frac{9}{4}\left[\theta - \frac{\sin 4\theta}{4}\right]_0^{\pi/2} = \frac{9\pi}{8}$$

The length of an arc of a curve

Consider a curve $y = f(x)$ and let PQ be a small section of length δs. Let A be a fixed point of the curve so that $AP = s$ measured along the curve.

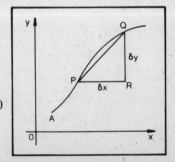

Hence if P and Q are the points (x, y) and $(x + \delta x, y + \delta y)$ respectively

$$(\delta y)^2 + (\delta x)^2 = PQ^2$$

by Pythagoras' theorem.
But if δx is small, $PQ \simeq \delta s$ *Figure 52*

$$(\delta s)^2 \simeq (\delta x)^2 + (\delta y)^2 \quad \text{or} \quad \left(\frac{\delta s}{\delta x}\right)^2 \simeq 1 + \left(\frac{\delta y}{\delta x}\right)^2 \tag{1}$$

In the limit as $\delta x \to 0$ $\quad \dfrac{ds}{dx} = \sqrt{1 + \left(\dfrac{dy}{dx}\right)^2}$

Integrating with respect to x gives $s = \displaystyle\int \sqrt{1 + \left(\frac{dy}{dx}\right)^2}\, dx$

Notice that by dividing the expression (1) by $(\delta y)^2$ we could have obtained an equivalent result in the form

$$s = \int \sqrt{1 + \left(\frac{dx}{dy}\right)^2}\, dy.$$

Example 44 Find the length of the curve $y^2 = 4x$ from the vertex to $x = 4$.

Now $\dfrac{dy}{dx} = \dfrac{2}{y} = \dfrac{2}{2\sqrt{x}}$, hence if s is the length of the curve

$$s = \int_0^4 \sqrt{1 + \frac{1}{x}}\, dx = \int_0^4 \sqrt{\frac{x+1}{x}}\, dx = \int_0^4 \frac{x+1}{\sqrt{x^2 + x}}\, dx$$

$$= \frac{1}{2} \int_0^4 \left(\frac{2x+1}{\sqrt{x^2+x}} + \frac{1}{\sqrt{x^2+x}}\right) dx$$

$$= \left[(x^2 + x)^{1/2}\right]_0^4 + \frac{1}{2} \int_0^4 \frac{1}{\sqrt{x^2+x}}\, dx$$

$$= \sqrt{20} + \frac{1}{2} \int_0^4 \frac{1}{\sqrt{(x + \frac{1}{2})^2 - \frac{1}{4}}}\, dx$$

$$= \sqrt{20} + \left[\tfrac{1}{2} \ln \left\{(x + \tfrac{1}{2}) + \sqrt{(x + \tfrac{1}{2})^2 - \tfrac{1}{4}}\right\}\right]_0^4$$

$$= \sqrt{20} + \tfrac{1}{2} \ln (9 + 2\sqrt{20})$$

The length of the arc is $2\sqrt{5} + \frac{1}{2} \ln (9 + 4\sqrt{5})$.

Area of a surface of revolution

Consider an arc of a curve of length δs being rotated about the x axis. The surface area of the disc formed is thus $2\pi y \delta s$.

∴ The surface area generated by one complete revolution about the x axis is found by summing all elements and taking the limit as $\delta s \to 0$.

i.e. Surface area $= \displaystyle\int 2\pi y\, ds$

Remembering that $(\delta s)^2 \simeq (\delta x)^2 + (\delta y)^2$ (see page 107)

$$\therefore \quad \frac{\delta s}{\delta x} \simeq \sqrt{1 + \left(\frac{\delta y}{\delta x}\right)^2} \Rightarrow \frac{ds}{dx} = \sqrt{1 + \left(\frac{dy}{dx}\right)^2}$$

Hence the integral above can usually be evaluated as follows

Surface area $= \displaystyle\int 2\pi y \sqrt{1 + \left(\frac{dy}{dx}\right)^2}\, dx = \int 2\pi y \sqrt{1 + \left(\frac{dx}{dy}\right)^2}\, dy$

A similar result holds for rotation about the y axis

Surface area $= \displaystyle\int 2\pi x \sqrt{1 + \left(\frac{dy}{dx}\right)^2}\, dx = \int 2\pi x \sqrt{1 + \left(\frac{dx}{dy}\right)^2}\, dy$

Example 45 Find the surface area of a sphere of radius a. Consider a circle centred on the origin whose equation is $x^2 + y^2 = a^2$. Let the part of the curve in the first quadrant be rotated about the x axis.

$\therefore$ Surface area of the hemisphere $=$

$$2\pi \int_{x=0}^{x=a} y\, ds = 2\pi \int_0^a y \sqrt{1 + \left(\frac{dy}{dx}\right)^2}\, dx$$

Now since $x^2 + y^2 = a^2$, $2x + 2y\dfrac{dy}{dx} = 0$, $\dfrac{dy}{dx} = -\dfrac{x}{y}$

Hence $\sqrt{1 + \left(\dfrac{dy}{dx}\right)^2} = \sqrt{1 + \left(\dfrac{-x}{y}\right)^2} = \sqrt{\dfrac{x^2 + y^2}{y^2}} = \sqrt{\dfrac{a^2}{y^2}} = \dfrac{a}{y}$

$\therefore$ Surface area $= 2\pi \displaystyle\int_0^a y \cdot \frac{a}{y}\, dx = 2\pi \int_0^a a\, dx$

$$= 2\pi \left[ax \right]_0^a = 2\pi a^2$$

Thus the surface area of the sphere $= 4\pi a^2$.

Theorems of Pappus

Two important theorems relating to volumes of revolution, areas of surfaces and centres of mass are due to Pappus.

1. If a plane area is revolved about an axis not intersecting it, the volume of revolution is equal to the product of the area and the length of the arc covered by its centre of mass.

Consider the area A as shown in fig. 53a, rotated about the x axis, and let the ordinate of its centre of mass be $\bar{y}$. Consider an element of area δA at P distant y from the axis.

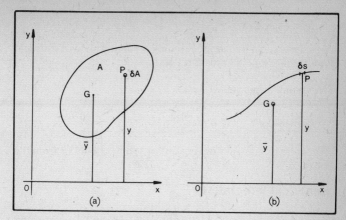

Figure 53

By definition $\bar{y} = \lim\limits_{\delta A \to 0} \left[\dfrac{\Sigma \, y \delta A}{\Sigma \, \delta A} \right]$ i.e. $A\bar{y} = \displaystyle\int y \, dA$ the integration ranging over the whole area.

The volume generated by δA in one revolution $\simeq 2\pi y \delta A$

$\therefore$ Total volume $= \displaystyle\int 2\pi \, y \, dA = 2\pi \bar{y} \, . \, A$

$\qquad\qquad\qquad$ = Area $\times$ distance moved by the centre of mass.

2. If an arc of a plane curve is revolved about a coplanar axis not intersecting it, the area of the surface of revolution is equal to the product of the length of the arc and the distance moved by the centre of mass.

Consider a plane curve of length s, (see fig. 53b) rotated about the x axis and let the ordinate of its centre of mass be $\bar{y}$. Consider an element of length δs distant y from the axis.

Then $\bar{y} = \lim\limits_{\delta s \to 0} \dfrac{\Sigma \, y \delta s}{\Sigma \, \delta s}$ i.e. $s\bar{y} = \displaystyle\int y \, ds$ where the integration extends over the whole length.

The surface area generated by δs in one revolution $\simeq 2\pi y \delta s$.

$\therefore$ Total area of surface $= 2\pi \displaystyle\int y \, ds = 2\pi \bar{y} s$

$\qquad\qquad\qquad$ = length of the arc $\times$ distance moved by G.

109

Example 46 Find the position of the centre of mass of a uniform semicircular lamina of radius a.

Consider a semicircular area, radius a, rotated about its straight edge through one complete revolution.

The volume generated $= \frac{4}{3}\pi a^3$

Using theorem 1 $\quad \frac{4}{3}\pi a^3 = \frac{1}{2}\pi a^2 . 2\pi\bar{y} \Rightarrow \bar{y} = \frac{4a}{3\pi}$

The centre of mass lies on a radius of symmetry $4a/3\pi$ from the centre.

Numerical integration

It may not always be possible to find the integral $\int_a^b f(x)\,dx$ by the methods described in this chapter. Indeed, it may not be possible to integrate a function by an analytical process at all and in this case we can use an approximate method.
1. the trapezium rule 2. Simpson's rule.

The trapezium rule

In this method the area is divided into a number of strips and the points where the ordinates meet the curve are joined by straight lines to form trapezia. See fig. 54.

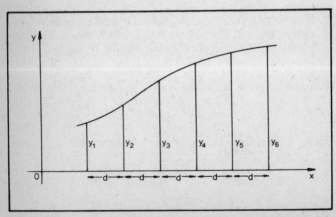

Figure 54

110

In this case we have chosen six ordinates $y_1, y_2, \ldots, y_6$ spaced at equal intervals d.

The total area is

$$\left(\frac{y_1 + y_2}{2}\right) d + \left(\frac{y_2 + y_3}{2}\right) d + \left(\frac{y_3 + y_4}{2}\right) d + \left(\frac{y_4 + y_5}{2}\right) d + \left(\frac{y_5 + y_6}{2}\right) d$$

$$= \tfrac{1}{2}d(y_1 + 2y_2 + 2y_3 + 2y_4 + 2y_5 + y_6)$$

This is the trapezium rule for six ordinates.

Example 47 Use the trapezium rule to estimate the value of $\int_1^2 \frac{2}{x}\, dx$.

Using six ordinates, area $= \tfrac{1}{2}d(y_1 + 2y_2 + 2y_3 + 2y_4 + 2y_5 + y_6)$

$$= \tfrac{1}{2}d[(y_1 + y_6) + 2(y_2 + y_3 + y_4 + y_5)]$$

With six ordinates $d = 0\cdot2$

x	1	1·2	1·4	1·6	1·8	2·0
y	2	1·6667	1·4286	1·2500	1·1111	1·0

Now $y_1 + y_6 = $ 1·0000 $\qquad\qquad y_2 = $ 1·6667

$\qquad\qquad\qquad\quad \underline{2\cdot0000}$ $\qquad\qquad\qquad\quad y_3 = $ 1·4286

$\qquad\qquad\qquad\quad 3\cdot0000$ $\qquad\qquad\qquad\quad y_4 = $ 1·2500

$\qquad\qquad\qquad\qquad\qquad\qquad\qquad\quad \underline{y_5 = \ 1\cdot1111}$

$\qquad\qquad\qquad\qquad\qquad\qquad\qquad\quad 5\cdot4564 \times 2$

$\qquad\quad \underline{10\cdot9128} \qquad\qquad\qquad\qquad = 10\cdot9128$

$\qquad\quad 13\cdot9128$

$$\therefore \quad \text{area} = \frac{0\cdot2}{2} \times 13\cdot9128 = 1\cdot3913$$

Note that $\int_1^2 \frac{2}{x}\, dx = 2\Big[\ln x\Big]_1^2 = 2\ln 2 = 1\cdot3862$

Simpson's rule

This method improves on the accuracy obtained by the trapezium rule by joining the tops of the ordinates with a smooth curve instead of straight lines. We assume a quadratic passes through three consecutive points on the curve. Let the quadratic be of the form $y = ax^2 + bx + c$ after translating so that the y axis lies along the centre ordinate.

Let the distance between successive ordinates be h. The coefficients a, b and c can be found since $(-h, y_1)$, $(0, y_2)$, (h, y_3) lie on the curve.

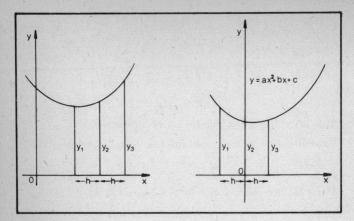

Figure 55

Thus

$$y_1 = ah^2 - bh + c \qquad (1)$$

$$y_2 = c \qquad (2)$$

$$y_3 = ah^2 + bh + c \qquad (3)$$

From equations (1) and (3) $y_1 + y_3 = 2ah^2 + 2c$ (4)

Now the area under the quadratic is

$$A = \int_{-h}^{h} (ax^2 + bx + c)\, dx = \left[\tfrac{1}{3}ax^3 + \tfrac{1}{2}bx^2 + cx \right]_{-h}^{h}$$

$$= \tfrac{2}{3}ah^3 + 2ch = \tfrac{1}{3}h(2ah^2 + 6c) = \tfrac{1}{3}h(y_1 + y_3 + 4y_2)$$

This is usually written as area $= \tfrac{1}{3}h(y_1 + 4y_2 + y_3)$.

The result holds for a quadratic arc in any position. This uses only three ordinates but, in practice, we normally require more than three ordinates. So provided there is an odd number of ordinates we can apply the rule a number of times. In fig. 56 seven ordinates are shown.

Clearly the total area is given by applying Simpson's rule three times.

Area $= \tfrac{1}{3}h(y_1 + 4y_2 + y_3) + \tfrac{1}{3}h(y_3 + 4y_4 + y_5) + \tfrac{1}{3}h(y_5 + 4y_6 + y_7)$

$\therefore$ Area $= \tfrac{1}{3}h(y_1 + 4y_2 + 2y_3 + 4y_4 + 2y_5 + 4y_6 + y_7)$

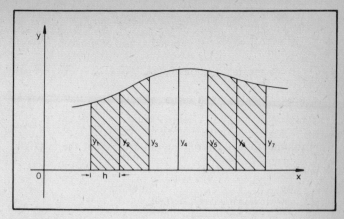

Figure 56

Learn the pattern of the result. The first and last terms have a coefficient of 1 and each of the other terms has a coefficient of 4 and 2 alternately.

i.e. for five ordinates area = $\frac{1}{3}h(y_1 + 4y_2 + 2y_3 + 4y_4 + y_5)$

Example 48 Use Simpson's rule with five ordinates to evaluate $\int_1^2 \frac{2}{x} \, dx$.

x	1	1·25	1·50	1·75	2
y	2	1·6000	1·3333	1·1428	1

$$
\begin{array}{ll}
y_1 = \ 2\cdot0000 & y_3 = 1\cdot3333 \quad y_2 = 1\cdot6000 \\
\underline{y_5 = \ 1\cdot0000} & \underline{\qquad\qquad \times 2} \quad \underline{y_4 = 1\cdot1428} \\
 3\cdot0000 & 2\cdot6666 \quad 2\cdot7428 \\
 2\cdot6666 & \times 4 \\
 10\cdot9712 & \underline{10\cdot9712} \\
 \underline{16\cdot6378} &
\end{array}
$$

Since $d = 0\cdot25 = \frac{1}{4}$, $\quad \frac{1}{3}d = \frac{1}{12}$

$$\text{Area} = \frac{1}{12} \times 16\cdot6378 = 1\cdot3865$$

This is a repeat of example 47 with a result nearer to $2\ln 2$.

Key terms

$$\int ax^n \, dx = \frac{ax^{n+1}}{n+1} + c \text{ for all rational values of } n \text{ except } n = -1.$$

If $F(x)$ is the integral function of $f(x)$, $\displaystyle\int_c^b f(x) \, dx = F(b) - F(a)$.

The **area** bounded by a curve $y = f(x)$ and the x axis is $\displaystyle\int_a^b f(x) \, dx$.

The **volume** formed by rotating an area about the x axis is $\pi \displaystyle\int_a^b y^2 \, dx$.

$$\int \frac{f'(x)}{f(x)} \, dx = \ln\{f(x)\} + c = \ln\{kf(x)\}.$$

There are two basic methods for **integration by substitution**,

(i) by using $u = h(x)$ whence $\displaystyle\int f(x) \, dx = \int f(x) \frac{dx}{du} \, du$

(ii) by using $x = f(u)$ whence $\displaystyle\int g(x) \, dx = \int g\{f(u)\} \frac{dx}{du} \, du$

Integration by parts. If $y = u(x) \,.\, v(x)$ then

$$\int u \frac{dv}{dx} \, dx = uv - \int v \frac{du}{dx} \, dx$$

The **mean value** of $f(x)$ over a range $a \leqslant x \leqslant b$ is

$$\frac{1}{(b-a)} \int_a^b f(x) \, dx.$$

The **area of a sector** in polar coordinates is $\displaystyle\int \frac{1}{2}r^2 \, d\theta$.

The **length of an arc** of a curve is $\displaystyle\int \sqrt{1 + \left(\frac{dy}{dx}\right)^2} \, dx$.

The **area of a surface** of revolution is found from $\displaystyle\int 2\pi y \, ds$.

The **trapezium rule** gives area $= \frac{1}{2}d\{y_1 + 2y_2 + 2y_3 + \cdots + y_n\}$.

Simpson's rule gives area $= \frac{1}{3}h\{y_1 + 4y_2 + y_3\}$.

This can be used with more ordinates, i.e. for five ordinates

$$\text{area} = \tfrac{1}{3}h\{y_1 + 4y_2 + 2y_3 + 4y_4 + y_5\}.$$

Chapter 5
Exponential and Logarithmic Functions

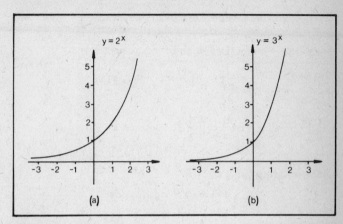

Figure 57

Figure 57 shows the graph of the function $y = 2^x$. From Chapter 1 we know $2^{-1} = \frac{1}{2}$, $2^0 = 1$ and $2^{1/2} = \sqrt{2} \simeq 1\cdot4142$ and these and other intermediate values enable a smooth curve to be drawn (2^k where k is irrational is defined to lie on the curve).

The gradient at $x = 0$ can be measured and found to be $\simeq 0.7$. We shall call this value M_2 (the gradient of 2^x when $x = 0$). If $y = 2^x$ then $y + \delta y = 2^{x+\delta x}$ if a small change δx in x corresponds to a small change δy in y.

$$\delta y = 2^{x+\delta x} - y = 2^{x+\delta x} - 2^x = 2^x(2^{\delta x} - 1)$$

$$\frac{dy}{dx} = \lim_{\delta x \to 0} \frac{\delta y}{\delta x} = \lim_{\delta x \to 0} \left(\frac{2^x(2^{\delta x} - 1)}{\delta x} \right)$$

$$x = 0 \Rightarrow \left(\frac{dy}{dx} \right)_{x=0} = M_2 = \lim_{\delta x \to 0} \left(\frac{2^0(2^{\delta x} - 1)}{\delta x} \right) = \lim_{\delta x \to 0} \left(\frac{2^{\delta x} - 1}{\delta x} \right)$$

In general $\dfrac{dy}{dx} = \lim_{\delta x \to 0} \dfrac{\delta y}{\delta x} = \lim \left(\dfrac{2^x(2^{\delta x} - 1)}{\delta x} \right) = 2^x M_2$

i.e. $\dfrac{dy}{dx} \simeq 0\cdot7 \,.\, 2^x$ if $y = 2^x$

115

Similarly (in fig. 57b) $y = 3^x \Rightarrow \dfrac{dy}{dx} = 3^x M_3$ where M_3 is the gradient of 3^x at $x = 0$. This can be measured and is $\simeq 1 \cdot 1$

$$y = 3^x \Rightarrow \frac{dy}{dx} \simeq 1 \cdot 1 \cdot 3^x$$

This suggests the existence of a function e^x where $\dfrac{dy}{dx} = 1 \cdot e^x$ and $2 < e < 3$. In fact $e \simeq 2.7$ ($e = 2 \cdot 71828$ to 5 decimal places). $y = e^x \Leftrightarrow x = \log_e y$ (Definition of logs, Chapter 1).

$$\frac{dx}{dy} = \frac{1}{\dfrac{dy}{dx}} = \frac{1}{e^x} = \frac{1}{y} \Leftrightarrow \frac{d}{dy}(\log_e y) = \frac{1}{y}$$

OR $\quad \dfrac{d}{dx}(\log_e x) = \dfrac{1}{x} \Leftrightarrow \displaystyle\int \frac{1}{x}\,dx = \log_e x + k$

$$= \log_e cx \text{ if } \log_e c = k$$

Differentiating $\log_{10} x$

$$y = \log_{10} x \Leftrightarrow y + \delta y = \log_{10}(x + \delta x)$$
$$\Leftrightarrow \quad \delta y = \log_{10}(x + \delta x) - y$$
$$= \log_{10}(x + \delta x) - \log_{10} x$$
$$= \log_{10}\left(\frac{x + \delta x}{x}\right)$$
$$\Rightarrow \quad \frac{\delta y}{\delta x} = \frac{1}{\delta x} \log_{10}\left(1 + \frac{\delta x}{x}\right)$$
$$\frac{dy}{dx} = \lim_{\delta x \to 0} \frac{\delta y}{\delta x} = \lim_{\delta x \to 0} \log_{10}\left(1 + \frac{\delta x}{x}\right)^{1/\delta x}$$

Let $\dfrac{\delta x}{x} = t$ then $\delta x \to 0 \Leftrightarrow t \to 0$ and $\dfrac{1}{\delta x} = \dfrac{1}{xt}$

$$\frac{dy}{dx} = \lim_{t \to 0} \log_{10}(1 + t)^{1/xt} = \lim_{t \to 0} \frac{1}{x} \log_{10}(1 + t)^{1/t} \qquad (1)$$

We need to evaluate the limit of $(1 + t)^{1/t}$ as $t \to 0$.
We do this by taking progressively smaller values of t.

i.e.
$$
\begin{array}{llll}
t = 0 \cdot 1 & (1 + t)^{1/t} = (1 \cdot 1)^{10} & = 2 \cdot 59374246 \\
t = 0 \cdot 01 & (1 + t)^{1/t} = (1 \cdot 01)^{100} & = 2 \cdot 70481386 \\
t = 0 \cdot 001 & (1 + t)^{1/t} = (1 \cdot 001)^{1000} & = 2 \cdot 71692302 \\
t = 0 \cdot 0001 & (1 + t)^{1/t} = (1 \cdot 0001)^{10000} & = 2 \cdot 71814591 \\
t = 0 \cdot 00001 & (1 + t)^{1/t} = (1 \cdot 00001)^{1000000} & = 2 \cdot 71828182
\end{array}
$$

It is clear that the limit is approaching a number which can be calculated to any degree of accuracy. This number we call e and correct to 9 decimal places $e = 2 \cdot 718281828$.

From the **Binomial theorem** (see Chapter 8)

$$(1+t)^{1/t} = 1 + \frac{1}{t} \cdot t + \left(\frac{1}{t}\right)\left(\frac{1}{t} - 1\right)\frac{t^2}{2} + \left(\frac{1}{t}\right)\left(\frac{1}{t} - 1\right)\left(\frac{1}{t} - 2\right)\frac{t^3}{2 \cdot 3} + \cdots$$

$$= 2 + \frac{1-t}{2} + \frac{(1-t)(1-2t)}{2 \cdot 3} + \frac{(1-t)(1-2t)(1-3t)}{2 \cdot 3 \cdot 4} + \cdots$$

$$t = 0 \text{ gives } \lim_{t \to 0}(1+t)^{1/t} = 2 + \frac{1}{2} + \frac{1}{6} + \frac{1}{24} + \cdots$$

$$= 2 \cdot 5 + 0 \cdot 1\dot{6} + 0 \cdot 041666 + 0 \cdot 008333 + 0 \cdot 001388 + \cdots$$

The sum of the first five terms $= 2 \cdot 7180555 \ldots$ and this will approach the value of e.

From (1) $\dfrac{dy}{dx} = \dfrac{1}{x}\log_{10} e$,

Consequently $y = \log_e x \Leftrightarrow \dfrac{dy}{dx} = \dfrac{1}{x}\log_e e = \dfrac{1}{x}$ since $\log_e e = 1$

OR $\displaystyle\int \frac{1}{x} dx = \log_e x + k = \log_e x + \log_e c = \log_e cx$

Integrating $f(x) = \dfrac{1}{x}$

$$\int x^m dx = \frac{x^{m+1}}{m+1} \quad \text{(from Chapter 4)}$$

and when $m = -1$ the formula gives $\displaystyle\int \frac{1}{x} dx = \frac{x^0}{0}$ which is infinite.

Define $F(a) = \displaystyle\int_1^a \frac{1}{x} dx$ so $F(1) = \displaystyle\int_1^1 \frac{1}{x} dx = 0$

$F(a) = $ area under graph from $x = 1$ to $x = a$ (horizontal shaded area in fig. 58b).

Apply the transformation represented by the matrix $\begin{pmatrix} b & 0 \\ 0 & 1/b \end{pmatrix}$

Under this transformation $(1, 0) \to (b, 0) \quad (1, 1) \to (b, 1/b)$

$\qquad\qquad\qquad (a, 0) \to (ab, 0) \quad (a, 1/a) \to (ab, 1/ab)$

The horizontal shaded area is transformed into the vertical shaded area.

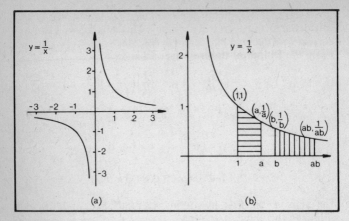

Figure 58

The determinant of $\begin{pmatrix} b & 0 \\ 0 & 1/b \end{pmatrix} = 1 \Rightarrow$ area is preserved under the transformation.

So $$\int_1^a \frac{1}{x}\, dx = \int_b^{ab} \frac{1}{x}\, dx = \int_1^{ab} \frac{1}{x}\, dx - \int_1^b \frac{1}{x}\, dx$$

i.e. $F(a) = F(ab) - F(b)$

or $F(a) + F(b) = F(ab)$ a fundamental law of logarithms.

Replacing b by b/a gives $F(a) + F(b/a) = F(b)$

so $F(b) - F(a) = F(b/a)$ another logarithmic law.

Writing $b = a$ gives $\displaystyle\int_1^a \frac{1}{x}\, dx = \int_a^{a^2} \frac{1}{x}\, dx$

$b = a^2$ gives $\displaystyle\int_1^a \frac{1}{x}\, dx = \int_{a^2}^{a^3} \frac{1}{x}\, dx$ and so on

$$F(a^n) = \int_1^{a^n} \frac{1}{x}\, dx = \int_1^a \frac{1}{x}\, dx + \int_a^{a^2} \frac{1}{x}\, dx + \int_{a^2}^{a^3} + \cdots + \int_{a^{n-1}}^{a^n} \frac{1}{x}\, dx$$

$$= n \int_1^a \frac{1}{x}\, dx = nF(a) \text{ another logarithmic law}$$

These three properties of $F(a)$ suggest F is a logarithmic function. Can we find its base (see fig. 59)?

$$F(2) = \int_1^2 \frac{1}{x}\, dx < \text{area trapezium } ABCD = \tfrac{1}{2}(1 \cdot 5) \times 1 = 0 \cdot 75$$

$$F(3) = \int_1^3 \frac{1}{x}\,dx > 4$$

shaded rectangles

$$= \frac{1}{3} + \frac{1}{4} + \frac{1}{5} + \frac{1}{6} = \frac{57}{60}$$

The area of 8 rectangles taken between $x = 1$ and $x = 3$ is $1 \cdot 019 > 1$.

Since $F(a)$ is continuously increasing there exists a number e such that

$$\int_1^e \frac{1}{x}\,dx = 1$$

and $2 < e < 3$

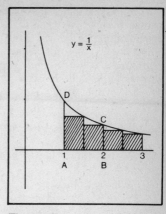

Figure 59

If $F(e) = 1$ and $F(a)$ is a logarithmic function then $F(e) = \log e = 1 \Leftrightarrow$ the base of the logarithms is e.

We now have two important functions

$$y = e^x \Leftrightarrow \frac{dy}{dx} = e^x \text{ and } x = \log_e y = \log_e e^x \qquad (1)$$

$$y = \log_e x \Leftrightarrow \frac{dy}{dx} = \frac{1}{x} \text{ and } x = e^y = e^{\log_e x} \qquad (2)$$

$\log_e x$ is the natural logarithm of x and is also denoted by $\ln x$. e^x is an exponential function (the exponent is x) and is also denoted by $\exp(x)$.

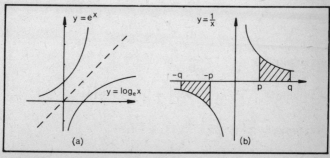

Figure 60

$y = e^x$ and $y = \log_e x$ are inverse functions of each other as is shown by relations (1) and (2). Their graphs are related by a reflection in the line $y = x$ (fig. 60a).

Since $\log(-x)$ is not defined care must be taken when integrating $1/x$ if the limits are negative. From the graph of $y = 1/x$ in fig. 60b, the shaded areas are equal in size since the graph is symmetrical.

$$\int_{-q}^{-p} \frac{1}{x}\, dx = -\int_{p}^{q} \frac{1}{x}\, dx \text{ since the area } A \text{ is below the } x\text{-axis.}$$

$$= -(\ln q - \ln p) = \ln p - \ln q$$

This is in fact $\left[\ln |x|\right]_{-q}^{-p}$

Consequently $\displaystyle\int_{q}^{p} \frac{1}{x}\, dx = \ln |x|$ if p, q are both positive or both negative.

$\displaystyle\int_{-2}^{+1} \frac{1}{x}\, dx$ consists of two infinite areas and cannot be evaluated.

Derivative of $e^{f(x)}$

$$y = e^{f(x)} \Leftrightarrow \log_e y = \log_e e^{f(x)} = f(x) \log_e e = f(x)$$

Differentiating $\dfrac{1}{y}\dfrac{dy}{dx} = \dfrac{df(x)}{dx} \Leftrightarrow \dfrac{dy}{dx} = \dfrac{df(x)}{dx}\, y = \dfrac{df(x)}{dx}\, e^{f(x)}$

Example 1 $y = e^{x^2} \Rightarrow \dfrac{dy}{dx} = 2x e^{x^2}$

Example 2 $y = e^{x^3 + x} \Rightarrow \dfrac{dy}{dx} = (3x^2 + 1) e^{x^3 + x}$

Example 3 $y = e^{\sin x} \Rightarrow \dfrac{dy}{dx} = \cos x . e^{\sin x}$

Example 4 $\displaystyle\int_{0}^{1} e^{2x}\, dx = \left[\tfrac{1}{2} e^{2x}\right]_{0}^{1} = \tfrac{1}{2} e^2 - \tfrac{1}{2} \simeq 3 \cdot 194$

Example 5 $\displaystyle\int x^2 e^{x^3}\, dx = \int \tfrac{1}{3} . 3x^2 e^{x^3}\, dx = \tfrac{1}{3} e^{x^3} + c$

Example 6 $\displaystyle\int \sin x e^{\cos x}\, dx = -e^{\cos x} + c$

120

Derivative of a^x and $a^{f(x)}$

$$y = a^x \Leftrightarrow \ln y = x \ln a \Rightarrow \frac{1}{y}\frac{dy}{dx} = \ln a \Leftrightarrow \frac{dy}{dx} = y \ln a = a^x \ln a$$

$$y = a^{f(x)} \Leftrightarrow \ln y = f(x) \ln a \Rightarrow \frac{1}{y}\frac{dy}{dx} = f'(x) \ln a$$

$$\Leftrightarrow \frac{dy}{dx} = f'(x) a^{f(x)} \ln a$$

Alternative proof of product rule for differentiation

If $y = uv$ where y, u and v are differentiable functions of x

$$\ln y = \ln(uv) = \ln u + \ln v \Rightarrow \frac{1}{y}\frac{dy}{dx} = \frac{1}{u}\frac{du}{dx} + \frac{1}{v}\frac{dv}{dx}$$

$$\Leftrightarrow \frac{dy}{dx} = \frac{y}{u}\frac{du}{dx} + \frac{y}{v}\frac{dv}{dx} = v\frac{du}{dx} + u\frac{dv}{dx}$$

This result can be extended to a product of three functions.

$$y = uvw$$

$$\Leftrightarrow \ln y = \ln u + \ln v + \ln w \Rightarrow \frac{dy}{dx} = vw\frac{du}{dx} + uw\frac{dv}{dx} + uv\frac{dw}{dx}$$

Derivative of $\log_e f(x)$

$$y = \log_e f(x) \Leftrightarrow e^y = f(x) \Rightarrow e^y \frac{dy}{dx} = f'(x) \Rightarrow \frac{dy}{dx} = \frac{f'(x)}{f(x)}$$

Example 7 $y = \log_e x^3 \Rightarrow \dfrac{dy}{dx} = \dfrac{1}{x^3} \cdot 3x^2 = \dfrac{3}{x}$

or $y = \log_e x^3 = 3\log_e x \Rightarrow \dfrac{dy}{dx} = 3 \cdot \dfrac{1}{x} = \dfrac{3}{x}$

Example 8 $y = \log_e \sin x \Rightarrow \dfrac{dy}{dx} = \dfrac{\cos x}{\sin x} = \cot x$

Example 9 $y = \log_e \sec x \Rightarrow \dfrac{dy}{dx} = \dfrac{\sec x \tan x}{\sec x} = \tan x$

Example 10 $y = \log_e(\sec x + \tan x)$

$$\frac{dy}{dx} = \frac{\sec x \tan x + \sec^2 x}{\sec x + \tan x}$$

$$= \frac{\sec x(\tan x + \sec x)}{\sec x + \tan x} = \sec x$$

Applications in integration

(Example 8 leads to the result)

$$\int \cot x \, dx = \int \frac{\cos x}{\sin x} \, dx = \log_e \sin x + \text{constant}$$

This is an example of an integral whose numerator is the derivative of the denominator, and this integral can be performed directly. Here is an example:

Example 11 $\int \frac{2x}{x^2 - 1} \, dx = \log(x^2 - 1) + c$

Alternatively by partial fractions $\frac{2x}{x^2 - 1} = \frac{1}{x - 1} + \frac{1}{x + 1}$

So $\int \frac{2x}{x^2 - 1} \, dx = \int \left(\frac{1}{x - 1} + \frac{1}{x + 1} \right) dx = \log_e (x - 1) + \log_e (x + 1)$

$$= \log_e (x^2 - 1) + c$$

In general $\int \frac{f'(x)}{f(x)} \, dx = \log_e f(x) + c$

Example 12 $\int \frac{x^2}{(x^3 - 2)} \, dx = \int \frac{1}{3} \frac{3x^2}{(x^3 - 2)} \, dx = \frac{1}{3} \log_e (x^3 - 2) + k$

This example may be more easily understood by using the method of integration by substitution.

$$\text{If } u = x^3 - 2 \text{ then } \frac{du}{dx} = 3x^2 \Leftrightarrow x^2 \, dx = \frac{du}{3}$$

$$\int \frac{x^2}{(x^3 - 2)} \, dx = \int \frac{1}{u} \frac{du}{3} = \frac{1}{3} \log_e u = \frac{1}{3} \log_e (x^3 - 2) + c$$

$$\text{If } u = f(x) \text{ then } \frac{du}{dx} = f'(x) \text{ so } \int \frac{f'(x)}{f(x)} \, dx = \int \frac{du}{u}$$

$$= \log_e u = \log_e f(x) + c$$

More complicated examples can be solved in the same way.

Example 13 $\int \frac{x^2}{(x^3 - 2)^3} \, dx = \int \frac{1}{3} \frac{3x^2}{(x^3 - 2)^3} \, dx$

$$= \int \frac{1}{3} 3x^2 (x^3 - 2)^{-3} \, dx$$

$$= \frac{1}{3} \frac{(x^3 - 2)^{-2}}{-2} = \frac{-1}{6(x^3 - 2)^2} + c$$

However, to avoid arithmetic errors, it may be better to use the method of substitution.

Let $u = x^3 - 2 \Rightarrow \dfrac{du}{dx} = 3x^2 \Rightarrow x^2 \, dx = \dfrac{du}{3}$

$$\int \frac{x^2}{(x^3-2)^3} = \int \frac{1}{u^3} \times \frac{du}{3} = \int \frac{1}{3} u^{-3} \, du = \left[\frac{1}{3} \frac{u^{-2}}{(-2)} \right] = \frac{-1}{6u^2}$$

$$= \frac{-1}{6(x^3-2)^2} + c$$

Example 14

$$\int \frac{1}{x^2 - a^2} \, dx = \int \frac{1}{(x+a)(x-a)} \, dx$$

$$= \int \left(\frac{1}{2a(x-a)} - \frac{1}{2a(x+a)} \right) dx \qquad \text{by partial fractions}$$

$$= \frac{1}{2a} \ln (x-a) - \frac{1}{2a} \ln (x+a) + c$$

$$= \frac{1}{2a} \ln \left| \frac{x-a}{x+a} \right| + c$$

$$\int \frac{1}{(a^2 - x^2)} \, dx = \int \left(\frac{1}{2a(a-x)} + \frac{1}{2a(a+x)} \right) dx$$

$$= -\frac{1}{2a} \ln (a-x) + \frac{1}{2a} \ln (a+x) + c$$

$$= \frac{1}{2a} \ln \left| \frac{a+x}{a-x} \right| + c$$

Hyperbolic functions

$$\cosh x = \frac{e^x + e^{-x}}{2} \qquad \sinh x = \frac{e^x - e^{-x}}{2}$$

pronounced *cosh*   pronounced *shine* or *sinch*

These definitions of cosh x and sinh x lead to some remarkable properties and a close analogy with the circular trigonometrical functions of Chapter 2.

$\cosh (-x) = \dfrac{e^{-x} + e^{x}}{2} = \cosh x$ so cosh x is an **even** function and its graph is symmetrical about $x = 0$

$\sinh (-x) = \dfrac{e^{-x} - e^{x}}{2} = -\sinh x$ so sinh x is an **odd** function and its graph has rotational symmetry (order 2) about the origin

The values of cosh x can be plotted for each x value by taking points half-way between the graphs of e^x and e^{-x}. The minimum value of cosh x is 1 when $x = 0$ (fig. 61a).

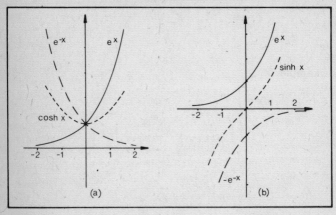

Figure 61

Sinh x lies half-way between the graphs of e^x and $-e^{-x}$ (fig. 61b). When x is large and positive both functions tend to $\frac{1}{2}e^x$ (since $e^{-x} \to 0$) in such a way that

$$\sinh x < \tfrac{1}{2}e^x < \cosh x$$

$$y = \cosh x \Rightarrow \frac{dy}{dx} = \frac{e^x - e^{-x}}{2} = \sinh x$$

$$y = \sinh x \Rightarrow \frac{dy}{dx} = \frac{e^x + e^{-x}}{2} = \cosh x$$

and the gradient of sinh x is 1 when $x = 0$.

$$\cosh a + \sinh a = e^a \text{ and } \cosh b + \sinh b = e^b$$

$$\cosh a - \sinh a = e^{-a} \text{ and } \cosh b - \sinh b = e^{-b}$$

so $e^{a+b} = (\cosh a + \sinh a)(\cosh b + \sinh b)$

$$= \cosh a \cosh b + \cosh a \sinh b$$
$$+ \sinh a \cosh b + \sinh a \sinh b \qquad (1)$$

and similarly

$$e^{-a-b} = \cosh a \cosh b - \cosh a \sinh b - \sinh a \cosh b$$
$$+ \sinh a \sinh b \qquad (2)$$

Adding (1) and (2) gives

$$e^{a+b} + e^{-a-b} = 2\cosh a \cosh b + 2\sinh a \sinh b$$

cosh $(a + b)$ = cosh a cosh b + sinh a sinh b   (3)

Subtracting (2) from (1) gives

$$e^{a+b} - e^{-a-b} = 2\sinh a \cosh b + 2\cosh a \sinh b$$

sinh $(a + b)$ = sinh a cosh b + cosh a sinh b   (4)

Substituting $-b$ for b and remembering that cosh x is **even** and sinh x is **odd** leads to

cosh $(a - b)$ = cosh a cosh b – sinh a sinh b   (5)

and **sinh $(a - b)$ = sinh a cosh b – cosh a sinh b**   (6)

Putting $a = b$ in (4) gives **sinh $2a$ = 2 sinh a cosh a**   (7)

Putting $a = b$ in (5) gives **1 = $\cosh^2 a - \sinh^2 a$**   (8)

$a = b$ in (7) gives **cosh $2a$ = $\cosh^2 a + \sinh^2 a$**   (9)

Using (8) **cosh $2a$ = 1 + 2 $\sinh^2 a$**   (10)

or **cosh $2a$ = 2 $\cosh^2 a$ – 1**   (11)

Equations (3) to (11) can be easily proved algebraically from the definitions but it is instructive and interesting to see how closely they follow the same lines as the circular function properties in Chapter 2.

The definitions can be extended to $\tanh x = \dfrac{\sinh x}{\cosh x} = \dfrac{e^x - e^{-x}}{e^x + e^{-x}}$

$$\operatorname{cosech} x = \frac{1}{\sinh x}; \quad \operatorname{sech} x = \frac{1}{\cosh x}; \quad \coth x = \frac{\cosh x}{\sinh x}$$

Dividing equation (8) by $\cosh^2 a$ gives **$\operatorname{sech}^2 a = 1 - \tanh^2 a$** and dividing (8) by $\sinh^2 a$ gives **$\operatorname{cosech}^2 a = \coth^2 a - 1$.**

$$\tanh (a + b) = \frac{\sinh (a + b)}{\cosh (a + b)} = \frac{\sinh a \cosh b + \cosh a \sinh b}{\cosh a \cosh b + \sinh a \sinh b}$$

$$= \frac{\tanh a + \tanh b}{1 + \tanh a \tanh b} \quad \begin{array}{l}\text{by dividing each term} \\ \text{by } \cosh a \cosh b\end{array}$$

Similar results can be deduced for $\tanh 2a$, $\sinh 3a$, $\cosh 3a$.

In Chapter 6 the equation of the hyperbola is given as $\dfrac{x^2}{a^2} - \dfrac{y^2}{b^2} = 1$.

$x = a \sec \theta$, $y = b \tan \theta$ satisfy this equation and can be regarded as the parametric equations of the hyperbola.

$$x = a \cosh \theta, \ y = b \sinh \theta \Rightarrow \frac{x^2}{a^2} - \frac{y^2}{b^2} = \frac{a^2 \cosh^2 \theta}{a^2} - \frac{b^2 \sinh^2 \theta}{b^2} = 1$$

and are more commonly used as the parametric form of the hyperbola. However, since $\cosh x \geqslant 1$ for all x this parametric form only describes points on one branch of the hyperbola. It is this property that leads to $\sinh x$ and $\cosh x$ being known as the **hyperbolic sine** and **cosine.**

Since $\sinh x$ and $\cosh x$ are the derivatives of each other it is straightforward to work out the derivatives of all the hyperbolic functions.

$$y = \tanh x = \frac{\sinh x}{\cosh x} \Rightarrow \frac{dy}{dx} = \frac{\cosh^2 x - \sinh^2 x}{\cosh^2 x} = \frac{1}{\cosh^2 x}$$
$$= \operatorname{sech}^2 x$$

$$y = \operatorname{sech} x = \frac{1}{\cosh x} \Rightarrow \frac{dy}{dx} = \frac{-\sinh x}{\cosh^2 x} = -\operatorname{sech} x \tanh x$$

$$y = \coth x = \frac{\cosh x}{\sinh x} \Rightarrow \frac{dy}{dx} = \frac{\sinh^2 x - \cosh^2 x}{\sinh^2 x} = \frac{-1}{\sinh^2 x}$$
$$= -\operatorname{cosech}^2 x$$

$$y = \operatorname{cosech} x = \frac{1}{\sinh x} \Rightarrow \frac{dy}{dx} = \frac{-\cosh x}{\sinh^2 x} = -\operatorname{cosech} x \coth x$$

$$y = p \cosh mx + q \sinh mx \Rightarrow \frac{d^2 y}{dx^2} = m^2 p \cosh x + m^2 q \sinh x$$
$$= m^2 y$$

and we have a result similar to that used for describing **Simple Harmonic Motion** in Mechanics.

Integration

$\sinh x$ and $\cosh x$ have important applications to integration.

You may remember $\int \dfrac{1}{\sqrt{a^2 - x^2}} \, dx = \sin^{-1}\left(\dfrac{x}{a}\right) + c$ and that this was achieved substituting $x = a \sin \theta$ so that the square root simplified (Chapter 4, page 83).

Similarly $\int \dfrac{1}{(a^2 + x^2)}\, dx =$

$\dfrac{1}{a} \tan^{-1} \dfrac{x}{a} + c$ by substituting $x = a \tan \theta$.

We can now integrate functions like $\dfrac{1}{\sqrt{x^2 - a^2}}$ and $\dfrac{1}{\sqrt{x^2 + a^2}}$.

Consider $\int \dfrac{1}{\sqrt{x^2 - a^2}}\, dx$ and let $x = a \cosh \theta \Rightarrow \dfrac{dx}{d\theta} = a \sinh \theta$.

$\int \dfrac{1}{\sqrt{x^2 - a^2}}\, dx = \int \dfrac{1}{\sqrt{a^2 \cosh^2 \theta - a^2}}\, a \sinh \theta\, d\theta = \int \dfrac{a \sinh \theta}{a \sinh \theta}\, d\theta$

since $\cosh^2 \theta - 1 = \sinh^2 \theta$

$$\int \dfrac{1}{\sqrt{x^2 - a^2}}\, dx = \int 1\, d\theta = \cosh^{-1}\left(\dfrac{x}{a}\right) + c$$

$\text{Cosh}^{-1} x$ is the inverse function of $\cosh x$ and so there is a need to define **inverse hyperbolic functions**.

Or, letting $x = a \sec \theta$ so $dx = a \sec \theta \tan \theta\, d\theta$

$\int \dfrac{dx}{\sqrt{x^2 - a^2}} = \int \dfrac{a \sec \theta \tan \theta\, d\theta}{\sqrt{a^2 \sec^2 \theta - a^2}} = \int \sec \theta\, d\theta$

$\qquad\qquad\qquad = \log_e (\sec \theta + \tan \theta) + c$

$\qquad\qquad\qquad = \log_e (x + \sqrt{x^2 - a^2}) + c$

Consider $\int \dfrac{1}{\sqrt{x^2 + a^2}}\, dx$ and let $x = a \sinh \theta \Rightarrow \dfrac{dx}{d\theta} = a \cosh \theta$

$\int \dfrac{dx}{\sqrt{x^2 + a^2}} = \int \dfrac{a \cosh \theta}{\sqrt{a^2 \sinh^2 \theta + a^2}}\, d\theta = \int 1\, d\theta = \sinh^{-1} \dfrac{x}{a} + c$

Or $\int \dfrac{1}{\sqrt{x^2 + a^2}}\, dx$; let $x = a \tan \theta \Rightarrow \dfrac{dx}{d\theta} = a \sec^2 \theta$

$\int \dfrac{dx}{\sqrt{x^2 + a}} = \int \dfrac{a \sec^2 \theta\, d\theta}{\sqrt{a^2 \tan^2 \theta + a^2}} = \int \sec \theta\, d\theta$

$\qquad\qquad = \ln (\sec \theta + \tan \theta) + c$

$\qquad\qquad = \ln \left(\dfrac{x}{a} + \sqrt{1 + \dfrac{x^2}{a^2}}\right) + c$

This suggests $\sinh^{-1} x = \ln (x + \sqrt{1 + x^2})$. This is derived on page 129.

Inverse hyperbolic functions

The inverse functions of $\sinh x$ and $\cosh x$ are denoted in the usual way by $\sinh^{-1} x$ and $\cosh^{-1} x$, although they both have alternative forms in terms of natural logarithms. Their graphs are easily obtained by reflection in the line $y = x$, (fig. 62).

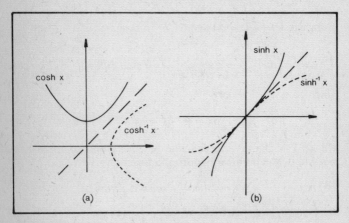

Figure 62

$\cosh^{-1} x$ is double-valued which means that for each value of x there are two possible values of $\cosh^{-1} x$ (or none). The graph of $\cosh^{-1} x$ has to be restricted to take only positive values.

$y = \cosh^{-1} x \Rightarrow x = \cosh y \Rightarrow \sinh y = \sqrt{x^2 - 1}$ since $\sinh^2 y = \cosh^2 y - 1$

$\cosh y + \sinh y = e^y = x \pm \sqrt{x^2 - 1} \Rightarrow y = \ln(x \pm \sqrt{x^2 - 1})$

The graph in fig. 62a suggests these two values are equal and opposite.

$\ln(x + \sqrt{x^2 + 1}) + \ln(x - \sqrt{x^2 - 1})$

$= \ln(x + \sqrt{x^2 - 1})(x - \sqrt{x^2 - 1})$

$= \ln[x^2 - (x^2 - 1)] = \ln 1 = 0$

Taking the positive sign as the principal value

$$\cosh^{-1} x = \ln(x + \sqrt{x^2 - 1})$$

Figure 62b shows the graph of $y = \sinh x$ (solid line).

Since $\dfrac{dy}{dx} = \cosh x$ the gradient of $\sinh x$ at $x = 0$ is 1. The inverse function $y = \sinh^{-1} x$ is shown by the dotted line in fig. 62b.

If $y = \sinh^{-1} x$ then $x = \sinh y$ and $\cosh^2 y = 1 + \sinh^2 y = 1 + x^2$

$$\cosh y + \sinh y = e^y \Rightarrow e^y = x + \sqrt{x^2 + 1}$$

since $e^y > 0$, the positive root is taken
$\Rightarrow y = \sinh^{-1} x = \ln(x + \sqrt{x^2 + 1})$.
Alternatively $x = \sinh y = \frac{1}{2}(e^y - e^{-y})$

$$\Rightarrow 2x = e^y - e^{-y}$$

$$\Rightarrow (e^y)^2 - 2xe^y - 1 = 0 \Rightarrow e^y = \frac{2x \pm \sqrt{4x^2 + 4}}{2} = x \pm \sqrt{x^2 + 1}$$

Taking the positive square root $y = \sinh^{-1} x = \ln(x + \sqrt{x^2 + 1})$.
Figure 63 shows the graph of $y = \tanh x$ and its inverse $y = \tanh^{-1} x$ (dotted).

$y = \tanh x = \dfrac{\sinh x}{\cosh x} = \dfrac{e^x - e^{-x}}{e^x + e^{-x}}$

$|\tanh x| < 1$ since $|\sinh x|$ $< \cosh x$.

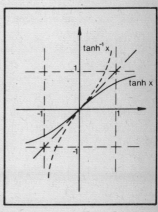

As x tends to $+\infty$, $\tanh x$ tends to $+1$. As x tends to $-\infty$, $\tanh x$ tends to -1. Consequently, $y = \tanh^{-1} x$ is only defined for $|x| < 1$, i.e. $-1 < x < 1$.

$y = \tanh x \Rightarrow \dfrac{dy}{dx} = \operatorname{sech}^2 x = 1$

when $x = 0$. So the gradient of both graphs is 1, when $x = 0$.

Figure 63

$y = \tanh^{-1} x \Rightarrow x = \tanh y$

$x = \dfrac{e^y - e^{-y}}{e^y + e^{-y}} \Rightarrow xe^y + xe^{-y} = e^y - e^{-y}$

$$\Rightarrow (x + 1)e^{-y} = e^y(1 - x) \Rightarrow e^{2y} = \frac{1+x}{1-x} \Rightarrow e^y = \sqrt{\frac{1+x}{1-x}}$$

$y = \ln \sqrt{\dfrac{1+x}{1-x}} = \tanh^{-1} x$ or $\tanh^{-1} x = \frac{1}{2} \ln\left(\dfrac{1+x}{1-x}\right)$

Example 15 Solve $3 \cosh x + 5 \sinh x = 12$.

It may be tempting to compare this with the auxiliary angle method for solving $a \cos x + b \sin x = c$ (Chapter 2, page 33) but it is more straightforward to substitute $\cosh x = \frac{1}{2}(e^x + e^{-x})$ and $\sinh x = \frac{1}{2}(e^x - e^{-x})$ and solve the resulting quadratic in e^x.

$$3 \times \tfrac{1}{2}(e^x + e^{-x}) + 5 \times \tfrac{1}{2}(e^x - e^{-x}) = 12 \Rightarrow 8e^x - 2e^{-x} = 24$$

$$4(e^x)^2 - 12e^x - 1 = 0 \Rightarrow e^x = \frac{12 \pm \sqrt{144 + 16}}{8} = \frac{3 \pm \sqrt{10}}{2}$$

$$x = \ln\left(\frac{3 + \sqrt{10}}{2}\right) \text{ since } \frac{3 - \sqrt{10}}{2} \text{ is negative.}$$

Key terms

$$y = \log_{10} x \Rightarrow \frac{dy}{dx} = \frac{1}{x} \log_{10} e \quad (e = 2\cdot7183\ldots)$$

$$y = \log_e x \Rightarrow \frac{dy}{dx} = \frac{1}{x} \quad \int \frac{1}{x}\, dx = \log_e x + k = \ln x + k$$

e^x and $\ln x$ are inverse functions; $e^{\ln x} = x$; $\ln e^x = x$.

$$y = e^{f(x)} \Rightarrow \frac{dy}{dx} = f'(x)\, e^{f(x)}$$

$$y = a^{f(x)} \Rightarrow \frac{dy}{dx} = f'(x)\, a^{f(x)} \ln x \quad \text{(a constant)}$$

Hyperbolic functions $\cosh x = \dfrac{e^x + e^{-x}}{2}$: $\sinh x = \dfrac{e^x - e^{-x}}{2}$

$$\frac{d}{dx}(\cosh x) = \sinh x; \frac{d}{dx}(\sinh x) = \cosh x$$

$$\int \frac{1}{\sqrt{x^2 - 1}}\, dx = \cosh^{-1} x + k = \ln(x + \sqrt{x^2 - 1}) + k$$

$$\sinh^{-1} x = \ln(x + \sqrt{x^2 + 1})$$

Chapter 6
Coordinate geometry

Introduction
Coordinate geometry is that branch of mathematics which studies the properties of lines, curves and planes using the techniques of algebra and calculus.

Coordinates
Figure 64a shows a plane in which two perpendicular lines have been drawn intersecting at O. The lines are called the axes of coordinates and O is the origin.

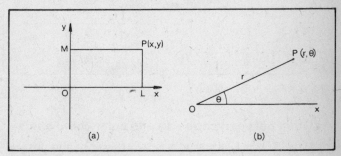

Figure 64

The position of point P is specified by two distances MP and LP, usually denoted by x and y respectively. As x and y take all possible values the point P moves to every position in the plane. The values x and y are called the **cartesian coordinates** of the point P and are written (x, y).

Definitions
The lengths of OL and OM are called the **abscissa** and the **ordinate** of P respectively.

An alternative system is to specify the position of a point in a plane by using a distance from a fixed point together with an angle measured from a given fixed line. See fig. 64b.

Let O be the fixed point, called the **pole,** and OX the fixed line called the **initial line.** If $OP = r$ and the angle $POX = \theta$ then r and θ are called the **polar coordinates** of the point P

and are written (r, θ). Note that θ is positive when measured anticlockwise. The same point may be described in different ways as shown in fig. 65a. The point A may be written as $(2, 240°)$ and the point B as $(3, 330°)$, $(3, -30°)$ or $(-3, 150°)$.

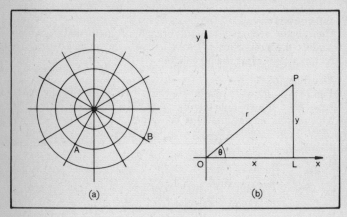

Figure 65

Relation between cartesian and polar coordinates

Consider a point P in a plane whose cartesian coordinates are given by (x, y) relative to axes OX and OY respectively. Let (r, θ) be the polar coordinates of the same point using OX as the initial line and O as the pole. See fig. 65b.

Now in $\triangle OPL$ $\cos \theta = \dfrac{x}{r}$ and $\sin \theta = \dfrac{y}{r}$.

$\therefore$ $x = r \cos \theta$ and $y = r \sin \theta$ which give x and y in terms of r and θ.

Squaring and adding these results we obtain

$$x^2 + y^2 = r^2 \cos^2 \theta + r^2 \sin^2 \theta = r^2$$

Dividing the two original equations gives $\dfrac{y}{x} = \dfrac{r \sin \theta}{r \cos \theta} = \tan \theta$.

Thus r and θ are given in terms of x and y by

$$r = \sqrt{(x^2 + y^2)} \text{ and } \theta = \tan^{-1}\left(\frac{y}{x}\right)$$

132

Example 1 Express the equation of the cardioid $r = \frac{1}{2}(1 + \cos \theta)$ in cartesian form.

Since $r = \sqrt{(x^2 + y^2)}$ and $\cos \theta = \dfrac{x}{r} = \dfrac{x}{\sqrt{x^2 + y^2}}$

we have $\sqrt{x^2 + y^2} = \dfrac{1}{2}\left\{1 + \dfrac{x}{\sqrt{x^2 + y^2}}\right\}$

$$\therefore \quad 2(x^2 + y^2) = \sqrt{x^2 + y^2} + x$$

Thus the cartesian equation is given by

$$\{2(x^2 + y^2) - x\}^2 = x^2 + y^2$$

Example 2 Find the polar equation of the curve whose cartesian equation is given by $y^2 = 8(2 - x)$.

Since $x = r \cos \theta$ and $y = r \sin \theta$ we have, by substituting in the equation, $(r \sin \theta)^2 = 8(2 - r \cos \theta)$

$$\therefore \quad r^2 \sin^2 \theta + 8r \cos \theta - 16 = 0$$
$$\therefore \quad r^2(1 - \cos^2 \theta) + 8r \cos \theta - 16 = 0$$
$$\therefore \quad r^2 = 16 - 8r \cos \theta + r^2 \cos^2 \theta$$
$$\therefore \quad r^2 = (4 - r \cos \theta)^2$$

Taking the positive square root we obtain $r = 4 - r \cos \theta$.
Hence $r(1 + \cos \theta) = 4$
Thus the polar equation is given by $r = 4/(1 + \cos \theta)$.

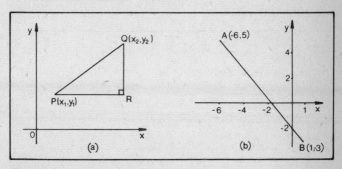

Figure 66

Consider the two points $P(x_1, y_1)$ and $Q(x_2, y_2)$ as shown in fig. 66a. By completing a right-angled triangle PQR with PR parallel to OX we have

$$RP = x_2 - x_1 \quad \text{and} \quad QR = y_2 - y_1.$$

Using the theorem of Pythagoras we have $PQ^2 = RP^2 + QR^2$.

$$\therefore \quad PQ^2 = (x_2 - x_1)^2 + (y_2 - y_1)^2$$

Thus the length of PQ is given by $PQ = \sqrt{\{(x_2 - x_1)^2 + (y_2 - y_1)^2\}}$. The positive square root is taken for the length of a line segment.

Example 3 Find the distance between the two points $A(-6, 5)$ and $B(1, -3)$.

We have $x_2 - x_1 = 1 - (-6) = 7$ units
and $y_2 - y_1 = -3 - (5) = -8$ units

$\therefore$ the length of AB is given by

$$AB = \sqrt{(x_2 - x_1)^2 + (y_2 - y_1)^2} = \sqrt{7^2 + (-8)^2}$$

$$= \sqrt{113} = 10 \cdot 6 \text{ to 1 decimal place}$$

Division of a line in a given ratio

There are two ways in which a line can be divided: **internally** and **externally**.

(i) (a) Divide a line segment AB **internally** in the ratio $3 : 2$. If P is the point of division then $AP : PB = 3 : 2$. See fig. 67a.

(b) Divide a line segment **internally** in the ratio of $2 : 3$. In this case $AP : PB = 2 : 3$ and P is nearer to A than B. See fig. 67b.

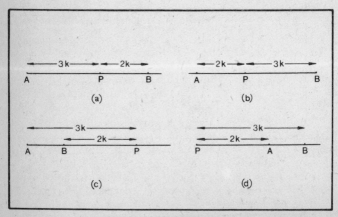

Figure 67

(ii) (a) Divide a line segment AB **externally** in the ratio of $3:2$. The same algebraic ratio still holds, i.e. $AP:PB = 3:2$ but for external division the point P lies on BA or AB produced, in this case to the right of B fig. 67c.

(b) Divide a line segment AB **externally** in the ratio of $2:3$. In this case P lies to the left of A so that $AP:PB = 2:3$. See fig. 67d.

Note that in external division the actual length of the line segment does not form part of the ratio.

General result To find the coordinates of the point D which divides the line joining the point $A(x_1, y_1)$ to the point $B(x_2, y_2)$ internally in the ratio $p:q$.

Let the coordinates of D be (x, y). It can be seen that $\triangle ADE$ is similar to $\triangle ABC$. See fig. 68.

$$\therefore \quad \frac{AD}{AB} = \frac{AE}{AC} = \frac{p}{p+q} \quad \text{which gives} \quad AE = \frac{p}{p+q} AC.$$

Therefore $AE = \dfrac{p}{p+q}(x_2 - x_1)$.

$\therefore$ since the x coordinate of D is $x = x_1 + AE$

we have $x = x_1 + \dfrac{p}{p+q}(x_2 - x_1) = \dfrac{qx_1 + px_2}{p+q}$

Similarly $\dfrac{CF}{CB} = \dfrac{AD}{AB} = \dfrac{p}{p+q} \Rightarrow CF = \dfrac{p}{p+q}(y_2 - y_1)$

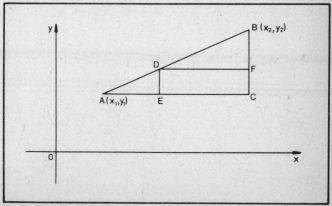

Figure 68

135

$\therefore$ the y coordinate of D is $y = y_1 + CF = \dfrac{qy_1 + py_2}{p + q}$

Therefore the coordinates of D are $\left(\dfrac{qx_1 + px_2}{p + q}, \dfrac{qy_1 + py_2}{p + q} \right)$.

If the line segment joining the point $A(-6, -3)$ to the point $B(4, 2)$ is divided internally in the ratio of $1:4$ we see that the coordinates of D are given by

$$x = \frac{4(-6) + 1(4)}{1 + 4} = \frac{-24 + 4}{5} = -4$$

$$y = \frac{4(-3) + 1(2)}{1 + 4} = \frac{-12 + 2}{5} = -2$$

$\therefore$ the coordinates of D are $(-4, -2)$.

Note that if $p = q = 1$, the mid point of the line AB will be found as $(\frac{1}{2}(x_1 + x_2), \frac{1}{2}(y_1 + y_2))$.

A similar result can be derived for the external division of a line in a given ratio.

The coordinate of the point E which divides externally the line joining the point $A(x_1, y_1)$ to the point $B(x_2, y_2)$ in the ratio of $p:q$ are $\left(\dfrac{px_2 - qx_1}{p - q}, \dfrac{py_2 - qy_1}{p - q} \right)$.

Example 4 Find the coordinates of a point E which divides a line joining $A(-1, -3)$ to the point $B(4, 1)$ externally in the ratio $3:2$.

Using the previous result with $p:q = 3:2$

A is the point $(-1, -3)$ i.e. $x_1 = -1$ and $y_1 = -3$
B is the point $(4, 1)$ i.e. $x_2 = 4$ and $y_2 = 1$

$\therefore$ the x coordinate of E is given by $\dfrac{3(4) - 2(-1)}{3 - 2} = 14$

$\therefore$ the y coordinate of E is given by $\dfrac{3(1) - 2(-3)}{3 - 2} = 9$

$\therefore$ the coordinates of E are $(14, 9)$.

The straight line

If we consider a point $P(x, y)$ in a plane where the values of x and y can vary, P could take any position in the plane. If, however, we define a relationship between x and y then we restrict the point P to a specific path. For example, if $y = x^2$ then the point P can only take positions where the y coordinate is the square of the x coordinate. If $y = x^2 - 4x + 3$ the possible points are shown in fig. 5a, page 19. If the relationship is linear the graph will be a straight line.

Example 5 For the relationship (i) $y = 2x - 1$, (ii) $y = 2 - x$ find the values of y corresponding to $x = -3, -2, -1, 0, 1, 2, 3$. Plot these pairs of values (x, y) using cartesian axes.

Tables are constructed as follows:

(i)

x	-3	-2	-1	0	1	2	3
$2x$	-6	-4	-2	0	2	4	6
-1	-1	-1	-1	-1	-1	-1	-1
$2x - 1$	-7	-5	-3	-1	1	3	5

(ii)

x	-3	-2	-1	0	1	2	3
$-x$	3	2	1	0	-1	-2	-3
2	2	2	2	2	2	2	2
$2 - x$	5	4	3	2	1	0	-1

These pairs of points are plotted on the same set of axes as in fig. 69.

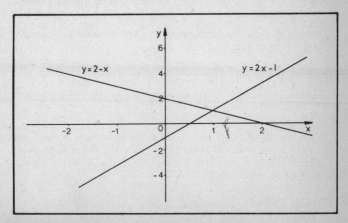

Figure 69

Gradient of a straight line

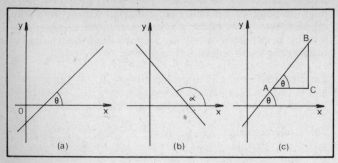

Figure 70

The gradient of a straight line is defined as the **tangent of the angle that it makes with the increasing direction of the x axis.** See fig. 70a. Note that in fig. 70b α will be obtuse and thus its tangent will be **negative.**

Check that the graphs of $y = 2x - 1$ and $y = 2 - x$ drawn in fig. 69 have gradients of 2 and -1 respectively.

The gradient can also be found by considering two points on the line and evaluating the ratio BC/AC which is equivalent to $\tan \theta$. See fig. 70c.

Remember to be careful when dealing with a line that has a negative gradient.

Example 6 Find the gradient of the line passing through the points $P(-6, 4)$ and $Q(4, -2)$.

Gradient of PQ is $\dfrac{(y \text{ coordinate of } Q) - (y \text{ coordinate of } P)}{(x \text{ coordinate of } Q) - (x \text{ coordinate of } P)}$

$$= \frac{-2 - (4)}{4 - (-6)} = -\frac{6}{10} = -\frac{3}{5}$$

Thus if the equation of the straight line is written in the form $y = mx + c$ the gradient is given by m and c defines the point at which the line cuts the y axis. This is referred to as the **intercept** on the y axis.

Special cases

(i) If $m = 0$ the line is parallel to the x axis and its equation reduces to $y = c$.

(ii) If the line is parallel to the y axis the equation $x = k$ defines the line.

Example 7 What is the equation of the straight line which cuts the y axis at the point $(0, -4)$ and whose gradient is $\frac{3}{2}$?

The general equation of a straight line is $y = mx + c$. In this case $m = \frac{3}{2}$ and $c = -4$ and the required equation is

$$y = \tfrac{3}{2}x - 4 \text{ or } 2y = 3x - 8.$$

Alternative methods for finding the equation of a straight line are illustrated by the following cases.

Method 1 Find the equation of the straight line through the point (x_1, y_1) whose gradient is m.

Let A be the given point whose coordinates are (x_1, y_1), and consider a general point P whose coordinates are (x, y).

The gradient of AP is $\dfrac{y - y_1}{x - x_1} = m$.

Rearranging this gives $y - y_1 = m(x - x_1)$.

This form of the equation is useful when the gradient and one point of the line are given.

Example 8 For a line whose gradient is -4, which passes through the point $(3, -2)$ the equation is given by $y - y_1 = m(x - x_1)$.

In this case $y - (-2) = -4(x - 3) \Rightarrow y + 2 = -4x + 12$.
$\therefore$ the equation is given by $y + 4x = 10$.

Method 2 Find the equation of the straight line joining the point $P(x_1, y_1)$ to the point $Q(x_2, y_2)$.

Consider a general point $R(x, y)$ on the line PQ. Since the points P, Q and R are collinear it follows that:

the gradient of PR = the gradient of QR

$\therefore$ the equation is $\dfrac{y - y_1}{x - x_1} = \dfrac{y - y_2}{x - x_2}$.

Example 9 Find the equation of the line joining the points $P(-2, 1)$ and $Q(3, 4)$.

We use the result from method 2.

The equation of the line is given by $\dfrac{y - y_1}{x - x_1} = \dfrac{y - y_2}{x - x_2}$.

i.e. $\dfrac{y - 1}{x - (-2)} = \dfrac{y - 4}{x - 3}$

Simplifying
$$(y - 1)(x - 3) = (x + 2)(y - 4)$$
$$xy - 3y - x + 3 = xy + 2y - 4x - 8$$
$$-5y + 3 = -3x - 8$$
$$\text{or} \quad 5y = 3x + 11$$

Perpendicular lines

The product of the gradients of two lines perpendicular to each other is equal to -1.

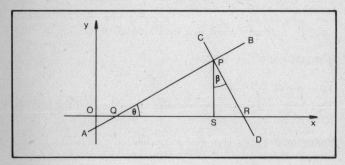

Figure 71

Consider two straight lines AB and CD intersecting at P. Let the lines AB and CD cut the x axis at Q and R respectively. PS is drawn perpendicular to the x axis. Let the gradient of AB be m.

$\therefore \quad m = \dfrac{SP}{QS} = \tan \theta$

Now the gradient of $CD = -SP/SR = -1/\tan \beta = -1/\tan \theta$
($\theta = \beta$ since the triangles PSR and QSP are similar.)
$\therefore$ the gradient of $CD = -1/m$.
i.e. gradient of $AB \times$ gradient of $CD = -1$

The angle between two lines

Consider two intersecting straight lines whose equations are $y = m_1x + c_1$ and $y = m_2x + c_2$ as shown in fig. 72.

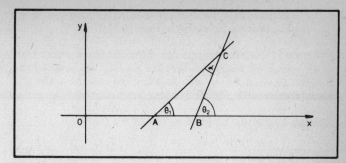

Figure 72

The angle between the two lines is the acute angle ACB. If the lines make angles of θ_1 and θ_2 with the x axis we have

angle $ACB = \theta_2 - \theta_1 = \alpha$ (say) $\quad \therefore \quad \tan \alpha = \tan(\theta_2 - \theta_1)$

$$= \frac{\tan \theta_2 - \tan \theta_1}{1 + \tan \theta_2 \tan \theta_1}$$

Now $\tan \theta_1 = m_1$ (the gradient of AC) and $\tan \theta_2 = m_2$ (the gradient of BC).

$$\therefore \qquad \tan \alpha = \frac{m_2 - m_1}{1 + m_1 m_2} \quad \Rightarrow \quad \alpha = \tan^{-1}\left(\frac{m_2 - m_1}{1 + m_1 m_2}\right)$$

Therefore the angle between two lines whose gradients are m_1 and m_2 is given by

$$\tan^{-1}\left(\frac{m_2 - m_1}{1 + m_1 m_2}\right)$$

Special cases

(i) If $\alpha = \dfrac{\pi}{2}$ then $\tan \dfrac{\pi}{2} = \dfrac{m_2 - m_1}{1 + m_1 m_2}$.

Thus it follows that $1 + m_1 m_2 = 0$, i.e. $m_1 m_2 = -1$ as before.
(ii) If $\alpha = 0$, $m_1 = m_2$ and the lines are parallel.

Example 10 Find the angle between the two lines $y = 2x + 3$ and $y = -\frac{1}{4}x$.

In this case $m_1 = 2$ and $m_2 = -\frac{1}{4}$. Using the above result

$$\alpha = \tan^{-1}\left(\frac{m_2 - m_1}{1 + m_1 m_2}\right) = \tan^{-1}\left(\frac{-\frac{1}{4} - 2}{1 + 2(-\frac{1}{4})}\right) = \tan^{-1}\left(-4\tfrac{1}{2}\right)$$

$$= 180° - 77 \cdot 47° = 102 \cdot 53°$$

141

$\therefore$ The obtuse angle between the lines is $102 \cdot 53°$.
$\therefore$ The acute angle between the lines is $77 \cdot 47°$.

Example 11 Find the coordinates of the point of intersection of the two perpendicular lines where one line has a gradient of 2 and cuts the x axis at $x = -1$ and the other line cuts the x axis at a point where $x = 19$.

The equation of the first line will be of the form
$$y - y_1 = m(x - x_1), \quad \text{i.e.} \quad y - 0 = 2\{x - (-1)\}$$
as it passes through the point $(-1, 0)$, i.e. $\quad y = 2x + 2$

The equation of the second line will be of the form
$$y - y_2 = m(x - x_2), \quad \text{i.e.} \quad y - 0 = -\tfrac{1}{2}(x - 19)$$
since the lines are perpendicular, $m = -\tfrac{1}{2}$ and it passes through $(19, 0)$. Thus the equation of the second line is $2y = -x + 19$.

If the lines represented by these equations intersect at $P(a, b)$
then $\quad b = 2a + 2$
$$2b = -a + 19$$

Solving gives $b = 8$ and $a = 3$

$\therefore$ The coordinates of the point of intersection are $(3, 8)$.

The **equation of a straight line** can be written in a number of different forms apart from $y = mx + c$ which we have considered.

1. **The form $ax + by + c = 0$**
   In this form it is assumed that a and b are not simultaneously zero. This is a rearrangement of the form $y = mx + c$.

2. Consider the equation of a line through the points $(a, 0)$ and $(0, b)$. Its gradient would be $-b/a$ and thus its equation is given by

$$y - b = -\frac{b}{a}(x - 0) \quad \text{or} \quad ay + bx = ab$$

Dividing by ab we obtain $\dfrac{x}{a} + \dfrac{y}{b} = 1$

This is known as the **intercept form** of the equation of a line.

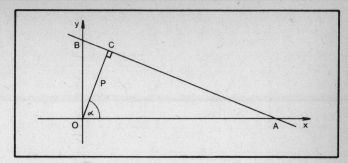

Figure 73

3. Consider fig. 73 in which the perpendicular from the origin to a given line is of length p and makes an angle α with the x axis as shown. Let the given line cut the x axis at A and the y axis at B. Let the perpendicular from the origin meet the line at C
Since $\triangle OCA$ is right-angled at C, $OA = p \sec \alpha$.
In $\triangle OBC$, angle $BOC = 90° - \alpha$ and hence angle $OBC = \alpha$.
∴ $OB = p \operatorname{cosec} \alpha$. Thus the coordinates of A are $(p \sec \alpha, 0)$ and of B $(0, p \operatorname{cosec} \alpha)$.

The equation of the line AB is $\dfrac{x}{p \sec \alpha} + \dfrac{y}{p \operatorname{cosec} \alpha} = 1$.

i.e. $x \cos \alpha + y \sin \alpha = p$.

This is the **perpendicular** or **normal** form for the equation of a straight line. p is taken to be positive.

Example 12 What is the perpendicular distance of the line $3x + 4y - 15 = 0$ from the origin?

If the equation is given in the form $x \cos \alpha + y \sin \alpha = p$ then p is the perpendicular distance of the line from the origin. Thus we require $\tan \alpha = 4/3$. i.e. $\cos \alpha = 3/5$ and $\sin \alpha = 4/5$. Divide the given equation by $\sqrt{4^2 + 3^2} = 5$.

Thus
$$\frac{3}{5}x + \frac{4}{5}y = \frac{15}{5} = 3$$

Therefore the line is 3 units from the origin.

The distance of a general point $P(x_1, y_1)$ from a line whose equation is $ax + by + c = 0$.

Since the gradient of the line AB is $-a/b$, the gradient of PQ (the perpendicular) must be b/a, i.e.

143

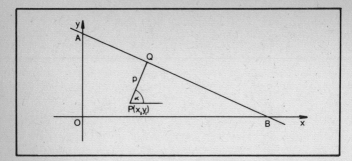

Figure 74

$\tan \alpha = \dfrac{b}{a}$ giving $\cos \alpha = \pm\dfrac{a}{\sqrt{a^2 + b^2}}$ and $\sin \alpha = \pm\dfrac{b}{\sqrt{a^2 + b^2}}$.

Then, if the length of the perpendicular $PQ = p$, the coordinates of Q will be $(x_1 + p \cos \alpha, y_1 + p \sin \alpha)$.
But Q lies on the given line $ax + by + c = 0$.

$\therefore \quad a(x_1 + p \cos \alpha) + b(y_1 + p \sin \alpha) + c = 0$

$\therefore \quad p(a \cos \alpha + b \sin \alpha) = -(ax_1 + by_1 + c)$

$\therefore \quad \dfrac{p(a^2 + b^2)}{\pm\sqrt{a^2 + b^2}} = -(ax_1 + by_1 + c)$

$\therefore \quad p = \pm\dfrac{ax_1 + by_1 + c}{\sqrt{a^2 + b^2}}$, the sign being chosen to make $p > 0$.

The sign, as evaluated, indicates whether the point (x_1, y_1) is on the same side of the line as the origin or not.

Example 13 Find the distances of the points $A(-3, 0)$ and $B(-5, 1)$ from the line $2x - 3y + 9 = 0$.
Substituting in the above result we have

$$p_A = \frac{2(-3) - 3(0) + 9}{\sqrt{2^2 + 3^2}} = \frac{3}{\sqrt{13}}$$

$$p_B = \frac{2(-5) - 3(1) + 9}{\sqrt{2^2 + 3^2}} = \frac{-4}{\sqrt{13}}$$

Hence the distances are $3/\sqrt{13}$ and $4/\sqrt{13}$ respectively.

It can be seen from a sketch that the two points are actually on opposite sides of the line as indicated by the difference in sign in the calculation.

The circle

The circle can be defined as a set of points equidistant from a fixed point, the centre of the circle.

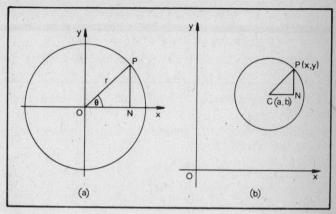

Figure 75

Figure 75a shows a circle, the centre of which has been chosen at the origin of coordinates.

Let $P(x, y)$ be any point on the circumference of the circle, radius r. Let PN be drawn perpendicular to the x axis. In triangle ONP, $x = r \cos \theta$ and $y = r \sin \theta$ where angle $PON = \theta$. Squaring and adding these two results we obtain

$$x^2 + y^2 = r^2(\cos^2 \theta + \sin^2 \theta)$$

i.e. $x^2 + y^2 = r^2$

This relationship between the coordinates of $P(x, y)$ is satisfied by all points on the circumference of the circle and hence it represents the equation of the circle whose centre is the origin and radius r.

Example 14 Find the equation of a circle whose centre is at the origin of coordinates which passes through the point $P(2, 3)$.

Since the point $(2, 3)$ lies on the circle it follows that these coordinates will satisfy the equation $x^2 + y^2 = r^2$.

$$\therefore \quad 2^2 + 3^2 = r^2, \quad \text{i.e.} \quad r^2 = 13$$

The equation of the circle is $x^2 + y^2 = 13$.

A circle of radius r whose centre is at $C(a, b)$

Refer to fig. 75b where $C(a, b)$ is the centre of the circle. Let $P(x, y)$ be a general point on the circumference of the circle. Construct the right-angled triangle CPN as shown.

In $\triangle CPN \qquad CN^2 + NP^2 = CP^2$ (Pythagoras' theorem)

$$\therefore \quad (x-a)^2 + (y-b)^2 = r^2 \qquad (1)$$

This is one form for the equation of a circle, radius r and centre (a, b).

Example 15 What is the equation of a circle, centre $(-1, 4)$ whose radius is 2.

Using equation (1) the equation becomes

$$[x - (-1)]^2 + (y-4)^2 = 2^2$$
i.e. $\qquad (x+1)^2 + (y-4)^2 = 4$

This can be written as $x^2 + y^2 + 2x - 8y + 13 = 0$.

We can expand equation (1) to give an alternative form,

$$(x-a)^2 + (y-b)^2 = r^2$$
$$x^2 - 2ax + a^2 + y^2 - 2by + b^2 = r^2$$
$$x^2 + y^2 - 2ax - 2by + a^2 + b^2 - r^2 = 0$$

This is usually written $x^2 + y^2 + 2gx + 2fy + c = 0$ (2)
where the values g, f and c are constants.

Note that in this form
 (i) the centre of the circle is at the point $(-g, -f)$,
(ii) the radius of the circle is r where $c = g^2 + f^2 - r^2$.

Example 16 What are the coordinates of the centre and the radius of the circle given by $x^2 + y^2 + 6x - 9y + 4 = 0$.

Comparing with equation (2) it follows that the coordinates of the centre are given by $(-3, +4\frac{1}{2})$.

The radius is given by $c = g^2 + f^2 - r^2$

$$\therefore \quad 4 = (3)^2 + (-4\frac{1}{2})^2 - r^2 \Rightarrow r^2 = 9 + \frac{81}{4} - 4 = \frac{101}{4}$$

$\therefore$ the radius is $\frac{1}{2}\sqrt{101}$.

Alternatively this problem can be solved by completing the

square in x and y in the given equation.

$$x^2 + y^2 + 6x - 9y + 4 = 0$$
$$(x^2 + 6x + 3^2) + (y^2 - 9y + (-4\tfrac{1}{2})^2) + 4 = 3^2 + (-4\tfrac{1}{2})^2$$
$$(x + 3)^2 + (y - 4\tfrac{1}{2})^2 = 101/4.$$

Hence the centre is $(-3, 4\tfrac{1}{2})$ and the radius $\tfrac{1}{2}\sqrt{101}$ as before.

Tangents to a circle

If the two points of intersection between a straight line and a circle are coincident then the line is said to be a tangent to the circle.

Example 17 What is the value of m if the line $y = mx$ is a tangent to the circle $x^2 + y^2 + 2x - 4y + 1 = 0$? What are the coordinates of the points at which the tangent touches the curve?

The points of intersection are found by solving simultaneously

$$x^2 + y^2 + 2x - 4y + 1 = 0 \quad \text{and} \quad y = mx$$

$\therefore \qquad x^2 + m^2x^2 + 2x - 4mx + 1 = 0$

$\therefore \qquad (1 + m^2)x^2 + (2 - 4m)x + 1 = 0 \qquad (1)$

Now since the general quadratic equation $ax^2 + bx + c = 0$ has equal roots if $b^2 = 4ac$ it follows that equation (1) has equal roots if

$$(2 - 4m)^2 = 4(1)(1 + m^2)$$
$$4 - 16m + 16m^2 = 4 + 4m^2$$

i.e. if $12m^2 - 16m = 0$ or $4m(3m - 4) = 0$ giving $m = 0$ or $4/3$

Thus if $m = 0$ or $m = 4/3$, $y = mx$ forms a tangent to the circle. The equations are $y = 0$ and $y = \tfrac{4}{3}x$.

The coordinates of the point of contact can be found from the fact that the x coordinates are given by the roots of equation (1), and since they are equal we can write

$$x_1 = x_2 = -\frac{b}{2a} = -\frac{(2 - 4m)}{2(1 + m^2)}$$

(from the sum of the roots of the quadratic equation $ax^2 + bx + c = 0$).

when $m = 0$ the x coordinate $= -1$

when $m = \tfrac{4}{3}$ the x coordinate $= -\dfrac{[2 - 4(\tfrac{4}{3})]}{2[1 + (\tfrac{4}{3})^2]} = \dfrac{3}{5}$

Using $y = 0$ and $y = \frac{4}{3}x$ the points of contact are given by $(-1, 0)$ and $(3/5, 4/5)$.

Parametric form

It is possible to express the position of a general point on a circle and hence its equation in terms of a third arbitrary variable. Such a variable is called a **parameter**.

Example 18 A point P has coordinates $(a \cos \theta, a \sin \theta)$ where a is a constant. If θ varies, describe the path of P.

Let the coordinates of P be (x, y).
Then $x = a \cos \theta$ and $y = a \sin \theta$. Squaring and adding gives

$$x^2 + y^2 = a^2 \cos^2 \theta + a^2 \sin^2 \theta = a^2(\cos^2 \theta + \sin^2 \theta) = a^2$$

Thus the relationship between the coordinates of P is $x^2 + y^2 = a^2$. We say that the locus of P is $x^2 + y^2 = a^2$. Thus as θ varies, this relationship holds for all positions of P and hence P moves on a circle, centre the origin, radius a.

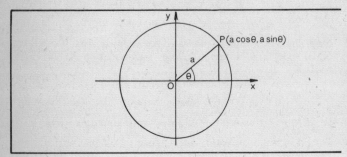

Figure 76

When the coordinates of P are given in the form $(a \cos \theta, a \sin \theta)$ θ is referred to as a parameter and the equation $\boldsymbol{x = a \cos \theta}$, $\boldsymbol{y = a \sin \theta}$ are the **parametric equations** of a circle, centre the origin and radius a.

In this example θ is interpreted geometrically as the angle that the radius vector from the origin to P makes with the x axis as shown in fig. 76.

Note that by choosing specific values of θ actual points will be defined.

For $\theta = 0$, P is the point $(a, 0)$.

For $\theta = \pi/4$, P is the point $(a \cos \pi/4, a \sin \pi/4) = (a/\sqrt{2}, a/\sqrt{2})$.

The conic sections

The circle, parabola, ellipse and hyperbola can be generated from the intersection of a right circular cone and an appropriate plane, and hence they are known as the conic sections.

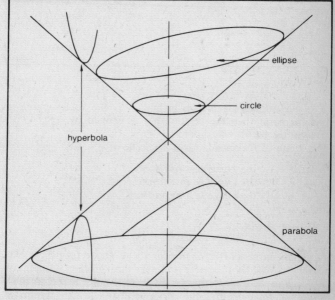

Figure 77

We have already considered the circle and we now examine the parabola, ellipse and hyperbola in terms of coordinate geometry. It will be noticed that many of the techniques can be applied to all three conics and that the parametric form is useful.

General definition The locus of a point which moves so that the ratio of its distance from a fixed point, S, to its distance from a fixed line is a constant, e, is a conic section.

The fixed point S is called the **focus**.
The fixed line is called the **directrix**.
The ratio e is called the **eccentricity**.

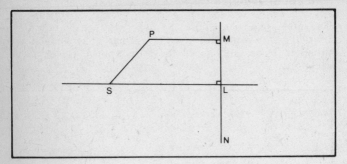

Figure 78

If in fig. 78 MN is the given line, S is the focus and P a
general point then the relationship is defined by $SP = ePM$.
Clearly the locus is symmetrical about SL.
The locus of P depends upon the value of e.

When $e = 1$, the curve is a **parabola**.
  $e < 1$, the curve is an **ellipse**.
  $e > 1$, the curve is a **hyperbola**.

The parabola

Since the parabola is symmetrical about SL we take the x axis
along LS. Clearly the curve will pass through the mid point of
LS and so we choose the y axis perpendicular to LS through
this point thus ensuring the curve will pass through the origin
of coordinates, $(0, 0)$. This is called the **vertex** of the parabola.
Let the distance of S from the origin be a, i.e. the focus is at the
point $(a, 0)$. Let $P(x, y)$ be any point of the curve. From P draw
PM parallel to SL to meet the line $x = -a$ (the directrix) at M.
Join PS, see fig. 79.

By definition $SP = PM$ (since $e = 1$) giving $SP^2 = PM^2$.

$\therefore \qquad\qquad (y - 0)^2 + (x - a)^2 = (x + a)^2$

$\therefore \qquad\qquad y^2 + x^2 - 2ax + a^2 = x^2 + 2ax + a^2$

$\therefore \qquad\qquad\qquad\qquad \boldsymbol{y^2 = 4ax}$

150

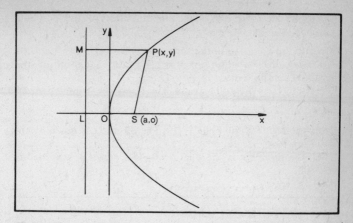

Figure 79

This is the **standard equation of a parabola.** When the equation is in this form it follows that

(i) the focus S is the point $(a, 0)$,
(ii) the directrix is the line $x = -a$,
(iii) the vertex is at the origin of coordinates,
(iv) the curve is symmetrical about the x axis,
(v) the y axis is a tangent at the vertex.

A line through the focus parallel to the y axis is called the **latus rectum.** If the equation is in the form $y^2 = 4ax$ then the length of the latus rectum is $4a$.

Example 19 Find the locus of the point $P(x, y)$ which moves such that its distance from the point $S(-3, 0)$ is equal to its distance from the fixed line $x = 3$.

Referring to fig. 80a

$$PS^2 = (x + 3)^2 + y^2 \quad \text{and} \quad PM^2 = (x - 3)^2$$

Since $PS = PM$ it follows that $PS^2 = PM^2$.

$\therefore$ $$(x + 3)^2 + y^2 = (x - 3)^2$$

$\therefore$ $$x^2 + 6x + 9 + y^2 = x^2 - 6x + 9$$

$\therefore$ $y^2 = -12x$ which is the locus of P.

If the equation is not already in the standard form it is possible to rearrange it.

Example 20 Find the focus and directrix of the parabola $y^2 = 8 - 2x$.

Rewrite the equation in the form $y^2 = 4(2 - \frac{1}{2}x)$.
By using the substitution $X = 2 - \frac{1}{2}x$ the equation of the parabola can be written as $y^2 = 4X$.
The focus is at the point where $X = 1$ and $y = 0$, i.e.

where $2 - \frac{1}{2}x = 1$ and $y = 0$ or $x = 2$ and $y = 0$

Thus the focus is the point $(2, 0)$. The directrix is given by the equation $X = -1$, i.e. $2 - \frac{1}{2}x = -1$.
Thus the directrix is the line $x = 6$. The curve is sketched in fig. 80b.

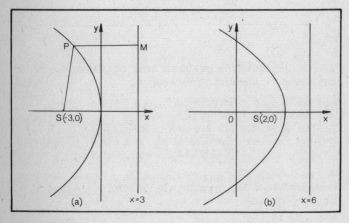

Figure 80

Example 21 Find the value of c which makes the straight line $y = mx + c$ a tangent to the parabola $y^2 = 4ax$.

Solving simultaneously we can form a quadratic equation in y.

$$y^2 = 4a(y - c)/m \quad \text{or} \quad my^2 = 4ay - 4ac$$

$$\therefore \qquad my^2 - 4ay + 4ac = 0$$

The roots of this equation will give the y coordinates of the points of intersection of the line and the curve. A repeated root will be obtained when $(-4a)^2 = 4m(4ac)$.

$$\therefore \qquad a^2 = mac \quad \text{or} \quad c = a/m$$

152

∴ The line $y = mx + a/m$ will be a tangent to the parabola. By using the substitution $t = 1/m$ this result can be written

$$y = \frac{x}{t} + at$$

This suggests a possible parametric form for the equation.

Any point on the parabola can be represented by the coordinates $(at^2, 2at)$. These values would satisfy the equation $y^2 = 4ax$.

The **parametric equations** of the parabola are **$x = at^2$, $y = 2at$.**

Note that t has to take all values from $-\infty$ to $+\infty$ if the complete curve is to be traced.

Much of the detailed work on the parabola is done using parameters.

Example 22 Find the equation of the tangent and the normal to the parabola $y^2 = 4ax$ at the point $P(at^2, 2at)$.

Differentiating implicitly with respect to x,

$$2y \frac{dy}{dx} = 4a \quad \text{giving} \quad \frac{dy}{dx} = \frac{2a}{2at} = \frac{1}{t} \quad \text{at the point } (at^2, 2at)$$

The gradient of the tangent is thus $1/t$ and since it passes through the point $(at^2, 2at)$ its equation is given by

$$y - 2at = \frac{1}{t}(x - at^2)$$

$$ty = x + at^2$$

Similarly the equation of the normal can be derived. If the gradient of the tangent at $(at^2, 2at)$ is $1/t$ the gradient of the normal is $-t$. The equation of the normal is

$$y - 2at = -t(x - at^2)$$

$$y + tx = 2at + at^3$$

Example 23 Find the equation of the chord joining the point $P(ap^2, 2ap)$ to the point $Q(aq^2, 2aq)$ on the parabola $y^2 = 4ax$.

The gradient of the chord $PQ = \dfrac{2ap - 2aq}{ap^2 - aq^2} = \dfrac{2a(p - q)}{a(p^2 - q^2)}$

153

$$= \frac{2(p-q)}{(p+q)(p-q)} = \frac{2}{p+q}$$

Since the chord passes through $(ap^2, 2ap)$ the equation is

$$y - 2ap = \frac{2}{(p+q)}(x - ap^2)$$

$$(p+q)y - 2ap^2 - 2apq = 2x - 2ap^2$$

$$\therefore \qquad \mathbf{(p+q)y = 2x + 2apq} \qquad\qquad (1)$$

From this equation of the chord we can deduce the equation of the tangent, for if $q \to p$ we have

$$2py = 2x + 2ap^2 \Rightarrow py = x + ap^2 \quad \text{as before.}$$

A focal chord is a chord passing through the focus $(a, 0)$. Since this point lies on the line, these values satisfy equation (1) which gives the equation of a chord joining two points.

$$\therefore \quad 0 = 2a + 2apq \Rightarrow pq = -1$$

Thus **the condition that a chord should be a focal chord is $pq = -1$.**

Non parametric method

It is possible to find equivalent equations to those derived for the tangent, normal and chord by considering a point (x_1, y_1) on the parabola.

Example 24 Find the equation of the tangent to the parabola $y^2 = 4ax$ at the point $P(x_1, y_1)$.

As before the gradient is given by $\dfrac{dy}{dx} = \dfrac{2a}{y} \Rightarrow \dfrac{dy}{dx} = \dfrac{2a}{y_1}$ at (x_1, y_1).

Hence the equation of the tangent is given by

$$y - y_1 = \frac{2a}{y_1}(x - x_1)$$

$$yy_1 - y_1^2 = 2ax - 2ax_1$$

But since the point (x_1, y_1) lies on the curve $y^2 = 4ax$, $y_1^2 = 4ax_1$. The equation of the tangent is **$yy_1 = 2a(x + x_1)$.**

Example 25 If tangents are drawn to the parabola $y^2 = 4ax$ at the ends of a focal chord show that they meet on the directrix.

Let P and Q be the points whose coordinates are given by $(ap^2, 2ap)$ and $(aq^2, 2aq)$ respectively.

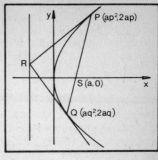

The equation of PQ is
$$(p + q)y = 2x + 2apq$$
(see equation (1) page 154).

Since PQ passes through the focus $S(a, 0)$ this equation gives

$$pq = -1 \qquad (1)$$

Now the equation of the tangent at $P(ap^2, 2ap)$ is

Figure 81

$$py = x + ap^2 \qquad (2)$$

Also the equation of the tangent at $Q(aq^2, 2aq)$ is

$$qy = x + aq^2 \qquad (3)$$

Solving equations (2) and (3) we find the point of intersection, R.

$$(p - q)y = ap^2 - aq^2 = a(p - q)(p + q) \Rightarrow y = a(p + q) \quad (4)$$

Substituting in equation (2)

$$pa(p + q) = x + ap^2 \Rightarrow x = apq$$

Thus the tangents intersect at $R(apq, a(p + q))$.

But from equation (1) $pq = -1$ and hence $x = -a$ which is the equation of the directrix. Thus the tangents always intersect at a point on the directrix.

Example 26 If the normal at a point $P(ap^2, 2ap)$ on the parabola $y^2 = 4ax$ meets the curve again at $Q(aq^2, 2aq)$ prove that p and q are related by the equation $p^2 + pq + 2 = 0$.

Since the equation of the normal at $P(ap^2, 2ap)$ is given by

$$y + px = 2ap + ap^3 \text{ (see example 22)}$$

We have, solving simultaneously with the equation of the curve,

$$y + \frac{py^2}{4a} = 2ap + ap^3$$

i.e. $$py^2 + 4ay - 4a(2ap + ap^3) = 0$$

The roots of this equation give the y coordinates of the points

155

of intersection, say y_1 and y_2.

Now $y_1 + y_2 = -4a/p$ (the sum of the roots of a quadratic equation). But $y_1 = 2ap$ and hence $y_2 = -2ap - 4a/p$.

$$\therefore \quad y_2 = 2aq = \frac{-4a - 2ap^2}{p} \Rightarrow q = \frac{-2 - p^2}{p}$$

i.e.
$$p^2 + pq + 2 = 0.$$

Example 27 Show that the normals drawn to the parabola $y^2 = 4ax$ at the points $P(ap^2, 2ap)$ and $Q(aq^2, 2aq)$ where PQ is a focal chord meet at a point $R(a(p^2 + q^2 + 1), a(p + q))$. Find the locus of R.

The equations of the normals at $P(ap^2, 2ap)$ and $Q(aq^2, 2aq)$ are

$$y + px = 2ap + ap^3 \quad \text{and} \quad y + qx = 2aq + aq^3$$

respectively. Subtracting these equations we obtain

$$(p - q)x = 2a(p - q) + a(p^3 - q^3)$$

i.e.
$$x = 2a + a(p^2 + pq + q^2) \quad \text{since } p - q \text{ is a factor.}$$

But since PQ is a focal chord $pq = -1$, thus $x = a(p^2 + q^2 + 1)$. Substituting in the equation of the normal at P we have

$$y + ap(p^2 + q^2 + 1) = 2ap + ap^3$$

$$y = 2ap - apq^2 - ap = a(p + q) \quad \text{since } pq = -1$$

Hence the point of intersection R is $(a(p^2 + q^2 + 1), a(p + q))$.

Let the coordinates of R be (x, y)

$$\therefore \quad x = a(p^2 + q^2 + 1) \tag{3}$$

and
$$y = a(p + q) \tag{4}$$

The relationship between x and y can be obtained by eliminating the parameters p and q.

Equation (3) gives $\quad x = a[(p + q)^2 - 2pq + 1]$

Since $pq = -1 \quad\quad x = a[(p + q)^2 + 3]$

But $y = a(p + q) \quad \therefore \quad x = a\left(\dfrac{y^2}{a^2} + 3\right)$

Thus the equation of the locus of R is $y^2 = ax - 3a^2$.

The ellipse

This curve has already been defined (see page 150) as the conic section whose eccentricity $e < 1$. We now derive its equation by choosing axes to give the most convenient form. The line through the focus, S, perpendicular to the directrix is chosen as the x axis. See fig. 82. Since, from fig. 77 on page 149, we can see that the ellipse is a closed curve let the curve cut the x axis at A and A'. We choose the y axis to be a line parallel to the directrix through the mid point of AA'.

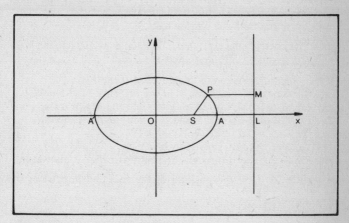

Figure 82

Let AA' be of length $2a$, i.e. A is the point $(a, 0)$ and A' is $(-a, 0)$. Let the focus be $S(s, 0)$ and the directrix $x = d$.
Since A and A' lie on the ellipse $SA = eAL$ and $SA' = eA'L$.

$$\therefore \quad a - s = e(d - a) \quad \text{and} \quad a + s = e(d + a)$$

Solving gives $s = ae$ and $d = a/e$.
Thus the focus is $S(ae, 0)$ and the directrix is the line $y = a/e$.
Consider a general point $P(x, y)$ on the ellipse.
By definition, $SP = ePM$, $SP^2 = e^2 PM^2$.

$$\therefore \qquad y^2 + (x - ae)^2 = e^2\left(\frac{a}{e} - x\right)^2$$

$$\therefore \qquad y^2 + x^2 - 2aex + a^2e^2 = a^2 - 2aex + e^2x^2$$

$$\therefore \qquad x^2(1 - e^2) + y^2 = a^2(1 - e^2)$$

By writing $b^2 = a^2(1 - e^2)$ and dividing by this factor we have

$$\frac{x^2}{a^2} + \frac{y^2}{b^2} = 1$$

This is the **standard equation of the ellipse.**

Note that we can write the equation as $\frac{y^2}{b^2} = 1 - \frac{x^2}{a^2}$, which gives two equal and opposite real values for y if $|x| < a$ but no real values if $x^2 > a^2$. Similarly if $y^2 > b^2$ it follows that there are no real values of x. Thus the curve is bounded by the lines $x = \pm a$ and $y = \pm b$. Hence the curve is closed and symmetrical about both the chosen axes.

This symmetry suggests that there is a second focus and directrix given by $S'(-ae, 0)$ and $x = -a/e$.

If the equation is given in the standard form then
  (i) the centre of the ellipse is at the origin of coordinates,
 (ii) the distance AA' is $2a$ and is called the **major axis,**
(iii) the distance BB' is $2b$ and is called the **minor axis,**
(iv) the foci are at $(\pm ae, 0)$,
 (v) the directrices are given by $x = \pm a/e$.
Note that any chord through the centre is called a **diameter.**

Example 28 Find the position of the foci and directrices of the ellipse $\frac{x^2}{16} + \frac{y^2}{9} = 1$.

Comparing the given equation with the standard equation we see that $a = 4$ and $b = 3$.

Now $b^2 = a^2(1 - e^2) \Rightarrow 9 = 16(1 - e^2) \Rightarrow 16e^2 = 7 \Rightarrow e = \sqrt{7}/4$
The foci are $(\pm ae, 0)$, i.e. $(\pm \sqrt{7}, 0)$.
The directrices are given by $x = \pm a/e$, i.e. $x = \pm 16/\sqrt{7}$.

Parametric form

The coordinates of any point on the ellipse can be represented by $(a \cos \theta, b \sin \theta)$.
The **parametric equations** of the ellipse are thus $x = a \cos \theta$ and $y = b \sin \theta$.

For the circle the geometrical interpretation of θ was simply the angle made by the radius vector to the given point with the x axis. A different interpretation is required for the ellipse.

Consider a general ellipse $\frac{x^2}{a^2} + \frac{y^2}{b^2} = 1$ and let a circle of radius a, centre the origin, be drawn as shown in fig. 83. Consider any point $P(a \cos \theta, b \sin \theta)$ on the ellipse and construct an ordinate through P to cut the circle at Q. The x coordinate of Q is $a \cos \theta$

and hence it follows from $\triangle OQN$ that angle $QON = \theta$.

θ is called the **eccentric angle** of the point P, and the circle, whose equation is $x^2 + y^2 = a^2$, is called the **auxiliary circle**.

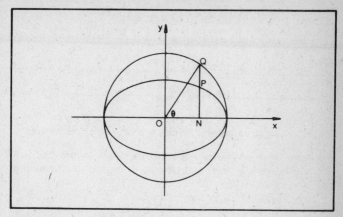

Figure 83

The techniques used for the parabola can be applied to the ellipse, again usually in terms of the parameter θ.

Example 29 Find the equation of the tangent to the ellipse given by $\dfrac{x^2}{a^2} + \dfrac{y^2}{b^2} = 1$ at the point $P(a \cos \theta, b \sin \theta)$.

Differentiating implicitly with respect to x

$$\frac{2x}{a^2} + \frac{2y}{b^2}\frac{dy}{dx} = 0 \quad \Rightarrow \quad \frac{dy}{dx} = -\frac{b \cos \theta}{a \sin \theta} \text{ at } P(a \cos \theta, b \sin \theta)$$

The equation of the tangent is

$$y - b \sin \theta = -\frac{b \cos \theta}{a \sin \theta}(x - a \cos \theta)$$

$\therefore \qquad ay \sin \theta - ab \sin^2 \theta = -bx \cos \theta + ab \cos^2 \theta$

$\therefore \qquad ay \sin \theta + bx \cos \theta = ab(\cos^2 \theta + \sin^2 \theta) = ab$

Dividing by ab the equation becomes

$$\frac{x}{a}\cos \theta + \frac{y}{b}\sin \theta = 1$$

159

Example 30 Find the equation of the normal to the ellipse $\frac{x^2}{a^2} + \frac{y^2}{b^2} = 1$ at the point $P(a \cos \theta, b \sin \theta)$.

The gradient of the tangent is $-\dfrac{b \cos \theta}{a \sin \theta}$.

Thus the gradient of the normal at P is $\dfrac{a \sin \theta}{b \cos \theta}$.

The equation of the normal is

$$y - b \sin \theta = \frac{a \sin \theta}{b \cos \theta} (x - a \cos \theta)$$

$\therefore$ $by \cos \theta - b^2 \cos \theta \sin \theta = ax \sin \theta - a^2 \cos \theta \sin \theta$
The equation is $\boldsymbol{ax \sin \theta - by \cos \theta = (a^2 - b^2) \cos \theta \sin \theta}$.

Example 31 Find the equation of the chord joining the two points $P(a \cos \theta, b \sin \theta)$ and $Q(a \cos \phi, b \sin \phi)$ on the ellipse $\frac{x^2}{a^2} + \frac{y^2}{b^2} = 1$.

The gradient of the chord is $\dfrac{b \sin \theta - b \sin \phi}{a \cos \theta - a \cos \phi}$

$$= -\frac{2b \cos \frac{1}{2}(\theta + \phi) \sin \frac{1}{2}(\theta - \phi)}{2a \sin \frac{1}{2}(\theta + \phi) \sin \frac{1}{2}(\theta - \phi)} = -\frac{b \cos \frac{1}{2}(\theta + \phi)}{a \sin \frac{1}{2}(\theta + \phi)}$$

Since the chord passes through $P(a \cos \theta, b \sin \theta)$ its equation is

$$y - b \sin \theta = -\frac{b \cos \frac{1}{2}(\theta + \phi)}{a \sin \frac{1}{2}(\theta + \phi)} (x - a \cos \theta)$$

$ay \sin \frac{1}{2}(\theta + \phi) + bx \cos \frac{1}{2}(\theta + \phi)$

$$= ab(\sin \theta \sin \frac{1}{2}(\theta + \phi) + \cos \theta \cos \frac{1}{2}(\theta + \phi)$$

$$= ab[\cos \frac{1}{2}(\theta + \phi) - \theta] = ab \cos \frac{1}{2}(\theta - \phi)$$

Thus the equation of the chord is

$$\boldsymbol{ay \sin \tfrac{1}{2}(\theta + \phi) + bx \cos \tfrac{1}{2}(\theta + \phi) = ab \cos \tfrac{1}{2}(\theta - \phi)}$$

Note that if $\phi \to \theta$, point Q approaches point P and the equation of the chord yields the equation of the tangent.
i.e. $\qquad ay \sin \theta + bx \cos \theta = ab$ as before.

Example 32 Find the condition that the line $y = mx + c$ is a tangent to the ellipse $\frac{x^2}{a^2} + \frac{y^2}{b^2} = 1$.

Solving simultaneously we have $\dfrac{x^2}{a^2} + \dfrac{(mx + c)^2}{b^2} = 1$

$$\therefore \qquad b^2x^2 + a^2m^2x^2 + 2a^2mcx + a^2c^2 = a^2b^2$$
$$\therefore \qquad (b^2 + a^2m^2)x^2 + 2a^2mcx + a^2(c^2 - b^2) = 0$$

For a tangent we need repeated roots,

$$\therefore \qquad (2a^2mc)^2 = 4(b^2 + a^2m^2)(c^2 - b^2)a^2$$
$$\therefore \qquad a^2m^2c^2 = (b^2 + a^2m^2)(c^2 - b^2)$$

Expanding and simplifying we obtain $c^2 = b^2 + a^2m^2$.
Thus the line $y = mx + c$ is a tangent to the given ellipse if $c^2 = b^2 + a^2m^2$.
The equations of the tangents are $y = mx \pm \sqrt{b^2 + a^2m^2}$.

The director circle

The locus of the point of intersection of two perpendicular tangents to an ellipse $\dfrac{x^2}{a^2} + \dfrac{y^2}{b^2} = 1$ is a circle whose equation is $x^2 + y^2 = a^2 + b^2$. This circle is called the **director circle**.

In example 32 the tangent to an ellipse was found to be of the form $y = mx \pm \sqrt{b^2 + a^2m^2}$ where m is the gradient.

i.e. $\quad y - mx = \pm\sqrt{b^2 + a^2m^2} \Rightarrow (y - mx)^2 = b^2 + a^2m^2$

$$\therefore \qquad y^2 - 2mxy + m^2x^2 = b^2 + a^2m^2$$

Writing this as a quadratic in m we have

$$m^2(a^2 - x^2) + 2xym + (b^2 - y^2) = 0 \qquad (1)$$

Since this is a quadratic with two roots m_1 and m_2 there are two possible tangents that can be drawn to the ellipse. If these are perpendicular then $m_1m_2 = -1$.

But from equation (1) $m_1m_2 = \dfrac{b^2 - y^2}{a^2 - x^2}$ (product of the roots)

Hence $\dfrac{b^2 - y^2}{a^2 - x^2} = -1 \Rightarrow b^2 - y^2 = -a^2 + x^2$

Thus the equation of the circle is $x^2 + y^2 = a^2 + b^2$.

Conjugate diameters
Two diameters of an ellipse are said to be **conjugate** when each one bisects all the chords parallel to the other.

Example 33 Show that for two conjugate diameters the product of their gradients is $-b^2/a^2$.

Consider fig. 84 where CD is a given diameter of gradient m and EF is a parallel chord.

Let E and F have parameters θ and ϕ respectively. Thus E is the point $(a \cos \theta, b \sin \theta)$ and F is $(a \cos \phi, b \sin \phi)$.

The gradient of EF is $-\dfrac{b \cos \frac{1}{2}(\theta + \phi)}{a \sin \frac{1}{2}(\theta + \phi)} = m$ (see example 31).

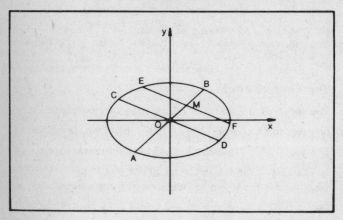

Figure 84

If $M(x, y)$ is the mid point of EF the gradient of OM is given by

$$\frac{y}{x} = \frac{\frac{1}{2}(b \sin \theta + b \sin \phi)}{\frac{1}{2}(a \cos \theta + a \cos \phi)}$$

$$\therefore \quad \frac{y}{x} = \frac{b \sin \frac{1}{2}(\theta + \phi) \cos \frac{1}{2}(\theta - \phi)}{a \cos \frac{1}{2}(\theta + \phi) \cos \frac{1}{2}(\theta - \phi)} = \frac{b \sin \frac{1}{2}(\theta + \phi)}{a \cos \frac{1}{2}(\theta + \phi)} = -\frac{b^2}{a^2 m}$$

$$\therefore \quad y = -\left(\frac{b^2}{a^2 m}\right) x$$

which is the equation of a straight line through the origin of gradient $-b^2/a^2 m$.

Thus the mid points of chords parallel to CD lie on a straight line through the centre of the ellipse. Hence AB and CD are conjugate diameters and if their gradients are m' and m then $mm' = -b^2/a^2$.

The hyperbola

This curve has already been defined (see page 150) as the conic section whose eccentricity $e > 1$. We derive its equation by choosing axes to give the most convenient form. The line through

the focus, S, perpendicular to the directrix is chosen as the x axis.
See fig. 85. Since the curve cuts this axis in two points, say A and
A', the mid point of AA' is chosen as the origin. Thus the y axis is
taken as the line parallel to the directrix through the mid point of
AA'.

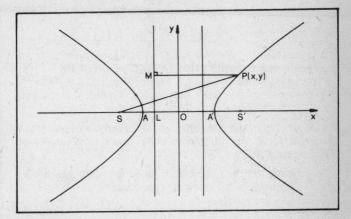

Figure 85

Let AA' be of length $2a$, i.e. A is the point $(-a, 0)$ and A' is $(a, 0)$.
Let the focus be $S(-s, 0)$ and the directrix $x = -d$.
By definition, since A and A' lie on the curve $SA = eAL$ and
$SA' = eA'L$.

$$\therefore \quad s - a = e(a - d) \quad \text{and} \quad s + a = e(a + d)$$
Solving gives $s = ae$ and $d = a/e$.

Thus the focus is $S(-ae, 0)$ and the directrix is the line
$y = -a/e$. Note that these are numerically the same as for the
ellipse but since $e > 1$ the focus is further from the origin than
the directrix.

Consider a general point $P(x, y)$ on the hyperbola.
By definition, $SP = ePM \Rightarrow SP^2 = e^2 PM^2$.

$$\therefore \qquad y^2 + (x + ae)^2 = e^2\left(\frac{a}{e} + x\right)^2$$

$$\therefore \qquad y^2 + x^2 + 2aex + a^2e^2 = a^2 + 2aex + e^2x^2$$

This can be simplified to give

$$\frac{x^2}{a^2} + \frac{y^2}{a^2(1-e^2)} = 1$$

In the hyperbola $e > 1$ and thus we write $b^2 = a^2(e^2 - 1)$. This gives the **standard equation of the hyperbola** as

$$\frac{x^2}{a^2} - \frac{y^2}{b^2} = 1$$

Note that by writing the equation in the form $y^2 = b^2\left(\frac{x^2}{a^2} - 1\right)$ it can be seen that for real y, $|x| \geqslant a$ and for $|x| > a$ there are two equal and opposite values of y. Similarly there are two equal and opposite values of x for every value of y. Thus the curve consists of two branches and is symmetrical about both axes.

As in the ellipse, the symmetry of the hyperbola suggests that a second focus and directrix can be taken as $S'(ae, 0)$ and $x = a/e$ respectively.

If the equation of the hyperbola is given in standard form then
(i) the centre is at the origin of coordinates,
(ii) the foci are at $(\pm ae, 0)$,
(iii) the directrices are $x = \pm a/e$,
(iv) AA' is of length $2a$ and is called the **transverse axis,**
(v) the **latus rectum** is of length $2b^2/a$.

We can investigate the behaviour of the curve as x becomes large by writing the equation as

$$y^2 = b^2\left(\frac{x^2}{a^2} - 1\right) \text{ giving } y = \pm b\sqrt{\frac{x^2}{a^2} - 1}.$$

As $x \to \infty$ the curve behaves as the line $y = \pm\frac{b}{a}x$. These lines form **asymptotes** to the curve.

This can be verified by considering the condition for a line $y = mx + c$ to be a tangent to the hyperbola $\frac{x^2}{a^2} - \frac{y^2}{b^2} = 1$.

Solving simultaneously $\frac{x^2}{a^2} - \frac{(mx + c)^2}{b^2} = 1$

$\therefore \qquad b^2x^2 - a^2m^2x^2 - 2a^2cmx - a^2c^2 = a^2b^2$

$\therefore \qquad (b^2 - a^2m^2)x^2 - 2a^2cmx - (a^2c^2 + a^2b^2) = 0 \qquad (1)$

This equation has repeated roots if

$$(-2a^2cm)^2 = -4(b^2 - a^2m^2)(a^2c^2 + a^2b^2)$$

$$\therefore \quad a^2c^2m^2 = -b^2c^2 + a^2m^2c^2 - b^4 + a^2m^2b^2$$

i.e. $$c^2 = a^2m^2 - b^2 \qquad (2)$$

The equations of the tangents are $y = mx \pm \sqrt{a^2m^2 - b^2}$ (3)
The x coordinate of the point of contact is given by $\frac{1}{2}(x_1 + x_2)$
where x_1 and x_2 are the roots of equation (1). Since $x_1 = x_2$ we
have $x = \frac{1}{2}\left(\frac{2a^2cm}{b^2 - a^2m^2}\right) = \frac{a^2mc}{(b^2 - a^2m^2)}$.

Now as $m \to \pm b/a$, $x \to \infty$ and $c \to 0$ (from equation (2)). In this
case equation (3) becomes $y = \pm \dfrac{b}{a}x$, which are the equations
of the asymptotes.

These can be written as $y - \dfrac{b}{a}x = 0$ and $y + \dfrac{b}{a}x = 0$.

$$\therefore \qquad \left(y - \frac{b}{a}x\right)\left(y + \frac{b}{a}x\right) = 0 \Rightarrow \frac{x^2}{a^2} - \frac{y^2}{b^2} = 0$$

This is the equation of a **line pair** (a pair of intersecting
straight lines) passing through the origin. The asymptotes are
shown in fig. 86.

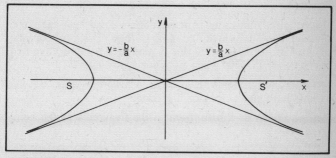

Figure 86

Parametric form

The general point $(a \sec \theta, b \tan \theta)$ will lie on the hyperbola since
the equation will be satisfied by the values **$x = a \sec \theta$, $y =$**
$b \tan \theta$. These equations giving x and y in terms of θ can be
taken as the **parametric equations** of the hyperbola.
The geometrical interpretation of θ needs the auxiliary circle but
the definition is not as easy as for the ellipse.

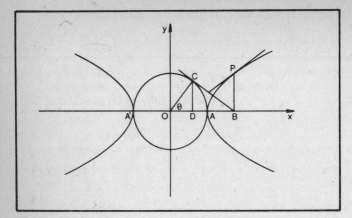

Figure 87

Let P be a point on the positive branch of the hyperbola and draw a perpendicular from P to meet AA' in B. Draw the tangent BC to the auxiliary circle $x^2 + y^2 = a^2$ and join OC.

Then θ is represented by the angle AOC.
In $\triangle OCB$ $OB = OC \sec \theta = a \sec \theta$.

Using the equation of the hyperbola, if $x = a \sec \theta$, $y = \pm b \tan \theta$. Clearly, if P is above the x axis then y is positive and $0° < \theta < 90°$. If P is below the x axis then y is negative and $270° < \theta < 360°$. Thus when $x = a \sec \theta$, $y = b \tan \theta$.

The methods used for the ellipse can, in general, be applied to the hyperbola to give corresponding results.

The equation of the tangent to the hyperbola at the point (x_1, y_1) is

$$\frac{xx_1}{a^2} - \frac{yy_1}{b^2} = 1.$$

The equation of the tangent to the hyperbola at the point $P(a \sec \theta, b \tan \theta)$ is

$$bx - ay \sin \theta = ab \cos \theta.$$

The rectangular hyperbola

There is a case of the general hyperbola that possesses some special properties. When the asymptotes are perpendicular the curve is called a **rectangular hyperbola**.

166

We have shown the asymptotes of the hyperbola $\dfrac{x^2}{a^2} - \dfrac{y^2}{b^2} = 1$

to be given by the equations $y = \pm \dfrac{b}{a} x$ which form a line pair.

Hence if the lines are to be perpendicular the product of their gradients must equal -1.

$$\text{i.e.} \quad \left(\frac{b}{a}\right)\left(-\frac{b}{a}\right) = -1 \Rightarrow b^2 = a^2$$

Thus the equation of the **hyperbola** becomes $x^2 - y^2 = a^2$ and its **asymptotes** are $x + y = 0$ and $x - y = 0$.

Since the asymptotes are perpendicular they could be used as axes and this, in fact, produces an easy form of the equation of a rectangular hyperbola. In effect this means a rotation of the axes about the origin through $-45°$. See fig. 88a.

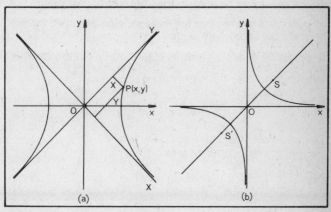

Figure 88

Consider a point $P(x, y)$ whose new coordinates relative to the rotated axes are (X, Y). This rotation is represented by

$$\begin{pmatrix} \cos(+45°) & -\sin(+45°) \\ \sin(+45°) & \cos(+45°) \end{pmatrix} \begin{pmatrix} x \\ y \end{pmatrix} = \begin{pmatrix} X \\ Y \end{pmatrix}$$

Thus $\begin{pmatrix} 1/\sqrt{2} & -1/\sqrt{2} \\ 1/\sqrt{2} & 1/\sqrt{2} \end{pmatrix} \begin{pmatrix} x \\ y \end{pmatrix} = \begin{pmatrix} X \\ Y \end{pmatrix}$ $\quad 1/\sqrt{2}(x - y) = X$

$\quad 1/\sqrt{2}(x + y) = Y$

Multiplying these two results we have $x^2 - y^2 = 2XY$. But the equation of the rectangular hyperbola is $x^2 - y^2 = a^2$. Thus in terms of the new axes its equation is $XY = \frac{1}{2}a^2$.

By writing $c^2 = \frac{1}{2}a^2$ we obtain an equation in the form **xy = c²** which is the **standard equation of a rectangular hyperbola** with its asymptotes as axes. The curve is shown in fig. 88b.

Note that since the eccentricity of $x^2 - y^2 = a^2$ is $\sqrt{2}$ the foci were at the points $(\pm a\sqrt{2}, 0)$. But $c^2 = \frac{1}{2}a^2$ and thus the foci are at points distant $2c$ along the major axis SS', i.e. at points whose coordinates are $(c\sqrt{2}, c\sqrt{2})$ and $(-c\sqrt{2}, -c\sqrt{2})$.

Parametric form

The point whose coordinates are $\left(ct, \dfrac{c}{t}\right)$ lies on the rectangular hyperbola for all values of t except $t = 0$. Thus the equations **x = ct, y = $\dfrac{c}{t}$** can be used as the **parametric equations.**

The methods used before for finding the equations of tangents normals and chords still apply.

Example 34 Find the equation of the tangent to the rectangular hyperbola $xy = c^2$ at the point $P\left(ct, \dfrac{c}{t}\right)$.

$$xy = c^2 \Rightarrow y = \frac{c^2}{x} \Rightarrow \frac{dy}{dx} = -\frac{c^2}{x^2}$$

The gradient of the curve at P is $-1/t^2$.
Hence the equation of the tangent is

$$y - \frac{c}{t} = -\frac{1}{t^2}(x - ct)$$

$$t^2 y - ct = -x + ct$$

Thus the **equation of the tangent is $t^2y + x = 2ct$.**

The **equation of the normal** to the rectangular hyperbola given by the equations $x = ct$, $y = c/t$ is

$$t^3x - ty + c - ct^4 = 0.$$

Example 35 Find the equation of the chord joining the points P and Q whose parameters are t_1 and t_2 respectively on the curve $xy = c^2$.

If P is the point $\left(ct_1, \dfrac{c}{t_1}\right)$ and Q is the point $\left(ct_2, \dfrac{c}{t_2}\right)$ then the

168

gradient of PQ is $\dfrac{(c/t_1) - (c/t_2)}{ct_1 - ct_2} = \dfrac{(t_2 - t_1)}{t_1 t_2 (t_1 - t_2)} = -\dfrac{1}{t_1 t_2}$

The equation of the chord is

$$y - \frac{c}{t_1} = -\frac{1}{t_1 t_2}(x - ct_1)$$

$$t_1 t_2 y - ct_2 = -x + ct_1$$

Thus the **equation of the chord PQ is $t_1 t_2 y + x = c(t_1 + t_2)$.**

Key terms

The distance between $P(x_1, y_1)$ and $Q(x_2, y_2)$ is
$\sqrt{(x_2 - x_1)^2 + (y_2 - y_1)^2}$

The angle between two lines of gradients m and m' is

$\tan^{-1}\left(\dfrac{m' - m}{1 + m'm}\right)$. If the lines are perpendicular the product of the gradients is -1.

$P(x_1, y_1)$ is a distance $\dfrac{ax_1 + by_1 + c}{\sqrt{a^2 + b^2}}$ from the line $ax + by + c = 0$.

Circle $(x - a)^2 + (y - b)^2 = r^2$.
The centre is (a, b) and radius r. The parametric equations are $x = a \cos \theta$, $y = a \sin \theta$.

Parabola $y^2 = 4ax$.
The equation of the **tangent** at $(at^2, 2at)$ is $ty = x + at^2$.
The equation of the **normal** at $(at^2, 2at)$ is $y + tx = 2at + at^3$.
The equation of the **chord** joining two points with parameters p and q is $(p + q)y = 2x + 2apq$.

Ellipse $\dfrac{x^2}{a^2} + \dfrac{y^2}{b^2} = 1$. General point is $P(a \cos \theta, b \sin \theta)$.

The equation of the **tangent** at P is $ay \sin \theta + bx \cos \theta = ab$.
The equation of the **normal** at P is
$ax \sin \theta - by \cos \theta = (a^2 - b^2) \cos \theta \sin \theta$

Hyperbola $\dfrac{x^2}{a^2} - \dfrac{y^2}{b^2} = 1$. General point is $P(a \sec \theta, b \tan \theta)$.

The equation of the **tangent** at P is $bx - ay \sin \theta = ab \cos \theta$.

Rectangular hyperbola $xy = c^2$. General point is $P\left(ct, \dfrac{c}{t}\right)$.

The equation of the **tangent** at P is $t^2 y + x = 2ct$.
The equation of the **normal** at P is $t^3 x - ty + c - ct^4 = 0$.

Chapter 7

Matrices and Determinants

$$
\begin{array}{c}
\text{Home} \qquad\qquad \text{Away} \\
\begin{array}{cccc}
 & W & D & L \\
\text{Liverpool} & \begin{pmatrix} 3 & 2 & 1 \\ 5 & 0 & 1 \end{pmatrix} & \\
\text{Arsenal} & & \\
 & \text{Matrix } H &
\end{array}
\end{array}
$$

$$
\begin{array}{cc}
\begin{array}{c}
\text{Home} \\
\begin{array}{ccc} W & D & L \end{array} \\
\begin{array}{l}\text{Liverpool} \\ \text{Arsenal}\end{array}
\begin{pmatrix} 3 & 2 & 1 \\ 5 & 0 & 1 \end{pmatrix} \\
\text{Matrix } H
\end{array}
&
\begin{array}{c}
\text{Away} \\
\begin{array}{ccc} W & D & L \end{array} \\
\begin{pmatrix} 2 & 4 & 0 \\ 2 & 1 & 2 \end{pmatrix} \\
\text{Matrix } A
\end{array}
\end{array}
$$

A **matrix** is a **rectangular array** of elements. In the example above the elements are numbers and by the headings of the rows and columns it can be seen that the matrices represent the home and away playing records of two football teams.

Matrix **addition** can be achieved by adding together the corresponding elements in each matrix.

$$
H + A = \begin{pmatrix} 3 & 2 & 1 \\ 5 & 0 & 1 \end{pmatrix} + \begin{pmatrix} 2 & 4 & 0 \\ 2 & 1 & 2 \end{pmatrix} = \begin{array}{c} \\ \text{L} \\ \text{A} \end{array}\!\!\begin{array}{c}\begin{array}{ccc}W & D & L\end{array}\\ \begin{pmatrix} 5 & 6 & 1 \\ 7 & 1 & 3 \end{pmatrix}\end{array} = \text{Matrix } T.
$$

T represents the total number of matches won, drawn or lost by each team.

Matrix **subtraction** is similar but in this case the resulting matrix has little meaning except to compare the teams' home and away records.

$$
H - A = \begin{pmatrix} 3 & 2 & 1 \\ 5 & 0 & 1 \end{pmatrix} - \begin{pmatrix} 2 & 4 & 0 \\ 2 & 1 & 2 \end{pmatrix} = \begin{pmatrix} 1 & -2 & 1 \\ 3 & -1 & -1 \end{pmatrix}
$$

The **order** of a matrix indicates its size or shape, and this is done by specifying the number of rows and columns.
Matrix H has order 2×3 (2 by 3) since it has 2 rows and 3 columns. Matrices can only be **added** or **subtracted** if they have the same order.

Matrix **multiplication** is more involved.
The points awarded for football results are 2 for a win, 1 for a draw and 0 for a defeat. Liverpool's total points are $5 \times 2 + 6 \times 1 + 1 \times 0 = 16$. Arsenal's points are $7 \times 2 + 1 \times 1 + 3 \times 0 = 15$

$$
\begin{array}{c}
\begin{array}{ccc} W & D & L \end{array}\ \ \text{Points} \\
\begin{array}{c}\text{L}\\\text{A}\end{array}\!\!\begin{pmatrix} 5 & 6 & 1 \\ 7 & 1 & 3 \end{pmatrix}
\end{array}
\begin{array}{c}
\\
\begin{array}{c}W\\D\\L\end{array}\!\!\begin{pmatrix} 2 \\ 1 \\ 0 \end{pmatrix}
\end{array}
= \begin{pmatrix} 5 \times 2 + 6 \times 1 + 1 \times 0 \\ 7 \times 2 + 1 \times 1 + 3 \times 0 \end{pmatrix}
\begin{array}{c}
\text{Points}\\
= \begin{array}{c}\text{L}\\\text{A}\end{array}\!\!\begin{pmatrix} 16 \\ 15 \end{pmatrix}
\end{array}
$$

$$
= \text{Matrix } P
$$

The points matrix has been written as a column matrix (order 3 by 1) and the product matrix P is a column matrix (order 2 by 1) giving the total number of points for the two teams. At first sight the arrangement of the points matrix as a column matrix seems to complicate the matrix multiplication process, but there are advantages to this system and it is the universal convention. If we had a different points system P_2, of 3 points for a win, 1 for a draw and 0 for a defeat, the total points are given by

$$
\begin{array}{c}
\begin{array}{ccc} W & D & L \end{array} \\
\begin{array}{c} L \\ A \end{array}\begin{pmatrix} 5 & 6 & 1 \\ 7 & 1 & 3 \end{pmatrix}
\end{array}
\begin{array}{c}
P_2 \\
\begin{array}{c} W \\ D \\ L \end{array}\begin{pmatrix} 3 \\ 1 \\ 0 \end{pmatrix}
\end{array}
=
\begin{array}{c}
P_2 \\
\begin{array}{c} L \\ A \end{array}\begin{pmatrix} 5\times3+6\times1+1\times0 \\ 7\times3+1\times1+3\times0 \end{pmatrix}
\end{array}
=
\begin{array}{c}
P_2 \\
\begin{array}{c} L \\ A \end{array}\begin{pmatrix} 21 \\ 22 \end{pmatrix}
\end{array}
$$

Both points systems can be incorporated in one matrix multiplication in the following way

$$
\begin{array}{c}
\begin{array}{ccc} W & D & L \end{array} \\
\begin{array}{c} L \\ A \end{array}\begin{pmatrix} 5 & 6 & 1 \\ 7 & 1 & 3 \end{pmatrix}
\end{array}
\begin{array}{c}
\begin{array}{cc} P_1 & P_2 \end{array} \\
\begin{array}{c} W \\ D \\ L \end{array}\begin{pmatrix} 2 & 3 \\ 1 & 1 \\ 0 & 0 \end{pmatrix}
\end{array}
$$

$$
\begin{array}{cc} T & P \end{array}
$$

$$
=
\begin{array}{c}
\begin{array}{cc} P_1 & \qquad\qquad P_2 \end{array} \\
\begin{array}{c} L \\ A \end{array}\begin{pmatrix} 5\times2+6\times1+1\times0 & 5\times3+6\times1+1\times0 \\ 7\times2+1\times1+3\times0 & 7\times3+1\times1+3\times0 \end{pmatrix}
\end{array}
=
\begin{array}{c}
\begin{array}{cc} P_1 & P_2 \end{array} \\
\begin{array}{c} L \\ A \end{array}\begin{pmatrix} 16 & 21 \\ 15 & 22 \end{pmatrix}
\end{array}
$$

$$
S
$$

In matrix multiplication each element of the product matrix comes from a row in the first matrix combined with a column in the second matrix. For example

$$
\begin{bmatrix} \text{row 1} \\ \text{in } T \end{bmatrix} \text{ combined with } \begin{bmatrix} \text{column 2} \\ \text{in } P \end{bmatrix} \text{ gives } \begin{bmatrix} \text{the element of} \\ \text{the product } S \\ \text{which is in the} \\ \text{first row and} \\ \text{second column.} \end{bmatrix}
$$

The number of columns in T must be equal to the number of rows in P.

Matrix	T	$\times$	P	$=$	S
Order	2 by 3		3 by 2		2 by 2

Consider two more points systems P_3 ($W = 1, D = 0, L = -1$) and P_4 ($W = 2, D = 0, L = -1$)

$$\begin{array}{c} \begin{array}{ccc} W & D & L \end{array} \\ \begin{array}{c} L \\ A \end{array}\!\begin{pmatrix} 5 & 6 & 1 \\ 7 & 1 & 3 \end{pmatrix} \end{array} \begin{array}{c} \begin{array}{cccc} P_1 & P_2 & P_3 & P_4 \end{array} \\ \begin{array}{c} W \\ D \\ L \end{array}\!\begin{pmatrix} 2 & 3 & 1 & 2 \\ 1 & 1 & 0 & 0 \\ 0 & 0 & -1 & -1 \end{pmatrix} \end{array} = \begin{array}{c} \begin{array}{cccc} P_1 & P_2 & P_3 & P_4 \end{array} \\ \begin{array}{c} L \\ A \end{array}\!\begin{pmatrix} 16 & 21 & 4 & 9 \\ 15 & 22 & 4 & 11 \end{pmatrix} \end{array}$$

$$\begin{array}{ccc} \text{2 by 3} & \text{3 by 4} & \text{2 by 4} \end{array}$$

First row × third column gives (the element in the first row, third column)

If two matrices can be multiplied together they are **compatible**. A matrix of order $a \times b$ multiplied by a matrix of order $b \times c$ gives a product matrix of order $a \times c$.

In common with standard algebraic practice it is understood that if two matrices are written with no sign between them, they are to be multiplied. Most of the applications of matrices in mathematics are confined to square matrices of order 2 by 2 and 3 by 3, but many of the results derived below can apply to matrices of any order. The rows and columns may have headings, but it is usually clear what the elements refer to from the context of the application.

Unit matrices

$$\begin{pmatrix} 1 & 0 \\ 0 & 1 \end{pmatrix}\begin{pmatrix} 1 & 2 \\ 3 & 4 \end{pmatrix} = \begin{pmatrix} 1 & 2 \\ 3 & 4 \end{pmatrix}; \quad \begin{pmatrix} 1 & 2 \\ 3 & 4 \end{pmatrix}\begin{pmatrix} 1 & 0 \\ 0 & 1 \end{pmatrix} = \begin{pmatrix} 1 & 2 \\ 3 & 4 \end{pmatrix}$$

The effect of premultiplying by $I_2 = \begin{pmatrix} 1 & 0 \\ 0 & 1 \end{pmatrix}$ is to leave the matrix $\begin{pmatrix} 1 & 2 \\ 3 & 4 \end{pmatrix}$ unchanged. I_2 is a unit matrix.

$$\begin{pmatrix} 1 & 0 & 0 \\ 0 & 1 & 0 \\ 0 & 0 & 1 \end{pmatrix}\begin{pmatrix} 1 & 2 \\ 3 & 4 \\ 5 & 6 \end{pmatrix} = \begin{pmatrix} 1 & 2 \\ 3 & 4 \\ 5 & 6 \end{pmatrix}; \quad I_3 = \begin{pmatrix} 1 & 0 & 0 \\ 0 & 1 & 0 \\ 0 & 0 & 1 \end{pmatrix} \text{ is a unit matrix.}$$

$\begin{pmatrix} 1 & 2 \\ 3 & 4 \\ 5 & 6 \end{pmatrix}\begin{pmatrix} 1 & 0 & 0 \\ 0 & 1 & 0 \\ 0 & 0 & 1 \end{pmatrix}$ is impossible, since the matrices are **incompatible.**

It is interesting to note the effect of multiplying by certain simple matrices.

$\begin{pmatrix} 2 & 0 \\ 0 & 2 \end{pmatrix}\begin{pmatrix} 1 & 2 \\ 3 & 4 \end{pmatrix} = \begin{pmatrix} 2 & 4 \\ 6 & 8 \end{pmatrix}; \quad \begin{pmatrix} 2 & 0 \\ 0 & 2 \end{pmatrix}$ has the effect of doubling each element.

172

$$\begin{pmatrix} 2 & 0 \\ 0 & 1 \end{pmatrix}\begin{pmatrix} 1 & 2 \\ 3 & 4 \end{pmatrix} = \begin{pmatrix} 2 & 4 \\ 3 & 4 \end{pmatrix}; \qquad \begin{pmatrix} 2 & 0 \\ 0 & 1 \end{pmatrix}$$ doubles the top line.

$$\begin{pmatrix} 1 & 1 \\ 0 & 1 \end{pmatrix}\begin{pmatrix} 1 & 2 \\ 3 & 4 \end{pmatrix} = \begin{pmatrix} 4 & 6 \\ 3 & 4 \end{pmatrix}; \qquad \begin{pmatrix} 1 & 1 \\ 0 & 1 \end{pmatrix}$$ adds the bottom row to the top row.

$$\begin{pmatrix} 0 & 1 \\ 1 & 0 \end{pmatrix}\begin{pmatrix} 1 & 2 \\ 3 & 4 \end{pmatrix} = \begin{pmatrix} 3 & 4 \\ 1 & 2 \end{pmatrix}; \qquad \begin{pmatrix} 0 & 1 \\ 1 & 0 \end{pmatrix}$$ exchanges the top and bottom rows.

$$\begin{pmatrix} 2 & 0 \\ 0 & 2 \end{pmatrix}\begin{pmatrix} 1 & 2 \\ 3 & 4 \end{pmatrix} = \begin{pmatrix} 2 & 4 \\ 6 & 8 \end{pmatrix} = 2 \times \begin{pmatrix} 1 & 2 \\ 3 & 4 \end{pmatrix}$$ To multiply a matrix by a number you must multiply each element by that number.

If $I = \begin{pmatrix} 1 & 0 \\ 0 & 1 \end{pmatrix}$, $A = \begin{pmatrix} 1 & 2 \\ 3 & 4 \end{pmatrix}$ and $B = \begin{pmatrix} 4 & 2 \\ 1 & 3 \end{pmatrix}$ then $IA = AI = A$ and $IB = BI = B$

The matrices I and A are **commutative** and so are I and B. However, this is not true for all matrices.

$$AB = \begin{pmatrix} 1 & 2 \\ 3 & 4 \end{pmatrix}\begin{pmatrix} 4 & 2 \\ 1 & 3 \end{pmatrix} = \begin{pmatrix} 6 & 8 \\ 16 & 18 \end{pmatrix} \quad BA = \begin{pmatrix} 4 & 2 \\ 1 & 3 \end{pmatrix}\begin{pmatrix} 1 & 2 \\ 3 & 4 \end{pmatrix} = \begin{pmatrix} 10 & 16 \\ 10 & 14 \end{pmatrix}$$

$AB \neq BA$ so **matrix multiplication is not commutative.**

If $C = \begin{pmatrix} 1 & 2 & 3 \\ 2 & 0 & -1 \end{pmatrix}$ and $D = \begin{pmatrix} 1 \\ 2 \\ 3 \end{pmatrix}$

$$(AC)D = \left[\begin{pmatrix} 1 & 2 \\ 3 & 4 \end{pmatrix}\begin{pmatrix} 1 & 2 & 3 \\ 2 & 0 & -1 \end{pmatrix}\right]\begin{pmatrix} 1 \\ 2 \\ 3 \end{pmatrix} = \begin{pmatrix} 5 & 2 & 1 \\ 11 & 6 & 5 \end{pmatrix}\begin{pmatrix} 1 \\ 2 \\ 3 \end{pmatrix} = \begin{pmatrix} 12 \\ 38 \end{pmatrix}$$

$$A(CD) = \begin{pmatrix} 1 & 2 \\ 3 & 4 \end{pmatrix}\left[\begin{pmatrix} 1 & 2 & 3 \\ 2 & 0 & -1 \end{pmatrix}\begin{pmatrix} 1 \\ 2 \\ 3 \end{pmatrix}\right] = \begin{pmatrix} 1 & 2 \\ 3 & 4 \end{pmatrix}\begin{pmatrix} 14 \\ -1 \end{pmatrix} = \begin{pmatrix} 12 \\ 38 \end{pmatrix}$$

$(AC)D = A(CD)$ and this result is true in general, as long as the matrices are compatible.

Matrix multiplication is associative.

Inverse matrices

$$\begin{pmatrix} 3 & 4 \\ 5 & 7 \end{pmatrix}\begin{pmatrix} 7 & -4 \\ -5 & 3 \end{pmatrix} = \begin{pmatrix} 1 & 0 \\ 0 & 1 \end{pmatrix} = \begin{pmatrix} 7 & -4 \\ -5 & 3 \end{pmatrix}\begin{pmatrix} 3 & 4 \\ 5 & 7 \end{pmatrix}$$

If $AB = BA = I$ then B is the **inverse** of matrix A and is denoted by A^{-1}. Does every matrix have an inverse?

Can we find the inverse of the matrix $G = \begin{pmatrix} a & b \\ c & d \end{pmatrix}$

Suppose $H = \begin{pmatrix} x & y \\ s & t \end{pmatrix}$ and $HG = I$ then

$$\begin{pmatrix} x & y \\ s & t \end{pmatrix}\begin{pmatrix} a & b \\ c & d \end{pmatrix} = \begin{pmatrix} xa+yc & xb+yd \\ sa+tc & sb+td \end{pmatrix} = \begin{pmatrix} 1 & 0 \\ 0 & 1 \end{pmatrix}$$

$xb + yd = 0 \Rightarrow \dfrac{x}{y} = \dfrac{-d}{b}$ so $x = \lambda d$ and $y = -\lambda b$

$xa + yc = 1 \Rightarrow \lambda da - \lambda bc = 1$ so $\lambda = \dfrac{1}{ad-bc}$ and $x = \dfrac{d}{ad-bc}$,

$y = \dfrac{-b}{ad-bc}$

Similarly $s = \dfrac{-c}{ad-bc}$, $t = \dfrac{a}{ad-bc}$

So $\begin{pmatrix} x & y \\ s & t \end{pmatrix} = G^{-1} = \begin{pmatrix} \dfrac{d}{ad-bc} & \dfrac{-b}{ad-bc} \\ \dfrac{-c}{ad-bc} & \dfrac{a}{ad-bc} \end{pmatrix} = \dfrac{1}{ad-bc}\begin{pmatrix} d & -b \\ -c & a \end{pmatrix}$

If $\begin{pmatrix} a & b \\ c & d \end{pmatrix} = G$ the quantity $\Delta = ad - bc = \begin{vmatrix} a & b \\ c & d \end{vmatrix}$ is the **determinant** of G.

An alternative method to find the inverse matrix follows.

$$\begin{pmatrix} x & y \\ s & t \end{pmatrix}\begin{pmatrix} a & b \\ c & d \end{pmatrix} = \begin{pmatrix} 1 & 0 \\ 0 & 1 \end{pmatrix}$$

Try to find x and y so that the first row multiplied by the second column is 0. To achieve this we let $x = d$ and $y = -b$.

$$\begin{pmatrix} d & -b \\ * & * \end{pmatrix}\begin{pmatrix} a & b \\ c & d \end{pmatrix} = \begin{pmatrix} ad-bc & 0 \\ * & * \end{pmatrix}$$

Let $s = -c$ and $t = a$ then

$$\begin{pmatrix} d & -b \\ -c & a \end{pmatrix}\begin{pmatrix} a & b \\ c & d \end{pmatrix} = \begin{pmatrix} ad-bc & 0 \\ 0 & -bc+ad \end{pmatrix}$$

If we divide the final matrix by $\Delta = ad - bc$, we obtain $\begin{pmatrix} 1 & 0 \\ 0 & 1 \end{pmatrix}$

If $G = \begin{pmatrix} a & b \\ c & d \end{pmatrix}$ $\qquad G^{-1} = \begin{pmatrix} d/\Delta & -b/\Delta \\ -c/\Delta & a/\Delta \end{pmatrix} = \dfrac{1}{(ad-bc)}\begin{pmatrix} d & -b \\ -c & a \end{pmatrix}$

This form of the inverse of a 2 by 2 matrix is easy to memorize. The elements on the main diagonal (top left to bottom right) are interchanged, the elements on the reverse diagonal (top right to bottom left) are reversed in sign and each element is divided by the determinant, $\Delta = ad - bc$.

Example 1 Find the inverse of $\begin{pmatrix} 1 & 2 \\ 3 & 4 \end{pmatrix}$.

$\Delta = 1 \times 4 - 3 \times 2 = 4 - 6 = -2$

Inverse $= -\dfrac{1}{2}\begin{pmatrix} 4 & -2 \\ -3 & 1 \end{pmatrix} = \begin{pmatrix} -2 & 1 \\ 1\frac{1}{2} & -\frac{1}{2} \end{pmatrix}$

If the determinant of a matrix is zero, the matrix has no inverse.

Example 2 $A = \begin{pmatrix} 1 & 2 \\ 3 & 6 \end{pmatrix}$. $\Delta = \det A = 0$. If $B = \begin{pmatrix} 6 & -2 \\ -3 & 1 \end{pmatrix}$

$BA = \begin{pmatrix} 6 & -2 \\ -3 & 1 \end{pmatrix}\begin{pmatrix} 1 & 2 \\ 3 & 6 \end{pmatrix} = \begin{pmatrix} 0 & 0 \\ 0 & 0 \end{pmatrix}$ Any attempt to produce zeros on the reverse diagonal produces zeros also on the main diagonal.

The product of B and A gives $\begin{pmatrix} 0 & 0 \\ 0 & 0 \end{pmatrix}$ the **zero matrix.**

To prove matrix multiplication is not commutative for 2 by 2 matrices

$$\begin{pmatrix} a & b \\ c & d \end{pmatrix}\begin{pmatrix} e & f \\ g & h \end{pmatrix} = \begin{pmatrix} ae + bg & af + bh \\ ce + dg & cf + dh \end{pmatrix}$$

$$\begin{pmatrix} e & f \\ g & h \end{pmatrix}\begin{pmatrix} a & b \\ c & d \end{pmatrix} = \begin{pmatrix} ae + cf & be + df \\ ag + ch & bg + dh \end{pmatrix}$$

The product matrices are in general not equal.

To prove matrix multiplication is **associative** for 2 by 2 matrices.

$$\left[\begin{pmatrix} a & b \\ c & d \end{pmatrix}\begin{pmatrix} e & f \\ g & h \end{pmatrix}\right]\begin{pmatrix} p & q \\ r & s \end{pmatrix} = \begin{pmatrix} ae + bg & af + bh \\ ce + dg & cf + dh \end{pmatrix}\begin{pmatrix} p & q \\ r & s \end{pmatrix}$$

$$= \begin{pmatrix} aep + bgp + afr + bhr & aeq + bgq + afs + bhs \\ cep + dgp + cfr + dhr & ceq + dgq + cfs + dhs \end{pmatrix}$$

$$\begin{pmatrix} a & b \\ c & d \end{pmatrix}\left[\begin{pmatrix} e & f \\ g & h \end{pmatrix}\begin{pmatrix} p & q \\ r & s \end{pmatrix}\right] = \begin{pmatrix} a & b \\ c & d \end{pmatrix}\begin{pmatrix} ep + fr & eq + fs \\ gp + hr & gq + hs \end{pmatrix}$$

$$= \begin{pmatrix} aep + afr + bgp + bhr & aeq + afs + bgq + bhs \\ cep + cfr + dgp + dhr & ceq + cfs + dgq + dhs \end{pmatrix}$$

These two matrices are identical term for term.

To prove **det(AB) = (det A) × (det B)** where A, B are two 2 by 2 matrices.

$$A = \begin{pmatrix} a & b \\ c & d \end{pmatrix} \quad B = \begin{pmatrix} e & f \\ g & h \end{pmatrix} \quad AB = \begin{pmatrix} ae+bg & af+bh \\ ce+dg & cf+dh \end{pmatrix}$$

$$\begin{aligned}
\det(AB) &= (ae+bg)(cf+dh) - (af+bh)(ce+dg) \\
&= aecf + aedh + bgcf + bgdh \\
&\quad - afce - afdg - bhce - bhdg \\
&= ad(eh-fg) - bc(he-gf) \\
&= (ad-bc)(he-fg) \\
&\equiv (\det A) \times (\det B)
\end{aligned}$$

This result is true for all square matrices. The determinant of a non-square matrix is not defined.

Simultaneous equations

Although the ideas of matrices and determinants are not necessary to solve simultaneous equations in two dimensions (2 equations and 2 unknowns) their study gives insight and understanding to simultaneous equations with 3 or more unknowns.

Consider first the equations $ax + by = 0$ (1)

$$cx + dy = 0 \tag{2}$$

These equations have the trivial solution $x = 0$, $y = 0$.

Geometrically the equations represent two straight lines through the origin. The only other solutions occur when each equation represents the same straight line. In this case equation (2) is a multiple of equation (1) or $ad - bc = 0$.

So $\begin{aligned} ax + by &= 0 \\ cx + dy &= 0 \end{aligned} \Leftrightarrow \begin{pmatrix} a & b \\ c & d \end{pmatrix}\begin{pmatrix} x \\ y \end{pmatrix} = 0$ which has non-trivial solutions when $ad - bc = 0$.

We shall use this idea in three dimensions to define a 3×3 determinant.

Consider now the equations $3x + 4y = 11$ (3)

$$x + 2y = 5 \tag{4}$$

Elimination method
Equation (3) $-2 \times$ equation (4) gives $x = 1$.
Substitution in either equation then gives $y = 2$.

Matrix method $\begin{pmatrix} 3 & 4 \\ 1 & 2 \end{pmatrix}\begin{pmatrix} x \\ y \end{pmatrix} = \begin{pmatrix} 11 \\ 5 \end{pmatrix}$

176

The inverse of the matrix $\begin{pmatrix} 3 & 4 \\ 1 & 2 \end{pmatrix}$ is $\frac{1}{2}\begin{pmatrix} 2 & -4 \\ -1 & 3 \end{pmatrix}$

Hence $\frac{1}{2}\begin{pmatrix} 2 & -4 \\ -1 & 3 \end{pmatrix}\begin{pmatrix} 3 & 4 \\ 1 & 2 \end{pmatrix}\begin{pmatrix} x \\ y \end{pmatrix} = \frac{1}{2}\begin{pmatrix} 2 & -4 \\ -1 & 3 \end{pmatrix}\begin{pmatrix} 11 \\ 5 \end{pmatrix}$

$$\begin{pmatrix} 1 & 0 \\ 0 & 1 \end{pmatrix}\begin{pmatrix} x \\ y \end{pmatrix} = \frac{1}{2}\begin{pmatrix} 2 \\ 4 \end{pmatrix}$$

$$\begin{pmatrix} x \\ y \end{pmatrix} = \begin{pmatrix} 1 \\ 2 \end{pmatrix}$$

The matrix method may look longer but the solution is obtained only by multiplying $\begin{pmatrix} 11 \\ 5 \end{pmatrix}$ by the inverse of $\begin{pmatrix} 3 & 4 \\ 1 & 2 \end{pmatrix}$

Matrices and transformations in two dimensions

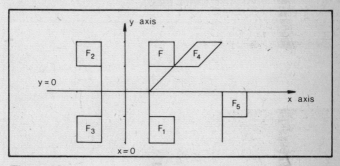

Figure 89

Figure 89 shows an object F and its images after certain transformations. F_1 is the image of F after reflection in the line $y = 0$. Under this reflection the point $(2, 1)$ is mapped onto the point $(2, -1)$ and in general (x, y) is mapped onto $(x, -y)$. Writing coordinates as column vectors, $\begin{pmatrix} x \\ y \end{pmatrix}$ is mapped onto $\begin{pmatrix} x \\ -y \end{pmatrix}$. This can be achieved by premultiplying by the matrix $\begin{pmatrix} 1 & 0 \\ 0 & -1 \end{pmatrix}$ so $\begin{pmatrix} 1 & 0 \\ 0 & -1 \end{pmatrix}\begin{pmatrix} x \\ y \end{pmatrix} = \begin{pmatrix} x \\ -y \end{pmatrix}$ and $M_1 = \begin{pmatrix} 1 & 0 \\ 0 & -1 \end{pmatrix}$ is the matrix of the transformation.

Similarly F to F_2 is given by $\begin{pmatrix} -1 & 0 \\ 0 & 1 \end{pmatrix}\begin{pmatrix} x \\ y \end{pmatrix} = \begin{pmatrix} -x \\ y \end{pmatrix}$ and $M_2 =$

177

$\begin{pmatrix} -1 & 0 \\ 0 & 1 \end{pmatrix}$ is the matrix representing a reflection in the line $x = 0$.

F to F_3 is given by $\begin{pmatrix} -1 & 0 \\ 0 & -1 \end{pmatrix}\begin{pmatrix} x \\ y \end{pmatrix} = \begin{pmatrix} -x \\ -y \end{pmatrix}$ and $M_3 = \begin{pmatrix} -1 & 0 \\ 0 & -1 \end{pmatrix}$ is the matrix representing a rotation of $180°$ about the origin $(0, 0)$.

Since $\begin{pmatrix} a & b \\ c & d \end{pmatrix}\begin{pmatrix} 1 \\ 0 \end{pmatrix} = \begin{pmatrix} a \\ c \end{pmatrix}$ and $\begin{pmatrix} a & b \\ c & d \end{pmatrix}\begin{pmatrix} 0 \\ 1 \end{pmatrix} = \begin{pmatrix} b \\ d \end{pmatrix}$ the first column of the transformation matrix is the image of $(1, 0)$ and the second column the image of $(0, 1)$.

F to F_4, a shear, takes $\begin{pmatrix} 1 \\ 0 \end{pmatrix}$ to $\begin{pmatrix} 1 \\ 0 \end{pmatrix}$ and $\begin{pmatrix} 0 \\ 1 \end{pmatrix}$ to $\begin{pmatrix} 1 \\ 1 \end{pmatrix}$.

So $M_4 = \begin{pmatrix} 1 & 1 \\ 0 & 1 \end{pmatrix}$.

Shearing and enlargement

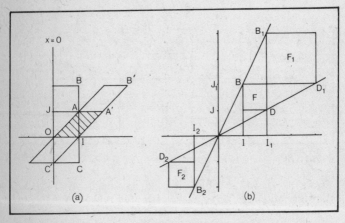

Figure 90

Figure 90a shows the effect of a **shear** with invariant line $y = 0$. A **shear** is a transformation in which all points move a distance which is proportional to their distance from an invariant line and parallel to that invariant line. In fig. 90a A moves 1 unit parallel to $y = 0$ and B moves 2 units parallel to $y = 0$.

Since $I(1, 0)$ is invariant and $J(0, 1)$ goes to $(1, 1)$ the transformation matrix is $\begin{pmatrix} 1 & 1 \\ 0 & 1 \end{pmatrix}$. The image of $C(1, -1)$ is

178

$C'(0, -1)$ since $\begin{pmatrix} 1 & 1 \\ 0 & 1 \end{pmatrix}\begin{pmatrix} 1 \\ -1 \end{pmatrix} = \begin{pmatrix} 0 \\ -1 \end{pmatrix}$. and points below the invariant line $y = 0$ move in the negative x direction.

Figure 90b shows two **enlargements**. F to F_1 is an enlargement of **scale factor** 2 (all lengths are doubled) and the centre of the enlargement is the origin (lines joining corresponding points BB_1, DD_2 etc. radiate from the origin).

$I(1, 0)$ goes $I_1(2, 0)$ and $J(0, 1)$ goes to $J_1(0, 2)$ the transformation matrix is $\begin{pmatrix} 2 & 0 \\ 0 & 2 \end{pmatrix}$.

$D(2, 1)$ goes to $D_1(4, 2)$ since $\begin{pmatrix} 2 & 0 \\ 0 & 2 \end{pmatrix}\begin{pmatrix} 2 \\ 1 \end{pmatrix} = \begin{pmatrix} 4 \\ 2 \end{pmatrix}$

An enlargement of scale factor -1 has a transformation matrix

$$\begin{pmatrix} -1 & 0 \\ 0 & -1 \end{pmatrix}$$

$I(1, 0)$ goes to $I_2(-1, 0)$ $B(1, 2)$ goes to $B_2(-1, -2)$ and $D(2, 1)$ goes to $D_2(-2, -1)$ and the image F_2 of F is shown in fig. 90b. The image is inverted and the enlargement of scale factor -1 is equivalent to a rotation of $180°$ with centre at the origin (this is not true in three dimensions).

F to F_3 can also be achieved by a reflection in $y = 0$ followed by a reflection in $x = 0$, see fig. 89.

$\begin{pmatrix} x \\ y \end{pmatrix}$ becomes $\begin{pmatrix} -1 & 0 \\ 0 & 1 \end{pmatrix}\left[\begin{pmatrix} 1 & 0 \\ 0 & -1 \end{pmatrix}\begin{pmatrix} x \\ y \end{pmatrix}\right] = \begin{pmatrix} -1 & 0 \\ 0 & 1 \end{pmatrix}\begin{pmatrix} x \\ -y \end{pmatrix} = \begin{pmatrix} -x \\ -y \end{pmatrix}$

Since matrix multiplication is associative

$\begin{pmatrix} -1 & 0 \\ 0 & 1 \end{pmatrix}\left[\begin{pmatrix} 1 & 0 \\ 0 & -1 \end{pmatrix}\begin{pmatrix} x \\ y \end{pmatrix}\right] = \left[\begin{pmatrix} -1 & 0 \\ 0 & 1 \end{pmatrix}\begin{pmatrix} 1 & 0 \\ 0 & -1 \end{pmatrix}\right]\begin{pmatrix} x \\ y \end{pmatrix}$

$$= \begin{pmatrix} -1 & 0 \\ 0 & -1 \end{pmatrix}\begin{pmatrix} x \\ y \end{pmatrix}$$

So $\begin{pmatrix} -1 & 0 \\ 0 & 1 \end{pmatrix}\begin{pmatrix} 1 & 0 \\ 0 & -1 \end{pmatrix} = \begin{pmatrix} -1 & 0 \\ 0 & -1 \end{pmatrix}$ or $M_2 M_1 = M_3$

$M_1 M_2$ gives the matrix of the transformation which is the result of a reflection in $x = 0$ (M_2) followed by a reflection in $y = 0$ (M_1).
In fact $M_1 M_2 = M_3 = M_2 M_1$ in this case.
A 2 by 2 matrix represents a transformation of the plane in which the origin $(0, 0)$ remains invariant.

The translation F to F_5 takes each point (x, y) to the point $(x + 3, y - 2)$.

This cannot be represented by a 2 by 2 matrix but simply by $\begin{pmatrix} x \\ y \end{pmatrix}$ becomes $\begin{pmatrix} x \\ y \end{pmatrix} + \begin{pmatrix} 3 \\ -2 \end{pmatrix}$

The matrix $M = \begin{pmatrix} a & b \\ c & d \end{pmatrix}$ transforms the square lattice of the xy plane into a lattice of parallelograms (fig. 91). The unit square $OABC$ becomes the parallelogram $OA'B'C'$.

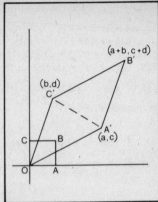

Area of $OA'B'C'$
$$= 2 \times \text{area of } \triangle OA'C'$$
$$= 2 \times \tfrac{1}{2}(ad - bc)$$
$$= \det M$$

Figure 91

The determinant of the transformation matrix is the ratio of the image area to the area of the object.

$\det M_3 = 1$ so area is unchanged by this transformation

$\det M_4 = 1$ so area is invariant under a shear

$\det M_1 = \det M_2 = -1$, so reflection preserves area and the negative sign distinguishes between a reflection and a rotation, and shows that a clockwise sense in the object becomes anti-clockwise in the image.

Transformations and invariant points in two dimensions

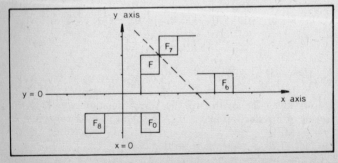

Figure 92

180

F to F_6 can be achieved by a rotation of $-90°$ (clockwise) about a centre of rotation other than the origin (see fig. 92). The complete transformation can be specified by first rotating $-90°$ about $(0, 0)$ to F_0 and then translating from F_0 to F_6.

$$\begin{pmatrix} x \\ y \end{pmatrix} \text{ becomes } \begin{pmatrix} 0 & 1 \\ -1 & 0 \end{pmatrix}\begin{pmatrix} x \\ y \end{pmatrix} + \begin{pmatrix} 4 \\ 2 \end{pmatrix} = \begin{pmatrix} y \\ -x \end{pmatrix} + \begin{pmatrix} 4 \\ 2 \end{pmatrix} = \begin{pmatrix} y + 4 \\ -x + 2 \end{pmatrix}$$

If (a, b) is an invariant point $\begin{pmatrix} b + 4 \\ -a + 2 \end{pmatrix} = \begin{pmatrix} a \\ b \end{pmatrix}$ and $\begin{matrix} a - b = 4 \\ a + b = 2 \end{matrix}$

So $a = 3$, $b = -1$ and $(3, -1)$ is the centre of the rotation.

F to F_7 is seen from the diagram to be a reflection in the line $x + y = 4$. Using F_8 as an intermediate position F to F_7 can be seen as a reflection in $x + y = 0$ (keeping the origin fixed) followed by a translation from F_8 to F_7.

$$\begin{pmatrix} x \\ y \end{pmatrix} \text{ becomes } \begin{pmatrix} 0 & -1 \\ -1 & 0 \end{pmatrix}\begin{pmatrix} x \\ y \end{pmatrix} + \begin{pmatrix} 4 \\ 4 \end{pmatrix} = \begin{pmatrix} -y + 4 \\ -x + 4 \end{pmatrix}$$

If (a, b) is an invariant point $\begin{pmatrix} -b + 4 \\ -a + 4 \end{pmatrix} = \begin{pmatrix} a \\ b \end{pmatrix}$ gives $\begin{matrix} a + b = 4 \\ a + b = 4 \end{matrix}$

The invariant points satisfy $x + y = 4$ so there is an invariant line, the reflection line $x + y = 4$. With most transformations the invariant points can be found by geometrical constructions and these constructions may be more direct and intuitive. Moreover, the constructions will work when the transformations are not conveniently described by coordinates.

Determinant of a 3 by 3 matrix

Consider the equations $a_1x + b_1y + c_1z = 0$ (1)

$$a_2x + b_2y + c_2z = 0 \qquad (2)$$

$$a_3x + b_3y + c_3z = 0 \qquad (3)$$

Solving (2) and (3) for x and y in terms of z

$$(a_2b_3 - a_3b_2)x = (b_2c_3 - b_3c_2)z \text{ and } x = \frac{(b_2c_3 - b_3c_2)z}{(a_2b_3 - a_3b_2)}$$

$$(a_2b_3 - a_3b_2)y = (a_3c_2 - a_2c_3)z \text{ and } y = \frac{(a_3c_2 - a_2c_3)z}{(a_2b_3 - a_3b_2)}$$

Substituting these values in (1) gives

$$a_1\frac{(b_2c_3 - b_3c_2)z}{(a_2b_3 - a_3b_2)} + b_1\frac{(a_3c_2 - a_2c_3)z}{(a_2b_3 - a_3b_2)} + c_1z = 0$$

181

For non-trivial solutions $z \neq 0$ so

$$a_1(b_2c_3 - b_3c_2) + b_1(a_3c_2 - a_2c_3) + c_1(a_2b_3 - a_3b_2) = 0$$

This expression is defined as the determinant of the matrix

$$\begin{pmatrix} a_1 & b_1 & c_1 \\ a_2 & b_2 & c_2 \\ a_3 & b_3 & c_3 \end{pmatrix}$$

It is usually written as

$$\Delta = a_1 \begin{vmatrix} b_2 & c_2 \\ b_3 & c_3 \end{vmatrix} - b_1 \begin{vmatrix} a_2 & c_2 \\ a_3 & c_3 \end{vmatrix} + c_1 \begin{vmatrix} a_2 & b_2 \\ a_3 & b_3 \end{vmatrix}$$

$$= a_1(b_2c_3 - b_3c_2) - b_1(a_2c_3 - a_3c_2) + c_1(a_2b_3 - a_3b_2)$$

$$= a_1b_2c_3 - a_1b_3c_2 - a_2b_1c_3 + a_3b_1c_2 + a_2b_3c_1 - a_3b_2c_1 \quad (4)$$

Each term contains an a, b and c and the suffixes are permutations of 1 2 3, the even permutations giving a positive sign and the odd permutations a negative sign.

Example 3
$$\begin{vmatrix} 1 & 2 & 6 \\ 3 & 5 & 7 \\ 4 & 8 & 9 \end{vmatrix} = 1 \begin{vmatrix} 5 & 7 \\ 8 & 9 \end{vmatrix} - 2 \begin{vmatrix} 3 & 7 \\ 4 & 9 \end{vmatrix} + 6 \begin{vmatrix} 3 & 5 \\ 4 & 8 \end{vmatrix}$$

$$= (45 - 56) - 2(27 - 28) + 6(24 - 20)$$

$$= -11 + 2 + 24$$

$$= 15$$

From (4)

$$\det M = -a_2(b_1c_3 - b_3c_1) + b_2(a_1c_3 - a_3c_1) - c_2(a_1b_3 - a_3b_1)$$

$$= -a_2 \begin{vmatrix} b_1 & c_1 \\ b_3 & c_3 \end{vmatrix} + b_2 \begin{vmatrix} a_1 & c_1 \\ a_3 & c_3 \end{vmatrix} - c_2 \begin{vmatrix} a_1 & b_1 \\ a_3 & b_3 \end{vmatrix}$$

and $\det M$ can be evaluated using the second row, or even using any column as long as the sign convention is maintained. Example 3 using the second row.

$$\begin{vmatrix} 1 & 2 & 6 \\ 3 & 5 & 7 \\ 4 & 8 & 9 \end{vmatrix} = -3 \begin{vmatrix} 2 & 6 \\ 8 & 9 \end{vmatrix} + 5 \begin{vmatrix} 1 & 6 \\ 4 & 9 \end{vmatrix} - 7 \begin{vmatrix} 1 & 2 \\ 4 & 8 \end{vmatrix}$$

$$= -3(-30) + 5(-15) - 7(0) = 15$$

Laws of determinants

Various laws and processes help in working with determinants which ease the arithmetic. The rules are true in general but we will prove them only for the third order determinant

$$M = \begin{pmatrix} a_1 & b_1 & c_1 \\ a_2 & b_2 & c_2 \\ a_3 & b_3 & c_3 \end{pmatrix} \qquad \det M = \Delta$$

Rule 1 Δ is changed in sign if two rows are interchanged.

$$\begin{vmatrix} a_2 & b_2 & c_2 \\ a_1 & b_1 & c_1 \\ a_3 & b_3 & c_3 \end{vmatrix} = a_2(b_1 c_3 - b_3 c_1) - b_2(a_1 c_3 - a_3 c_1) + c_2(a_1 b_3 - a_3 b_1)$$
$$= -a_1(b_2 c_3 - b_3 c_2) + b_1(a_2 c_3 - a_3 c_2)$$
$$- c_1(a_2 b_3 - a_3 b_2) = -\Delta$$

The same result is true if two columns are interchanged.

Rule 2 $\Delta = 0$ if two rows are identical.

From rule 1, interchanging the rows changes the sign of Δ, but this produces the same determinant so $\Delta = -\Delta \Rightarrow \Delta = 0$.

Rule 3 If each element of one row, or column is multiplied by k, then Δ is multiplied by k.

$$\begin{vmatrix} ka_1 & kb_2 & kc_3 \\ a_2 & b_2 & c_2 \\ a_3 & b_3 & c_3 \end{vmatrix} = k\Delta \quad \text{since there will be a factor of } k \text{ in each term}$$

An important consequence is that if every element of a row or column has a factor k, then Δ has a factor k.

Rule 4 Δ is unchanged if a multiple of a row is added to another row.

$$\begin{vmatrix} a_1 + ka_2 & b_1 + kb_2 & c_1 + kc_2 \\ a_2 & b_2 & c_2 \\ a_3 & b_3 & c_3 \end{vmatrix}$$

$$= (a_1 + ka_2)\begin{vmatrix} b_2 & c_2 \\ b_3 & c_3 \end{vmatrix} - (b_1 + kb_2)\begin{vmatrix} a_2 & c_2 \\ a_3 & c_3 \end{vmatrix} + (c_1 + kc_2)\begin{vmatrix} a_2 & b_2 \\ a_3 & b_3 \end{vmatrix}$$

$$= a_1\begin{vmatrix} b_2 & c_2 \\ b_3 & c_3 \end{vmatrix} - b_1\begin{vmatrix} a_2 & c_2 \\ a_3 & c_3 \end{vmatrix} + c_1\begin{vmatrix} a_2 & b_2 \\ a_3 & b_3 \end{vmatrix}$$

$$+ ka_2\begin{vmatrix} b_2 & c_2 \\ b_3 & c_3 \end{vmatrix} - kb_2\begin{vmatrix} a_2 & c_2 \\ a_3 & c_3 \end{vmatrix} + kc_2\begin{vmatrix} a_2 & b_2 \\ a_3 & b_3 \end{vmatrix}$$

$$= \begin{vmatrix} a_1 & b_1 & c_1 \\ a_2 & b_2 & c_2 \\ a_3 & b_3 & c_3 \end{vmatrix} + k\begin{vmatrix} a_2 & b_2 & c_2 \\ a_2 & b_2 & c_2 \\ a_3 & b_3 & c_3 \end{vmatrix} = \Delta$$

since the second determinant has two identical rows and is zero.

Rule 4 gives a way of adding or subtracting determinants

$$\begin{vmatrix} a_1+a & b_1+b & c_1+c \\ a_2 & b_2 & c_2 \\ a_3 & b_3 & c_3 \end{vmatrix} = \begin{vmatrix} a_1 & b_1 & c_1 \\ a_2 & b_2 & c_2 \\ a_3 & b_3 & c_3 \end{vmatrix} + \begin{vmatrix} a & b & c \\ a_2 & b_2 & c_2 \\ a_3 & b_3 & c_3 \end{vmatrix}$$

Notice that two rows in the determinants are identical.

Examples using these rules

$$\begin{vmatrix} 20 & 17 & 16 \\ 21 & 19 & 12 \\ 23 & 20 & 14 \end{vmatrix} = \begin{vmatrix} 3 & 17 & 16 \\ 2 & 19 & 12 \\ 3 & 20 & 14 \end{vmatrix} \quad \text{subtracting column 2 from column 1}$$

$$= \begin{vmatrix} 3 & 1 & 16 \\ 2 & 7 & 12 \\ 3 & 6 & 14 \end{vmatrix} \quad \text{subtracting column 3 from column 2}$$

$$= \begin{vmatrix} 1 & -6 & 4 \\ 2 & 7 & 12 \\ 1 & -1 & 2 \end{vmatrix} \quad \text{subtracting row 2 from rows 1 and 3}$$

$$= 1(14+12) + 6(4-12) + 4(-2-7)$$

$$= 26 - 48 - 36 = -58$$

Cofactors

$$\Delta = \begin{vmatrix} a_1 & b_1 & c_1 \\ a_2 & b_2 & c_2 \\ a_3 & b_3 & c_3 \end{vmatrix} = a_1 \begin{vmatrix} b_2 & c_2 \\ b_3 & c_3 \end{vmatrix} - b_1 \begin{vmatrix} a_2 & c_2 \\ a_3 & c_3 \end{vmatrix} + c_1 \begin{vmatrix} a_2 & b_2 \\ a_3 & b_3 \end{vmatrix}$$

$$= a_1(b_2c_3 - b_3c_2) - b_1(a_2c_3 - a_3c_2) + c_1(a_2b_3 - a_3b_2)$$

$$= a_1A_1 + b_1B_1 + c_1C_1$$

where $A_1 = \begin{vmatrix} b_2 & c_2 \\ b_3 & c_3 \end{vmatrix}$, $B_1 = -\begin{vmatrix} a_2 & c_2 \\ a_3 & c_3 \end{vmatrix}$, $C_1 = \begin{vmatrix} a_2 & b_2 \\ a_3 & b_3 \end{vmatrix}$

A_1, B_1, ... are the cofactors of a_1, b_1, ... and their numerical values are the determinants left by omitting the row and column containing the respective elements. The signs of the cofactors are alternately + and − starting with A_1 positive and proceeding row by row in turn.

So $A_2 = -\begin{vmatrix} b_1 & c_1 \\ b_3 & c_3 \end{vmatrix}$, $B_2 = \begin{vmatrix} a_1 & c_1 \\ a_3 & c_3 \end{vmatrix}$, $C_2 = -\begin{vmatrix} a_1 & b_1 \\ a_3 & b_3 \end{vmatrix}$

$A_3 = \begin{vmatrix} b_1 & c_1 \\ b_2 & c_2 \end{vmatrix}$, $B_3 = -\begin{vmatrix} a_1 & c_1 \\ a_2 & c_2 \end{vmatrix}$, $C_3 = \begin{vmatrix} a_1 & b_1 \\ a_2 & b_2 \end{vmatrix}$

$$\Delta = -a_2 \begin{vmatrix} b_1 & c_1 \\ b_3 & c_3 \end{vmatrix} + b_2 \begin{vmatrix} a_1 & c_1 \\ a_3 & c_3 \end{vmatrix} - c_2 \begin{vmatrix} a_1 & b_1 \\ a_3 & b_3 \end{vmatrix}$$

by expanding from the second row

$\Delta = a_2 A_2 + b_2 B_2 + c_2 C_2$

$\Delta = a_3 A_3 + b_3 B_3 + c_3 C_3$ by expanding from the third row

$\Delta = a_1 A_1 + a_2 A_2 + a_3 A_3$ by expanding from the first column

$\Delta =$ the sum of the products of any row or column with the corresponding cofactors.

If the products are taken with the cofactors of a different row or column the result is zero.

$$a_1 A_2 + b_1 B_2 + c_1 C_2 = -a_1 \begin{vmatrix} b_1 & c_1 \\ b_3 & c_3 \end{vmatrix} + b_1 \begin{vmatrix} a_1 & c_1 \\ a_3 & c_3 \end{vmatrix} - c_1 \begin{vmatrix} a_1 & b_1 \\ a_3 & b_3 \end{vmatrix}$$

$$= \begin{vmatrix} a_1 & b_1 & c_1 \\ a_1 & b_1 & c_1 \\ a_3 & b_3 & c_3 \end{vmatrix} = 0$$

since 2 rows are identical, from rule 2

The results for cofactors can be summarized as;

by rows $a_1 A_1 + b_1 B_1 + c_1 C_1 = \Delta$
$\quad\quad\quad a_1 A_2 + b_1 B_2 + c_1 C_2 = 0$
by columns $a_1 A_1 + a_2 A_2 + a_3 A_3 = \Delta$
$\quad\quad\quad\quad a_1 C_1 + a_2 C_2 + a_3 C_3 = 0$

The definitions can be extended to larger determinants and the results are still true.

Equations in three unknowns

$$\begin{pmatrix} a_1 & b_1 & c_1 \\ a_2 & b_2 & c_2 \\ a_3 & b_3 & c_3 \end{pmatrix} \begin{pmatrix} x \\ y \\ z \end{pmatrix} = \begin{pmatrix} d_1 \\ d_2 \\ d_3 \end{pmatrix} \qquad \begin{matrix} a_1 x + b_1 y + c_1 z = d_1 & (1) \\ a_2 x + b_2 y + c_2 z = d_2 & (2) \\ a_3 x + b_3 y + c_3 z = d_3 & (3) \end{matrix}$$

Multiply (1) by A_1, (2) by A_2 and (3) by A_3 and add to give

$$(a_1 A_1 + a_2 A_2 + a_3 A_3)x + (b_1 A_1 + b_2 A_2 + b_3 A_3)y$$
$$+ (c_1 A_1 + c_2 A_2 + c_3 A_3)z = d_1 A_1 + d_2 A_2 + d_3 A_3$$

But $a_1A_1 + a_2A_2 + a_3A_3 = \Delta$

and $b_1A_1 + b_2A_2 + b_3A_3 = c_1A_1 + c_2A_2 + c_3A_3 = 0$

So $\Delta x = d_1A_1 + d_2A_2 + d_3A_3$

$$= d_1 \begin{vmatrix} b_2 & c_2 \\ b_3 & c_3 \end{vmatrix} - d_2 \begin{vmatrix} b_1 & c_1 \\ b_3 & c_3 \end{vmatrix} + d_3 \begin{vmatrix} b_1 & c_1 \\ b_2 & c_2 \end{vmatrix}$$

$$= \begin{vmatrix} d_1 & b_1 & c_1 \\ d_2 & b_2 & c_2 \\ d_3 & b_3 & c_3 \end{vmatrix}$$

Similarly $\Delta y = d_1B_1 + d_2B_2 + d_3B_3 = \begin{vmatrix} a_1 & d_1 & c_1 \\ a_2 & d_2 & c_2 \\ a_3 & d_3 & c_3 \end{vmatrix}$

and $\quad \Delta z = d_1C_1 + d_2C_2 + d_3C_3 = \begin{vmatrix} a_1 & b_1 & d_1 \\ a_2 & b_2 & d_2 \\ a_3 & b_3 & d_3 \end{vmatrix}$

These results can be summarized as

$$\frac{x}{\begin{vmatrix} d_1 & b_1 & c_1 \\ d_2 & b_2 & c_2 \\ d_3 & b_3 & c_3 \end{vmatrix}} = \frac{y}{\begin{vmatrix} a_1 & d_1 & c_1 \\ a_2 & d_2 & c_2 \\ a_3 & d_3 & c_3 \end{vmatrix}} = \frac{z}{\begin{vmatrix} a_1 & b_1 & d_1 \\ a_2 & b_2 & d_2 \\ a_3 & b_3 & d_3 \end{vmatrix}} = \frac{1}{\begin{vmatrix} a_1 & b_1 & c_1 \\ a_2 & b_2 & c_2 \\ a_3 & b_3 & c_3 \end{vmatrix}}$$

where the determinants are taken from Δ with the d_1, d_2, d_3 column replacing the first column under x, the second column under y and the third column under z. In this way the results are easy to remember.

Example 4 Solve $x + 2y + 3z = 14$

$\qquad\qquad\qquad 2x + 5y - z = 9$

$\qquad\qquad\qquad 3x + 8y - 4z = 7$

$$\frac{x}{\begin{vmatrix} 14 & 2 & 3 \\ 9 & 5 & -1 \\ 7 & 8 & -4 \end{vmatrix}} = \frac{y}{\begin{vmatrix} 1 & 14 & 3 \\ 2 & 9 & -1 \\ 3 & 7 & -4 \end{vmatrix}} = \frac{z}{\begin{vmatrix} 1 & 2 & 14 \\ 2 & 5 & 9 \\ 3 & 8 & 7 \end{vmatrix}} = \frac{1}{\begin{vmatrix} 1 & 2 & 3 \\ 2 & 5 & -1 \\ 3 & 8 & -4 \end{vmatrix}}$$

$$\begin{vmatrix} 14 & 2 & 3 \\ 9 & 5 & -1 \\ 7 & 8 & -4 \end{vmatrix} = \begin{vmatrix} 41 & 17 & 0 \\ 9 & 5 & -1 \\ -29 & -12 & 0 \end{vmatrix} = \begin{vmatrix} 12 & 5 & 0 \\ 9 & 5 & -1 \\ -29 & -12 & 0 \end{vmatrix}$$

$$= \begin{vmatrix} 12 & 5 & 0 \\ 9 & 5 & -1 \\ 7 & 3 & 0 \end{vmatrix} = 1(36-35) = 1$$

$$\begin{vmatrix} 1 & 14 & 3 \\ 2 & 9 & -1 \\ 3 & 7 & -4 \end{vmatrix} = \begin{vmatrix} 1 & 14 & 3 \\ 0 & -19 & -7 \\ 0 & -35 & -13 \end{vmatrix} = \begin{vmatrix} 1 & 14 & 3 \\ 0 & 19 & -7 \\ 0 & 3 & 1 \end{vmatrix}$$

$$= 1(-19+21) = 2$$

$$\begin{vmatrix} 1 & 2 & 14 \\ 2 & 5 & 9 \\ 3 & 8 & 7 \end{vmatrix} = \begin{vmatrix} 1 & 2 & 14 \\ 0 & 1 & -19 \\ 0 & 2 & -35 \end{vmatrix} = 1(-35+38) = 3$$

$$\begin{vmatrix} 1 & 2 & 3 \\ 2 & 5 & -1 \\ 3 & 8 & -4 \end{vmatrix} = \begin{vmatrix} 1 & 2 & 3 \\ 0 & 1 & -7 \\ 0 & 2 & -13 \end{vmatrix} = 1(-13+14) = 1$$

$$\frac{x}{1} = \frac{y}{2} = \frac{z}{3} = \frac{1}{1} \Rightarrow x = 1, \ y = 2, \ z = 3$$

This method involves working out four determinants. We shall compare this method with others.

The inverse of a 3 by 3 matrix

$$\begin{pmatrix} a_1 & b_1 & c_1 \\ a_2 & b_2 & c_2 \\ a_3 & b_3 & c_3 \end{pmatrix} \begin{pmatrix} x \\ y \\ z \end{pmatrix} = \begin{pmatrix} d_1 \\ d_2 \\ d_3 \end{pmatrix} \Rightarrow \begin{aligned} \Delta x &= d_1 A_1 + d_2 A_2 + d_3 A_3 \\ \Delta y &= d_1 B_1 + d_2 B_2 + d_3 B_3 \\ \Delta z &= d_1 C_1 + d_2 C_2 + d_3 C_3 \end{aligned}$$

$$\Delta \begin{pmatrix} x \\ y \\ z \end{pmatrix} = \begin{pmatrix} A_1 & A_2 & A_3 \\ B_1 & B_2 & B_3 \\ C_1 & C_2 & C_3 \end{pmatrix} \begin{pmatrix} d_1 \\ d_2 \\ d_3 \end{pmatrix} \text{ or } \begin{pmatrix} x \\ y \\ z \end{pmatrix} = \frac{1}{\Delta} \begin{pmatrix} A_1 & A_2 & A_3 \\ B_1 & B_2 & B_3 \\ C_1 & C_2 & C_3 \end{pmatrix} \begin{pmatrix} d_1 \\ d_2 \\ d_3 \end{pmatrix}$$

Writing $M = \begin{pmatrix} a_1 & b_1 & c_1 \\ a_2 & b_2 & c_2 \\ a_3 & b_3 & c_3 \end{pmatrix}$, det $M = \Delta$, $\begin{pmatrix} x \\ y \\ z \end{pmatrix} = \mathbf{r}$ and $\begin{pmatrix} d_1 \\ d_2 \\ d_3 \end{pmatrix} = \mathbf{d}$

$$M\mathbf{r} = \mathbf{d} \Leftrightarrow \mathbf{r} = M^{-1}\mathbf{d}$$

where M^{-1} is the inverse of the matrix M, i.e. $M^{-1}M = MM^{-1} = I_3$.

$$M^{-1} = \frac{1}{\Delta} \begin{pmatrix} A_1 & A_2 & A_3 \\ B_1 & B_2 & B_3 \\ C_1 & C_2 & C_3 \end{pmatrix} = \frac{1}{\Delta} M_c^{1\prime}$$

where $M_c^{1\prime}$ is the transpose of M_c, the matrix of cofactors of M.

187

As a check

$$M^{-1}M = \frac{1}{\Delta}\begin{pmatrix} A_1 & A_2 & A_3 \\ B_1 & B_2 & B_3 \\ C_1 & C_2 & C_3 \end{pmatrix}\begin{pmatrix} a_1 & b_1 & c_1 \\ a_2 & b_2 & c_2 \\ a_3 & b_3 & c_3 \end{pmatrix}$$

$$= \Delta\begin{pmatrix} \Delta & 0 & 0 \\ 0 & \Delta & 0 \\ 0 & 0 & \Delta \end{pmatrix} = \begin{pmatrix} 1 & 0 & 0 \\ 0 & 1 & 0 \\ 0 & 0 & 1 \end{pmatrix} = I_3$$

Example 4 repeated

$$\begin{matrix} x + 2y + 3z = 14 \\ 2x + 5y - 1z = 9 \\ 3x + 8y - 4z = 7 \end{matrix} \Leftrightarrow \begin{pmatrix} 1 & 2 & 3 \\ 2 & 5 & -1 \\ 3 & 8 & -4 \end{pmatrix}\begin{pmatrix} x \\ y \\ z \end{pmatrix} = \begin{pmatrix} 14 \\ 9 \\ 7 \end{pmatrix} \Leftrightarrow M\mathbf{r} = \begin{pmatrix} 14 \\ 9 \\ 7 \end{pmatrix}$$

$$\Delta = \det M = \begin{vmatrix} 1 & 2 & 3 \\ 2 & 5 & -1 \\ 3 & 8 & -4 \end{vmatrix} = \begin{vmatrix} 1 & 2 & 3 \\ 0 & 1 & -7 \\ 0 & 2 & -13 \end{vmatrix} = 1(-13 + 14) = 1$$

$$M^{-1} = \frac{1}{\Delta}\begin{pmatrix} A_1 & A_2 & A_3 \\ B_1 & B_2 & B_3 \\ C_1 & C_2 & C_3 \end{pmatrix} = \begin{pmatrix} -12 & 32 & -17 \\ 5 & -13 & 7 \\ 1 & -2 & 1 \end{pmatrix}$$

$$\begin{pmatrix} x \\ y \\ z \end{pmatrix} = \begin{pmatrix} -12 & 32 & -17 \\ 5 & -13 & 7 \\ 1 & -2 & 1 \end{pmatrix}\begin{pmatrix} 14 \\ 9 \\ 7 \end{pmatrix} = \begin{pmatrix} 1 \\ 2 \\ 3 \end{pmatrix}$$

This method is difficult arithmetically with opportunities for arithmetic errors. The following method of elimination leads to a simple matrix method for setting out the solution which is very instructive.

Coefficient matrix

$$\begin{matrix} x + 2y + 3z = & 14 & (1) \\ 2x + 5y - z = & 9 & (2) \\ 3x + 8y - 4z = & 7 & (3) \end{matrix} \qquad \begin{pmatrix} 1 & 2 & 3 \\ 2 & 5 & -1 \\ 3 & 8 & -4 \end{pmatrix}\begin{pmatrix} 14 \\ 9 \\ 7 \end{pmatrix}$$

Stage 1

(1)	is	$x + 2y + 3z = 14$	(1)	
(2) − 2(1)	gives	$y - 7z = -19$	(2′)	$\begin{pmatrix} 1 & 2 & 3 \\ 0 & 1 & -7 \\ 0 & 2 & -13 \end{pmatrix}\begin{pmatrix} 14 \\ -19 \\ -35 \end{pmatrix}$
(3) − 3(1)	gives	$2y - 13z = -35$	(3′)	

Stage 2

(1)	is	$x + 2y + 3z = 14$	(1)	
(2′)	is	$y - 7z = -19$	(2′)	$\begin{pmatrix} 1 & 2 & 3 \\ 0 & 1 & -7 \\ 0 & 0 & 3 \end{pmatrix}\begin{pmatrix} 14 \\ -19 \\ 3 \end{pmatrix}$
(3′) − 2(2′)	gives	$z = 3$	(3″)	

188

Stage 3

(1) − 3(3″) gives $\qquad x + 2y = 5 \qquad$ (1′)

(2′) + 7(3″) gives $\qquad \quad y = 2 \qquad$ (2″)

$\qquad\qquad\qquad\qquad\qquad z = 3 \qquad$ (3″)

$$\begin{pmatrix} 1 & 2 & 0 \\ 0 & 1 & 0 \\ 0 & 0 & 1 \end{pmatrix} \begin{pmatrix} 5 \\ 2 \\ 3 \end{pmatrix}$$

Stage 4

(1′) − 2(2″) gives $\qquad x = 1$

2″ is $\qquad\qquad\qquad\ y = 2$

3″ is $\qquad\qquad\qquad\ z = 3$

$$\begin{pmatrix} 1 & 0 & 0 \\ 0 & 1 & 0 \\ 0 & 0 & 1 \end{pmatrix} \begin{pmatrix} 1 \\ 2 \\ 3 \end{pmatrix}$$

Each instruction on a pair of equations is equivalent to pre-multiplying the matrices by a simple 3×3 matrix. The first two lines of working are achieved by premultiplying by

$$\begin{pmatrix} 1 & 0 & 0 \\ -2 & 1 & 0 \\ -3 & 0 & 1 \end{pmatrix} \text{ so } \begin{pmatrix} 1 & 0 & 0 \\ -2 & 1 & 0 \\ -3 & 0 & 1 \end{pmatrix} \begin{pmatrix} 1 & 2 & 3 \\ 2 & 5 & -1 \\ 3 & 8 & -4 \end{pmatrix} = \begin{pmatrix} 1 & 2 & 3 \\ 0 & 1 & -7 \\ 0 & 2 & -13 \end{pmatrix}$$

and $\begin{pmatrix} 1 & 0 & 0 \\ -2 & 1 & 0 \\ -3 & 0 & 1 \end{pmatrix} \begin{pmatrix} 14 \\ 9 \\ 7 \end{pmatrix} = \begin{pmatrix} 14 \\ -19 \\ -35 \end{pmatrix}$

To solve $\begin{pmatrix} 1 & 2 & 3 \\ 2 & 5 & -1 \\ 3 & 8 & -4 \end{pmatrix} \begin{pmatrix} x \\ y \\ z \end{pmatrix} = \begin{pmatrix} 14 \\ 9 \\ 7 \end{pmatrix}$

Stage 1

Multiply both sides by $M_1 = \begin{pmatrix} 1 & 0 & 0 \\ -2 & 1 & 0 \\ -3 & 0 & 1 \end{pmatrix}$

$$\begin{pmatrix} 1 & 0 & 0 \\ -2 & 1 & 0 \\ -3 & 0 & 1 \end{pmatrix} \begin{pmatrix} 1 & 2 & 3 \\ 2 & 5 & -1 \\ 3 & 8 & -4 \end{pmatrix} \begin{pmatrix} x \\ y \\ z \end{pmatrix} = \begin{pmatrix} 1 & 0 & 0 \\ -2 & 1 & 0 \\ -3 & 0 & 1 \end{pmatrix} \begin{pmatrix} 14 \\ 9 \\ 7 \end{pmatrix}$$

to give $\begin{pmatrix} 1 & 2 & 3 \\ 0 & 1 & -7 \\ 0 & 2 & -13 \end{pmatrix} \begin{pmatrix} x \\ y \\ z \end{pmatrix} = \begin{pmatrix} 14 \\ -19 \\ -35 \end{pmatrix}$

Stage 2

Multiply both sides by $M_2 = \begin{pmatrix} 1 & 0 & 0 \\ 0 & 1 & 0 \\ 0 & -2 & 1 \end{pmatrix}$

$$\begin{pmatrix} 1 & 0 & 0 \\ 0 & 1 & 0 \\ 0 & -2 & 1 \end{pmatrix} \begin{pmatrix} 1 & 2 & 3 \\ 0 & 1 & -7 \\ 0 & 2 & -13 \end{pmatrix} \begin{pmatrix} x \\ y \\ z \end{pmatrix} = \begin{pmatrix} 1 & 0 & 0 \\ 0 & 1 & 0 \\ 0 & -2 & 1 \end{pmatrix} \begin{pmatrix} 14 \\ -19 \\ -35 \end{pmatrix}$$

to give
$$\begin{pmatrix} 1 & 2 & 3 \\ 0 & 1 & -7 \\ 0 & 0 & 1 \end{pmatrix} \begin{pmatrix} x \\ y \\ z \end{pmatrix} = \begin{pmatrix} 14 \\ -19 \\ 3 \end{pmatrix}$$

Stage 3

Multiply both sides by $M_3 = \begin{pmatrix} 1 & 0 & -3 \\ 0 & 1 & 7 \\ 0 & 0 & 1 \end{pmatrix}$

$$\begin{pmatrix} 1 & 0 & -3 \\ 0 & 1 & 7 \\ 0 & 0 & 1 \end{pmatrix} \begin{pmatrix} 1 & 2 & 3 \\ 0 & 1 & -7 \\ 0 & 0 & 1 \end{pmatrix} \begin{pmatrix} x \\ y \\ z \end{pmatrix} = \begin{pmatrix} 1 & 0 & -3 \\ 0 & 1 & 7 \\ 0 & 0 & 1 \end{pmatrix} \begin{pmatrix} 14 \\ -19 \\ 3 \end{pmatrix}$$

to give
$$\begin{pmatrix} 1 & 2 & 0 \\ 0 & 1 & 0 \\ 0 & 0 & 1 \end{pmatrix} \begin{pmatrix} x \\ y \\ z \end{pmatrix} = \begin{pmatrix} 5 \\ 2 \\ 3 \end{pmatrix}$$

Stage 4

Multiply both sides by $M_4 = \begin{pmatrix} 1 & -2 & 0 \\ 0 & 1 & 0 \\ 0 & 0 & 1 \end{pmatrix}$

$$\begin{pmatrix} 1 & -2 & 0 \\ 0 & 1 & 0 \\ 0 & 0 & 1 \end{pmatrix} \begin{pmatrix} 1 & 2 & 0 \\ 0 & 1 & 0 \\ 0 & 0 & 1 \end{pmatrix} \begin{pmatrix} x \\ y \\ z \end{pmatrix} = \begin{pmatrix} 1 & -2 & 0 \\ 0 & 1 & 0 \\ 0 & 0 & 1 \end{pmatrix} \begin{pmatrix} 5 \\ 2 \\ 3 \end{pmatrix}$$

to give
$$\begin{pmatrix} 1 & 0 & 0 \\ 0 & 1 & 0 \\ 0 & 0 & 1 \end{pmatrix} \begin{pmatrix} x \\ y \\ z \end{pmatrix} = \begin{pmatrix} 1 \\ 2 \\ 3 \end{pmatrix} \text{ or } \begin{pmatrix} x \\ y \\ z \end{pmatrix} = \begin{pmatrix} 1 \\ 2 \\ 3 \end{pmatrix}$$

The numbers in this example were specially chosen to make the arithmetic simple as you may have noticed. After stage 1 the second row, second column element is conveniently 1. If this does not happen an extra stage can be inserted to make it 1, or it can be incorporated in the matrix M_1. A later example will illustrate this point.

The setting out of the matrix process on the previous page shows that the effect of premultiplying by M_1 then M_2, M_3 and M_4 is to reduce the coefficient matrix M to the identity.

So $M_4\{M_3[M_2(M_1M)]\} = I_3$ or $(M_4M_3M_2M_1)M = I$

The brackets can be rearranged since matrix multiplication is associative. Remembering that $M^{-1}M = I$ where M^{-1} is the inverse matrix of M we have $M_4M_3M_2M_1 = M^{-1}$ and a process for finding the inverse of a 3 by 3 matrix.

Since $M_4M_3M_2M_1 = M_4M_3M_2M_1 I$ if the matrices M_1, M_2, M_3, M_4 are multiplied with I then we will end up with the inverse M^{-1}. It is not even necessary to formulate M_1, M_2, M_3, M_4 because these only have the effect of row operations and it is sufficient only to devise the row operations. The calculations are set out below.

Row operation	Coefficient matrix	Vector	Identity
	$\begin{pmatrix} 1 & 2 & 3 \\ 2 & 5 & -1 \\ 3 & 8 & -4 \end{pmatrix}$	$\begin{pmatrix} 14 \\ 9 \\ 7 \end{pmatrix}$	$\begin{pmatrix} 1 & 0 & 0 \\ 0 & 1 & 0 \\ 0 & 0 & 1 \end{pmatrix}$
Row $2 - 2 \times$ Row 1 Row $3 - 3 \times$ Row 1 gives	$\begin{pmatrix} 1 & 2 & 3 \\ 0 & 1 & -7 \\ 0 & 2 & -13 \end{pmatrix}$	$\begin{pmatrix} 14 \\ -19 \\ -35 \end{pmatrix}$	$\begin{pmatrix} 1 & 0 & 0 \\ -2 & 1 & 0 \\ -3 & 0 & 1 \end{pmatrix}$
Row $3 - 2 \times$ Row 2 gives	$\begin{pmatrix} 1 & 2 & 3 \\ 0 & 1 & -7 \\ 0 & 0 & 1 \end{pmatrix}$	$\begin{pmatrix} 14 \\ -19 \\ 3 \end{pmatrix}$	$\begin{pmatrix} 1 & 0 & 0 \\ -2 & 1 & 0 \\ 1 & -2 & 1 \end{pmatrix}$
Row $1 - 3 \times$ Row 3 Row $2 + 7 \times$ Row 3 gives	$\begin{pmatrix} 1 & 2 & 0 \\ 0 & 1 & 0 \\ 0 & 0 & 1 \end{pmatrix}$	$\begin{pmatrix} 5 \\ 2 \\ 3 \end{pmatrix}$	$\begin{pmatrix} -2 & 6 & -3 \\ 5 & -13 & 7 \\ 1 & -2 & 1 \end{pmatrix}$
Row $1 - 2 \times$ Row 2 gives	$\begin{pmatrix} 1 & 0 & 0 \\ 0 & 1 & 0 \\ 0 & 0 & 1 \end{pmatrix}$	$\begin{pmatrix} 1 \\ 2 \\ 3 \end{pmatrix}$	$\begin{pmatrix} -12 & 32 & -17 \\ 5 & -13 & 7 \\ 1 & -2 & 1 \end{pmatrix}$
	Identity	Solution	Inverse

In four stages to produce the identity we have calculated the solution and the inverse matrix if we need it. As a check on the inverse

$$\begin{pmatrix} -12 & 32 & -17 \\ 5 & -13 & 7 \\ 1 & -2 & 1 \end{pmatrix} \begin{pmatrix} 1 & 2 & 3 \\ 2 & 5 & -1 \\ 3 & 8 & -4 \end{pmatrix} = \begin{pmatrix} 1 & 0 & 0 \\ 0 & 1 & 0 \\ 0 & 0 & 1 \end{pmatrix}$$

$$= \begin{pmatrix} 1 & 2 & 3 \\ 2 & 5 & -1 \\ 3 & 8 & -4 \end{pmatrix} \begin{pmatrix} -12 & 32 & -17 \\ 5 & -13 & 7 \\ 1 & -2 & 1 \end{pmatrix}$$

You will have noticed the same calculations turning up whichever method you use to solve the three equations. It may happen that with certain equations it is more convenient to use one method in preference to another. Perhaps the easiest way is to use the elimination method and not bother with determinants or matrices at all. If you are only concerned with solving three equations in three unknowns the best way is to learn one method and stick to it. However, if you want some insight into mathematics, the methods of matrices and determinants give valuable mathematical tools with which to explore and understand other branches and topics of mathematics. (See the **Vectors** chapter in which each equation represents a plane and the solution the intersection of those planes.)

Transformations in three dimensions

Three dimensional transformations are described with reference to a set of three mutually perpendicular axes Ox, Oy and Oz. Points are described by three coordinates (x, y, z). The transformation can be described by a 3×3 matrix.
If $P(x, y, z)$ is transformed to $P_1(x_1, y_1, z_1)$

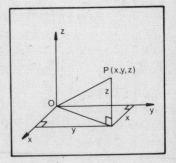

Figure 93

$$\begin{pmatrix} x_1 \\ y_1 \\ z_1 \end{pmatrix} = \begin{pmatrix} a_1 & b_1 & c_1 \\ a_2 & b_2 & c_2 \\ a_3 & b_3 & c_3 \end{pmatrix} \begin{pmatrix} x \\ y \\ z \end{pmatrix} \text{ where } M = \begin{pmatrix} a_1 & b_1 & c_1 \\ a_2 & b_2 & c_2 \\ a_3 & b_3 & c_3 \end{pmatrix}$$

is the transformation matrix.

The mirror for a reflection in three dimensions is a plane, (instead of a line in two dimensions) and a rotation takes place about a line (instead of a point as in two dimensions).

Example 5 Reflection in the plane $z = 0$.
The point (x, y, z) is reflected to $(x, y, -z)$.
$$\begin{pmatrix} x \\ y \\ z \end{pmatrix} \text{ becomes } \begin{pmatrix} 1 & 0 & 0 \\ 0 & 1 & 0 \\ 0 & 0 & -1 \end{pmatrix} \begin{pmatrix} x \\ y \\ z \end{pmatrix} = \begin{pmatrix} x \\ y \\ -z \end{pmatrix} \text{ and } M_1 = \begin{pmatrix} 1 & 0 & 0 \\ 0 & 1 & 0 \\ 0 & 0 & -1 \end{pmatrix}$$

is the matrix for a reflection in the plane $z = 0$.

Example 6 Reflection in the plane $x = y$.

(x, y, z) is reflected to (y, x, z).

$$\begin{pmatrix} x \\ y \\ z \end{pmatrix} \text{ becomes } \begin{pmatrix} 0 & 1 & 0 \\ 1 & 0 & 0 \\ 0 & 0 & 1 \end{pmatrix} \begin{pmatrix} x \\ y \\ z \end{pmatrix} = \begin{pmatrix} y \\ x \\ z \end{pmatrix} \text{ and } M_2 = \begin{pmatrix} 0 & 1 & 0 \\ 1 & 0 & 0 \\ 0 & 0 & 1 \end{pmatrix}$$

represents reflection in $y = x$.

As in two dimensions the easiest way to find a transformation matrix is to find the images of the unit vectors along the axes.

$$\begin{pmatrix} a_1 & b_1 & c_1 \\ a_2 & b_2 & c_2 \\ a_3 & b_3 & c_3 \end{pmatrix} \begin{pmatrix} 1 \\ 0 \\ 0 \end{pmatrix} = \begin{pmatrix} a_1 \\ a_2 \\ a_3 \end{pmatrix}$$

and $(1, 0, 0)$ is transformed to (a_1, a_2, a_3).

Similarly $(0, 1, 0)$ goes to (b_1, b_2, b_3) and $(0, 0, 1)$ goes to (c_1, c_2, c_3).

Example 7 Rotation of $180°$ about Oz.

$(1, 0, 0)$ rotates to $(-1, 0, 0)$
$(0, 1, 0)$ rotates to $(0, -1, 0)$
$(0, 0, 1)$ is invariant

$$\text{so } M_3 = \begin{pmatrix} -1 & 0 & 0 \\ 0 & -1 & 0 \\ 0 & 0 & 1 \end{pmatrix}$$

is the matrix for a rotation of $180°$ about Oz.

A **translation** necessarily moves the origin. Consequently since the image of $(0, 0, 0)$ is always $(0, 0, 0)$ with a 3×3 matrix transformation, a translation cannot be represented by a 3×3 matrix alone. It is sufficient to say

$$\begin{pmatrix} x \\ y \\ z \end{pmatrix} \text{ is translated to } \begin{pmatrix} x + 1 \\ y + 2 \\ z + 3 \end{pmatrix} = \begin{pmatrix} x \\ y \\ z \end{pmatrix} + \begin{pmatrix} 1 \\ 2 \\ 3 \end{pmatrix}$$

for the translation represented by the vector $\begin{pmatrix} 1 \\ 2 \\ 3 \end{pmatrix}$.

An **enlargement** of scale factor 3, centre the origin, will take $(1, 0, 0)$ to $(3, 0, 0)$, $(0, 1, 0)$ to $(0, 3, 0)$ and $(0, 0, 1)$ to $(0, 0, 3)$.

$$\begin{pmatrix} x \\ y \\ z \end{pmatrix} \text{ becomes } \begin{pmatrix} 3 & 0 & 0 \\ 0 & 3 & 0 \\ 0 & 0 & 3 \end{pmatrix} \begin{pmatrix} x \\ y \\ z \end{pmatrix} = \begin{pmatrix} 3x \\ 3y \\ 3z \end{pmatrix}$$

so $M_4 = \begin{pmatrix} 3 & 0 & 0 \\ 0 & 3 & 0 \\ 0 & 0 & 3 \end{pmatrix}$ is the matrix for an enlargement of scale

factor 3, centre $(0, 0, 0)$.

An enlargement of scale factor

-1 has matrix $\begin{pmatrix} -1 & 0 & 0 \\ 0 & -1 & 0 \\ 0 & 0 & -1 \end{pmatrix}$

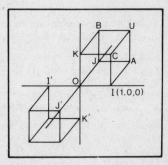

Its effect on the unit cube is shown in fig. 94. det $M_6 = -1$ so the transformation is not a rotation and it is not a reflection as the lines joining corresponding points are not parallel.

Figure 94

A **shear** in the direction of the x axis with **invariant plane** $z = 0$ so that points in the plane $z = 1$ (i.e. distance 1 unit from invariant plane) move 3 units will take $(0, 0, 1)$ to $(3, 0, 1)$ and leave $(1, 0, 0)$ and $(0, 1, 0)$ invariant.

$$\begin{pmatrix} x \\ y \\ z \end{pmatrix} \text{ becomes } \begin{pmatrix} 1 & 0 & 3 \\ 0 & 1 & 0 \\ 0 & 0 & 1 \end{pmatrix} \begin{pmatrix} x \\ y \\ z \end{pmatrix} = \begin{pmatrix} x + 3z \\ y \\ z \end{pmatrix} \text{ so } \begin{pmatrix} 1 & 0 & 3 \\ 0 & 1 & 0 \\ 0 & 0 & 1 \end{pmatrix}$$

is the matrix for this shear.

The 3×3 matrices considered above are quite simple. More complicated transformations can be represented by a 3×3 matrix but it is difficult to interpret these in terms of simple transformations. It is most helpful to consider their effect on the unit cube whose vertices are $0(0, 0, 0)$, $I(1, 0, 0)$, $J(0, 1, 0)$, $K(0, 0, 1)$, $A(1, 1, 0)$, $B(0, 1, 1)$, $C(1, 0, 1)$, $U(1, 1, 1)$. (See fig. 94). You can check that the determinant of $M_1 = -1$ and det $M_2 = -1$ while det $M_3 = 1$. These results are similar to those for reflections and rotations in two dimensions and in fact the determinant of a 3×3 transformation matrix represents the ratio of the volume of the object to the volume of the image.

det $M_4 = 27$ which means that the image volume is 27 times the object volume in an enlargement scale factor 3.

det $M_5 = 1$ which confirms the important shearing property that volume is invariant.

Volume scale factor

The determinant of the

$$\text{matrix } M = \begin{pmatrix} a_1 & b_1 & c_1 \\ a_2 & b_2 & c_2 \\ a_3 & b_3 & c_3 \end{pmatrix}$$

is the volume scale factor of the transformation, i.e. the ratio of image volume to object volume.

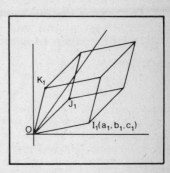

Figure 95

Figure 95 shows the image of the unit cube which is a parallelepiped (squashed cuboid) generated by the vectors $\overrightarrow{OI_1}$, $\overrightarrow{OJ_1}$, $\overrightarrow{OK_1}$. Opposite faces are parallel and the parallelepiped has three other edges parallel to OI_1, three parallel to OJ_1 and three parallel to OK_1. It can be proved (see Chapter 11 on Vectors page 299) that the volume of this parallelepiped is equal to

$$\det M = \begin{pmatrix} a_1 & b_1 & c_1 \\ a_2 & b_2 & c_2 \\ a_3 & b_3 & c_3 \end{pmatrix}$$

Key terms

A matrix of order $p \times q$ multiplied with a matrix of order $q \times r$ gives a matrix of order $p \times r$.

Matrix multiplication is associative but not commutative.

For a matrix M, the identity I satisfies $IM = MI = M$.

The inverse, M^{-1}, of the matrix M satisfies $MM^{-1} = M^{-1}M = I$.

The inverse of $M = \begin{pmatrix} a & b \\ c & d \end{pmatrix}$ is $\frac{1}{\Delta} \begin{pmatrix} d & -b \\ -c & a \end{pmatrix}$ where $\Delta = \det M = ad - bc$.

The inverse of a 3×3 matrix N is the transpose of the matrix of cofactors of N, divided by det N.

Chapter 8
Series

Sequences

A **sequence** of numbers can be obtained by writing a set of values in a given order such that there is a definite rule by which each term may be calculated.

For example,   (a) $1, 3, 5, 7, \ldots$ (c) $1, -1, 1, -1, 1, \ldots$
    (b) $1, 4, 9, 16, \ldots$ (d) $2, 5, 8, 11, 14, 17, \ldots$

In general we write the terms of a sequence as $u_1, u_2, u_3, \ldots u_n, \ldots$. There is a more convenient way of defining a sequence which saves having to write out the terms as we have above. We can use an **inductive definition** which gives an initial term and a rule for determining each term from a previous one. So (a) above could be defined as

$$u_1 = 1, \quad u_{k+1} = u_k + 2$$

This gives all the terms starting with $u_1 = 1$, $u_2 = u_1 + 2 = 3$, $u_3 = u_2 + 2 = 5$ and so on.

Alternatively we can define a sequence by giving an **algebraic formula** for a general term. So the sequence (a) above could be written as $u_n = 2n - 1$ for $n = 1, 2, 3, \ldots$.

Two important sequences are formed when (i) each term is formed by adding a constant to the previous term and (ii) each term is formed by multiplying the previous term by a constant. The first case gives an **arithmetic progression** (A.P.) and the constant may be positive, negative or fractional.

The second case gives a **geometric progression** (G.P.) and the constant may be a positive or negative integer or fraction.

Examples of arithmetic progressions are

  (i) $u_n = 2n - 5$ for $n = 0, 1, 2, \ldots$
 (ii) $-2, -4, -6, -8, \ldots$
(iii) $u_1 = -3, u_{k+1} = u_k + 4$

Examples of geometric progressions are

  (i) $u_n = 3^n$ for $n = 0, 1, 2, 3, \ldots$
 (ii) $1, -\frac{1}{2}, +\frac{1}{4}, -\frac{1}{8}, +\frac{1}{16}, \ldots$
(iii) $u_1 = 5, u_{k+1} = 2u_k$

We may consider the terms of a sequence as a sum, in which

case we write it as $u_1 + u_2 + u_3 + \cdots + u_n$.
This is called a **series** and may be **finite** or **infinite**.

i.e. $3 + 7 + 11 + 15 + 19$ is a finite series
but $1 + 2 + 3 + 4 + 5 + \ldots$ is infinite.

The greek letter Σ (sigma) is used to denote a summation and we can write the general series above as $\displaystyle\sum_{r=1}^{n} u_r$. The arithmetic series would be written as $\displaystyle\sum_{r=1}^{7} (2r - 1)$ which is equivalent to $1 + 3 + 5 + 7 + 9 + 11 + 13$.
This is a finite series. If we have an infinite series we write
$$\sum_{r=2}^{\infty} r^2 = 2^2 + 3^2 + 4^2 + 5^2 + \cdots$$

Example 1 Find the terms of the series $\displaystyle\sum_{r=1}^{4} \frac{1}{r(r+1)}$.

Clearly $\displaystyle\sum_{r=1}^{4} \frac{1}{r(r+1)} = \frac{1}{1.2} + \frac{1}{2.3} + \frac{1}{3.4} + \frac{1}{4.5}$.

Example 2 Write the series $1 - 2 + 4 - 8 + 16 - 32$ in the Σ notation. We must find an algebraic expression for the general term. In this case each term can be obtained from $(-1)^r 2^r$ by substituting $r = 0, 1, 2, \ldots, 5$.
Hence $1 - 2 + 4 - 8 + 16 - 32 = \displaystyle\sum_{r=0}^{5} (-1)^r 2^r$.

The arithmetic series

An A.P. is formed by adding a positive or negative constant to each term to obtain the next term of the series. If a is the first term and d is the value to be added we can write the series as

$$a + (a + d) + (a + 2d) + (a + 3d) + \cdots$$

If we take n terms the nth term will be $a + (n - 1)d$. Thus if the sum to n terms is S_n we have

$$S_n = a + (a + d) + (a + 2d) + \cdots [a + (n - 1)d]$$

or by rewriting in the opposite order

$$S_n = [a + (n - 1)d] + [a + (n - 2)d] + \cdots + a$$

adding $2S_n = [2a + (n - 1)d] \times n$
$\therefore$ The **sum of an A.P.** is given by $S_n = \frac{1}{2}n[2a + (n - 1)d]$ where a is the first term, d is the **common difference** and n is the number of terms.

The **value of the *n*th term** is given by $a + (n - 1)d$.

Example 3 Find the value of the 15th term and the sum of the first 10 terms of the series $8 + 5 + 2 + \cdots$.

The value of the 15th term $= a + 14d = 8 + 14(-3) = -34$.

The sum of the first 10 terms $S_{10} = \frac{1}{2} \times 10[(2 \times 8) + 14(-3)]$
$$= 5 \times (-26) = -130$$

Example 4 Find the number of terms in the arithmetic progression $2\cdot8 + 3\cdot3 + 3\cdot8 + \cdots + 17\cdot8$.

Now $a = 2\cdot8$ and the value of the *n*th term is $17\cdot8$.

$\therefore \quad 17\cdot8 = 2\cdot8 + (n - 1) \times (0\cdot5)$

$\therefore \quad \frac{1}{2}(n - 1) = 15 \Rightarrow n - 1 = 30 \Rightarrow n = 31$

$\therefore$ there are 31 terms in the series

Note that if we denote the last term by (l) then the sum to n terms can be written

$$S_n = \frac{1}{2}n[2a + (n - 1)d] = \frac{1}{2}n\{a + [a + (n - 1)d]\} = \frac{1}{2}n(a + l)$$

This can be a useful alternative when the first and last terms of the series are given.

Geometric series

A G.P. is formed by multiplying each term by a constant factor to produce the next term. If a is the first term and r is the given factor the series can be written as

$$a + ar + ar^2 + ar^3 + \cdots$$

If we take n terms, the *n*th term is ar^{n-1}. Thus the sum to n terms is given by

$$S_n = a + ar + ar^2 + ar^3 + \cdots + ar^{n-1} \tag{1}$$

Multiplying by r

$$rS_n = ar + ar^2 + ar^3 + \cdots + ar^{n-1} + ar^n \tag{2}$$

Subtracting (2) from (1) we have

$$(1 - r)S_n = a - ar^n \Rightarrow S_n = \frac{a(1 - r^n)}{1 - r}$$

If $r > 1$ then this is best expressed as $S_n = \dfrac{a(r^n - 1)}{r - 1}$.

r is called the **common ratio** of the series.

Hence the **value of the nth term** $= ar^{n-1}$

The **sum to n terms** $= \dfrac{a(1 - r^n)}{1 - r} = \dfrac{a(r^n - 1)}{r - 1}$

Example 5 Find the sum of the first 10 terms of the geometric progression $3 + 6 + 12 + 24 + \cdots$.

The sum of the G.P. is $S_{10} = \dfrac{3(2^{10} - 1)}{2 - 1} = 3(2^{10} - 1)$

$$= 3 \times 1023 = 3069$$

It will be noted that if more terms of the series in Example 5 were taken the sum would be correspondingly larger. In fact, the value of S_n increases without limit and we say that $S_n \to \infty$ and $n \to \infty$. However, this is not always true and we have the idea of a limiting value.

A limit

Consider the function $f(r) = 1 + \dfrac{1}{r}$. As r takes values of $1, 10, 100, 1000, \ldots$ and so on $f(r)$ takes the set of values $2, 1 \cdot 1, 1 \cdot 01, 1 \cdot 001, \ldots$ and so on. Although $f(r)$ decreases as r tends to infinity it is approaching the value of 1 more closely. In fact, we can make $f(r)$ as near to 1 as we like by choosing r to be sufficiently large.

We say that $f(r) \to 1$ as $r \to \infty$ or $\lim\limits_{r \to \infty} f(r) = 1$.

We can define this mathematically by saying that any function S_n of n will tend to a limit S (which is independent of n) if for all $n \geqslant N$, $|S_n - S|$ is less than some arbitrary small positive value, ϵ.

i.e. there exists an N such that $|S_n - S| < \epsilon$ for all $n \geqslant N$.

If in an infinite series, $S_n = u_1 + u_2 + u_3 + \cdots + u_n$ is the sum of the first n terms of $\sum\limits_{r=1}^{\infty} u_r$, and if as n tends to infinity, S_n tends to a finite limit S, then the infinite series $\sum\limits_{r=1}^{\infty} u_r$ is said to be **convergent** and S is called its **sum to infinity**. So for a G.P. we have already shown that the sum to n terms is given by

$$S_n = \frac{a(1 - r^n)}{1 - r}$$

If $-1 < r < 1, r^n \to 0$ as $n \to \infty$ and hence S_n tends to a finite limit

$$S = \frac{a}{1-r}$$

Hence an infinite G.P. converges if $|r| < 1$.

Note that if $r > 1, r^n \to \infty$ as $n \to \infty$ and hence S_n does not tend to a finite limit and we say that the series **diverges.**

Also if $r < -1, r^n$ takes alternately positive and negative values and hence as $n \to \infty, r^n \to \pm \infty$ and we say that r^n **oscillates infinitely** and hence S_n **oscillates infinitely** and the series cannot converge.

Example 6 Find the sum of the first n terms of the series $1 - \frac{1}{2} + \frac{1}{4} - \frac{1}{8} + \cdots$ and deduce its sum to infinity.

Since $a = 1$ and $r = -\frac{1}{2}$

$$S_n = \frac{a(1-r^n)}{1-r} = \frac{1[1-(-\frac{1}{2})^n]}{1-(-\frac{1}{2})} = \frac{2}{3}[1-(-\frac{1}{2})^n]$$

Now as $n \to \infty, (-\frac{1}{2})^n \to 0.$ $\therefore$ sum to infinity $= \frac{2}{3}$.

Other finite series

Many other finite series exist apart from the arithmetic and geometric progressions. The methods of finding the sums of these series differ but there are a number of recognized techniques.

The arithmetico–geometrical series which is of the form $a + (a+d)x + (a+2d)x^2 + (a+3d)x^3 + \cdots$ can be summed by multiplying by x and subtracting the result from the original series.

Example 7 Find the sum of the arithmetico–geometrical series $2 + 5x + 8x^2 + 11x^3 + \cdots + 29x^9$.

There are 10 terms in the series so let

$$S_{10} = 2 + 5x + 8x^2 + 11x^3 + \cdots + 29x^9 \qquad (1)$$

Multiply by x

$$\therefore \quad xS_{10} = 2x + 5x^2 + 8x^3 + \cdots + 26x^9 + 29x^{10} \qquad (2)$$

Subtracting (2) from (1) we have

$$(1-x)S_{10} = 2 + 3x + 3x^2 + 3x^3 + \cdots + 3x^9 - 29x^{10}$$

$$= 2 + \frac{3x(1-x^9)}{1-x} - 29x^{10}$$

$$\therefore \qquad S_{10} = \frac{2-29x^{10}}{1-x} + \frac{3x(1-x^9)}{(1-x)^2}$$

The difference method can be applied to evaluate the sum of a series $\sum_{r=1}^{n} u_r = u_1 + u_2 + u_3 + \cdots + u_r + \cdots + u_n$.

If u_r can be expressed as the difference of two functions, say, $f(r) - f(r-1)$, then clearly

$$u_n = f(n) - f(n-1)$$
$$u_{n-1} = f(n-1) - f(n-2)$$
$$u_{n-2} = f(n-2) - f(n-3)$$
$$\vdots$$
$$u_2 = f(2) - f(1)$$
$$u_1 = f(1) - f(0)$$

By adding these terms it follows that

$$u_1 + u_2 + u_3 + \cdots + u_{n-1} + u_n = f(n) - f(0)$$

Example 8 Find the sum of the series $\sum_{r=1}^{n} r(r+1)(r+2)$.

Now $u_r = r(r+1)(r+2)$. If we let $f(r) = r(r+1)(r+2)(r+3)$ then

$$f(r) - f(r-1) = r(r+1)(r+2)(r+3) - r(r-1)(r+1)(r+2)$$
$$= r(r+1)(r+2)(r+3-r+1)$$
$$= 4r(r+1)(r+2) = 4u_r$$
$$\therefore \quad u_r = \tfrac{1}{4}[f(r) - f(r-1)]$$

Hence
$$\sum_{r=1}^{n} u_r = \sum_{r=1}^{n} \tfrac{1}{4}[f(r) - f(r-1)]$$
$$= \tfrac{1}{4}[f(n) - f(0)]$$
$$= \tfrac{1}{4}n(n+1)(n+2)(n+3)$$

The method of differences is useful in series where each term is a reciprocal of a product of a constant number of factors in A.P., the first factors of each term being in the same A.P.

Example 9 Find the sum to n terms of the series

$$\frac{1}{1.2} + \frac{1}{2.3} + \frac{1}{3.4} + \cdots$$

Clearly the series can be written $\displaystyle\sum_{r=1}^{n} \frac{1}{r(r+1)}$

i.e. $u_r = \dfrac{1}{r(r+1)} = \dfrac{1}{r} - \dfrac{1}{r+1}.$ Hence

$$\sum_{r=1}^{n} u_r = \left(\frac{1}{1} - \frac{1}{2}\right) + \left(\frac{1}{2} - \frac{1}{3}\right) + \cdots + \left(\frac{1}{n-1} - \frac{1}{n}\right) + \left(\frac{1}{n} - \frac{1}{n+1}\right)$$

adding $\displaystyle\sum_{r=1}^{n} \frac{1}{r(r+1)} = 1 - \frac{1}{n+1}$

Or we could have defined $f(r)$ as $\dfrac{1}{r+1}$ in which case $u_r = -[f(r) - f(r-1)]$.

In this case $\displaystyle\sum_{r=1}^{n} u_r = \sum_{r=1}^{n} -[f(r) - f(r-1)]$

$$= -[f(n) - f(0)] = 1 - \frac{1}{n+1} \text{ as before.}$$

This method can be extended to cover more values in the denominator.

Series of natural numbers

Three very useful results, which should be memorized, are (i) the sum of the natural numbers, (ii) the sum of the squares of the natural numbers and (iii) the sum of the cubes of the natural numbers.

(i) Let $S_n = 1 + 2 + 3 + 4 + \cdots + n$.
   Since this is an A.P., $S_n = \frac{1}{2}n(n+1)$ (see page 197).

(ii) Let $S_n = 1^2 + 2^2 + 3^2 + 4^2 + \cdots + (n-1)^2 + n^2$.

   Let $f(r) = r(r+1)(2r+1) \Rightarrow f(r-1) = (r-1)r(2r-1)$

   Hence $f(r) - f(r-1) = r[(r+1)(2r+1) - (r-1)(2r-1)]$

   $\therefore \quad f(r) - f(r-1) = r[2r^2 + 3r + 1 - 2r^2 + 3r - 1] = 6r^2$

   Now $u_r = r^2$ hence $u_r = \frac{1}{6}[f(r) - f(r-1)]$

   Thus by the method of differences $S_n = \frac{1}{6}n(n+1)(2n+1)$.

(iii) Let $S_n = 1^3 + 2^3 + 3^3 + 4^3 + \cdots + n^3$.

   Let $f(r) = [r(r+1)]^2 \Rightarrow f(r-1) = [(r-1)r]^2$

   $\therefore \quad f(r) - f(r-1) = r^2(r+1)^2 - r^2(r-1)^2$

   $$= (r^2 + r)^2 - (r^2 - r)^2 = (2r^2)(2r) = 4r^3$$

Thus $r^3 = \frac{1}{4}[f(r) - f(r-1)]$.

$\therefore$ by the method of differences $S_n = \frac{1}{4}n^2(n+1)^2$.

These three results are written again and should be learnt

$$1 + 2 + 3 + 4 + \cdots + n = \frac{1}{2}n(n+1)$$

$$1^2 + 2^2 + 2^3 + 4^2 + \cdots + n^2 = \frac{1}{6}n(n+1)(2n+1)$$

$$1^3 + 2^3 + 3^3 + 4^3 + \cdots + n^3 = \frac{1}{4}n^2(n+1)^2 = [\frac{1}{2}n(n+1)]^2$$

Note that this is the square of the sum of the natural numbers.

We can use these results to sum series as the following example shows.

Example 10 Find the sum of the series $\sum_{r=1}^{n} (r+1)(r+3)$.

This series is $2.4 + 3.5 + 4.6 + 5.7 + \cdots + (n+1)(n+3)$ and the general term is $r^2 + 4r + 3$.

$$\therefore \sum_{r=1}^{n} (r^2 + 4r + 3) = \sum_{r=1}^{n} r^2 + \sum_{r=1}^{n} 4r + \sum_{r=1}^{n} 3$$

$$= \frac{1}{6}n(n+1)(2n+1) + 4 \times \frac{1}{2}n(n+1) + 3n$$

$$= \frac{1}{6}n[(n+1)(2n+1) + 12(n+1) + 18]$$

$$= \frac{1}{6}n[2n^2 + 3n + 1 + 12n + 12 + 18]$$

$$= \frac{1}{6}n(2n^2 + 15n + 31)$$

Proof by induction

It sometimes happens that the sum of a series can be found by experiment or by a process that does not form a proof. In this situation we can prove the truth of the result by a special process called induction. There are two basic steps.

(i) Show that if it is true for some value of n then it is also true for the next integral value of n.

(ii) Show that it is true for the first value.

To illustrate the method we consider the series of the squares of the natural numbers.

i.e. $\quad 1^2 + 2^2 + 3^2 + 4^2 + \cdots + n^2 = \frac{1}{6}n(n+1)(2n+1)$ \hfill (1)

Suppose this result is true for some value of n, say $n = k$, then

$$1^2 + 2^2 + 3^2 + \cdots + k^2 = \frac{1}{6}k(k+1)(2k+1)$$

Adding the next term of the series we obtain

$$1^2 + 2^2 + 3^2 + \cdots + k^2 + (k+1)^2 = \tfrac{1}{6}k(k+1)(2k+1) + (k+1)^2$$
$$= \tfrac{1}{6}(k+1)[k(2k+1) + 6(k+1)]$$
$$= \tfrac{1}{6}(k+1)(2k^2 + 7k + 6)$$
$$= \tfrac{1}{6}(k+1)(k+2)(2k+3)$$

Now this is clearly the result for $n = k+1$ and thus if the result is true for $n = k$ it is also true for $n = k+1$.
Now when $n = 1$, considering (1)

LHS $= 1^2 = 1$ and RHS $= \tfrac{1}{6} \cdot 1 \cdot 2 \cdot 3 = 1$

$\therefore$ The result is true for $n = 1$ and hence for $n = 2$. From this it follows that it is also true for $n = 3$ and so on for all positive integral value of n.

This can be written out in a slightly different manner as follows.

(i) Proof that $k \in T \Rightarrow k+1 \in T$ where T is the set of values for which the statement is true.

Now $k \in T \Rightarrow S_k = \tfrac{1}{6}k(k+1)(2k+1)$
$$\Rightarrow S_{k+1} = S_k + (k+1)^2 = \tfrac{1}{6}k(k+1)(2k+1) + (k+1)^2$$
$$\Rightarrow S_{k+1} = \tfrac{1}{6}(k+1)(k+2)(2k+3)$$

This is the result with $n = k+1$. Hence we have shown that $k \in T \Rightarrow k+1 \in T$.

(ii) Proof that $1 \in T$.

Now $S_1 = 1^2 = 1$. When $n = 1$, $\tfrac{1}{6}n(n+1)(2n+1) = 1 \Rightarrow 1 \in T$. So both parts of the induction principle are satisfied and it follows that the result is true for all integral values of n.

Remember that to use induction we must have the answer first.

The induction principle can also be used in a number of other contexts.

Example 11 Prove that $10^n + 3 \cdot 4^{n+2} + 5$ is divisible by 9 for all positive integral values of n.

Let $f(n) = 10^n + 3 \cdot 4^{n+2} + 5$ and consider the difference $f(n+1) - f(n)$.
Now $f(n+1) - f(n) = (10^{n+1} + 3 \cdot 4^{n+3} + 5) - (10^n + 3 \cdot 4^{n+2} + 5)$
$$= 10^n(10 - 1) + 3 \cdot 4^{n+2}(4 - 1)$$
$$= 9(10^n + 4^{n+2})$$

Thus if $f(n)$ is divisible by 9 so is $f(n+1)$.

Now $f(1) = 10 + 3 \cdot 4^3 + 5 = 207 = 9 \times 23$.

Since when $n = 1$ the expression is divisible by 9, it is also divisible when $n = 2$ and when $n = 3$. Hence it is divisible by 9 for all positive integral values of n.

Permutations and combinations

We now consider certain problems in selection and arrangement. A **permutation** is an arrangement of items in a particular order.

Example 12 How many numbers can be made by using all of the digits 2, 3, 6 and 7?

The 1st digit chosen can be any one of 4 digits.

The 2nd digit must then be chosen from 3 digits, one having already been chosen.

The 3rd digit is then restricted to a choice of 2 digits and the 4th digit is then determined.

The total is obtained by multiplying these results since for each of the first 4 choices there are 3 possibilities for the next, and so on.

i.e. The number of different arrangements is $4 \times 3 \times 2 \times 1 = 24$.

i.e.

2367	3267	6327	7326
2376	3276	6372	7362
2637	3627	6273	7263
2673	3672	6237	7236
2763	3726	6732	7632
2736	3762	6723	7623

It follows that if we had to place 8 counters of different colours in order the total number of arrangements is given by $8 \times 7 \times 6 \times 5 \times 4 \times 3 \times 2 \times 1$.

We can use a dot to signify products, i.e. $8 \cdot 7 \cdot 6 \cdot 5 \cdot 4 \cdot 3 \cdot 2 \cdot 1$.

But an alternative notation is adopted, namely 8!, which is read **factorial** eight. Remember that it includes each integer from the highest down to 1 inclusive.

i.e. $n! = n(n-1)(n-2)(n-3)\ldots4 \cdot 3 \cdot 2 \cdot 1$

If some values are missing it is sometimes possible to write the product in factorial notation.

Example 13 Express $\dfrac{10 \times 9}{2 \times 1}$ in factorial notation.

$$\frac{10 \times 9}{2 \times 1} = \frac{10 \times 9 \times 8!}{2 \times 1 \times 8!} = \frac{10!}{2!8!}$$

Example 14 Simplify $\frac{15!}{11!4!} + \frac{15!}{12!3!}$.

Now $\frac{15!}{11!4!} + \frac{15!}{12!3!} = \frac{12 \times 15!}{12! \times 4!} + \frac{4 \times 15!}{12!4!}$

$$= \frac{16 \times 15!}{12!4!} = \frac{16!}{12!4!}$$

This notation means that there are $n!$ permutations of n unlike objects.

If we only select **r objects from n unlike objects**
the 1st is chosen in n ways,
the 2nd is chosen in $(n-1)$ ways,
the 3rd is chosen in $(n-2)$ ways, and so on until we arrive at
the last object which can be chosen in $(n-r+1)$ ways.
∴ Total number of arrangements

$$= n(n-1)(n-2) \cdots (n-r+1)$$

$$= \frac{n!}{(n-r)!}$$

We use the notation $^nP_r = \dfrac{n!}{(n-r)}$ **(r ≤ n).**

The following examples illustrate the methods which can be used.

Example 15 How many three letter arrangements can be formed from the first eight letters of the alphabet!

Since we are choosing 3 objects from 8 the number of different arrangements is $^8P_3 = \dfrac{8!}{5!} = 8 . 7 . 6 = 336$.

When some items are repeated some modification is necessary.

Example 16 In how many ways can the letters of the word EXPERIMENT be arranged?

Since there are three E's call them E_1, E_2, E_3. So treating the E's as different there are 10! ways of arranging the letters. But in every distinct arrangement the three E's can be rearranged amongst themselves without altering the other letters. So, TNIEXEPREM, for example, has been included 3! times in the total 10! arrangements. Hence the number of distinct arrangements of the letters of the word EXPERIMENT $= \dfrac{10!}{3!}$.

Sometimes it pays to find the exact opposite of what is asked in the question and subtract this from the number of unrestricted arrangements.

Example 17 How many arrangements of eight stories in a book can be made if the longest and the shortest must not come together.

If there is no restriction, the number of arrangements $= 8!$

If we consider the longest and the shortest to be together so that they cannot be separated then we effectively have seven stories to arrange. This can be done in 7! ways but the longest and shortest can be placed in either order. Thus the total number of arrangements when the longest and the shortest are together is $2 \times 7!$

$\therefore$ The number of arrangements when they are not together is clearly $8! - 2 \times 7! = 7!(8 - 2) = 6 \times 7!$

A combination is a selection of objects where the order is unimportant.

We have already seen that we can arrange r objects from n in $\dfrac{n!}{(n-r)!}$ different ways. Clearly for each of these arrangements we can arrange the r objects amongst themselves in $r!$ ways, and hence if the order is unimportant we can form $\dfrac{n!}{(n-r)!r!}$ different groups of the r objects.

We use the notation $^nC_r = \dfrac{n!}{(n-r)!r!}$.

This is sometimes written as $_nC_r$ or $\begin{pmatrix} n \\ r \end{pmatrix}$.

Example 18 In how many ways can a committee of 4 men and 3 ladies be formed from 10 men and 8 ladies.

Since the order is not important,
4 men can be chosen from 10 men in $^{10}C_4$ ways
3 ladies can be chosen from 8 ladies in $^8C_3$ ways

Since with each selection of men we could put each selection of the ladies, the total ways of forming the committee is obtained by

207

multiplying these results, i.e.

$$^{10}C_4 \times {}^8C_3 = \frac{10!}{6!4!} \times \frac{8!}{5!3!}$$

$$= \frac{10.9.8.7}{4.3.2.1} \times \frac{8.7.6}{3.2.1}$$

$$= 210 \times 56 = 11760$$

Two useful relationships should be noted: (i) $^nC_r = {}^nC_{n-r}$ since

$$^nC_r = \frac{n!}{(n-r)!r!} \quad \text{and} \quad {}^nC_{n-r} = \frac{n!}{[n-(n-r)]!(n-r)!} = \frac{n!}{r!(n-r)!}$$

(ii) $^nC_r + {}^nC_{r-1} = {}^{n+1}C_r$

Considering the LHS

$$^nC_r + {}^nC_{r-1} = \frac{n!}{(n-r)!r!} + \frac{n!}{(n-r+1)!(r-1)!}$$

$$= \frac{(n-r+1)n!}{(n-r+1)!r!} + \frac{n!r}{(n-r+1)!r!}$$

$$= \frac{n!}{(n-r+1)!r!}[n-r+1+r] = \frac{(n+1)!}{(n-r+1)!r!} = {}^{n+1}C_r$$

The Binomial theorem for a positive integral index

Consider the expansions $(1+x)^2 = 1 + 2x + x^2$

$$(1+x)^3 = (1+x)(1+x)^2 = 1 + 3x + 3x^2 + x^3$$
$$(1+x)^4 = (1+x)(1+x)^3 = 1 + 4x + 6x^2 + 4x^3 + x^4 \quad (1)$$

Now in the expansion of $(1+x)^4$ we have

$$(1+x)^4 = (1+x)(1+x)(1+x)(1+x)$$

Each term in the expansion (1) is obtained by selecting one element from each bracket so that the complete expansion consists of the sum of all such combinations.

i.e. the term $4x^3$ is obtained by selecting three x's from four which can be done in $^4C_3$ ways. Thus the coefficient is $\frac{4!}{3!1!} = 4$.

Extending this process suggests that

$$(1+x)^n = 1 + {}^nC_1 x + {}^nC_2 x^2 + \cdots {}^nC_r x^r + \cdots + x^n$$

We can prove this by induction.
Assume that the result is true for some value of $n = k$,
i.e. $(1+x)^k = 1 + {}^kC_1 x + {}^kC_2 x^2 + \cdots {}^kC_r x^r + \cdots + x^k$
Now $(1+x)^{k+1} = (1+x)(1+x)^k$

Multiplying and collecting like terms in powers of x,

$$(1+x)^{k+1} = 1 + ({}^kC_1 + 1)x + ({}^kC_2 + {}^kC_1)x^2 + \cdots$$
$$+ ({}^kC_r + {}^kC_{r-1})x^r + \cdots + x^{k+1}$$

We have shown on page 208 that ${}^kC_r + {}^kC_{r-1} = {}^{k+1}C_r$.

$$\therefore \quad (1+x)^{k+1} = 1 + {}^{k+1}C_1 x + {}^{k+1}C_2 x^2 + \cdots$$
$$+ {}^{k+1}C_r x^r + \cdots + x^{k+1}$$

This is the correct form of the result with $n = k + 1$. Hence, if it holds for $n = k$ it also holds for $n = k + 1$.

Now for $n = 1, 2$ it has been shown to be true by direct expansion and hence it is true for $n = 3$ and $n = 4$, and so on.

$\therefore$ The result is true for all positive integral values of n.

In general, the expansion of $(a + x)^n$ can be obtained by writing it in the form $a^n\left(1 + \dfrac{x}{a}\right)^n$.

$$\therefore \quad (a+x)^n = a^n\left\{ 1 + {}^nC_1\left(\frac{x}{a}\right) + {}^nC_2\left(\frac{x}{a}\right)^2 + {}^nC_3\left(\frac{x}{a}\right)^3 + \cdots \right.$$
$$\left. + {}^nC_r\left(\frac{x}{a}\right)^r + \cdots + \left(\frac{x}{a}\right)^n \right\}$$
$$= a^n + {}^nC_1 a^{n-1}x + {}^nC_2 a^{n-2}x^2 + \cdots$$
$$+ {}^nC_r a^{n-r}x^r + \cdots + x^n$$

Note that   (i) there are $n + 1$ terms in this finite expansion,

           (ii) the sum of the indices of a and x in each term is equal to n,

          (iii) the term in x^r is the $(r + 1)$th term in the expansion in ascending powers of x.

The coefficients in the binomial expansion using different values of n form a pattern and it is useful to remember this for fairly small values. They can be written in the form of a triangle and it is known as Pascal's triangle.

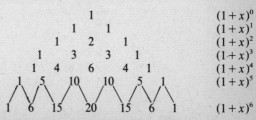

$$\begin{array}{ccccccccccccc}
 & & & & & & 1 & & & & & & & (1+x)^0 \\
 & & & & & 1 & & 1 & & & & & & (1+x)^1 \\
 & & & & 1 & & 2 & & 1 & & & & & (1+x)^2 \\
 & & & 1 & & 3 & & 3 & & 1 & & & & (1+x)^3 \\
 & & 1 & & 4 & & 6 & & 4 & & 1 & & & (1+x)^4 \\
 & 1 & & 5 & & 10 & & 10 & & 5 & & 1 & & (1+x)^5 \\
1 & & 6 & & 15 & & 20 & & 15 & & 6 & & 1 & (1+x)^6
\end{array}$$

Apart from the first and last terms which are always unity each term of a line is formed by adding the two terms on either side of it in the row above as shown.

i.e. $5 + 10 = 15$ and $10 + 10 = 20$.

Example 19 Expand $(1 + 2x)^7$ in ascending powers of x.

Making use of Pascal's triangle the coefficients required will be given by the next line in the above table,

i.e. 1 7 21 35 35 21 7 1

$$\therefore \quad (1 + 2x)^7 = 1 + 7(2x) + 21(2x)^2 + 35(2x)^3 + 35(2x)^4 + 21(2x)^5$$
$$+ 7(2x)^6 + (2x)^7$$
$$= 1 + 14x + 84x^2 + 280x^3 + 560x^4 + 672x^5$$
$$+ 448x^6 + 128x^7$$

Example 20 Find the term in x^6 in the expansion of $\left(x - \dfrac{2}{x}\right)^8$. It is not necessary to write out the whole series. The term in x^r is given by $^nC_r a^{n-r} x^r$ in the expansion of $(a + x)^n$.

In this case $n = 8$, $a = x$ and x is replaced by $\left(-\dfrac{2}{x}\right)$.

$\therefore$ The term in x^6 is formed when $n - r - r = 6$,

i.e. $8 - 2r = 6$ giving $r = 1$

$\therefore$ The term in x^6 is given by $^8C_1 x^7 \left(-\dfrac{2}{x}\right)^1 = -16x^6$

Example 21 Find the first four terms of the expansion of $(1 - x)^{12}$ in ascending powers of x and hence find the value of $(0.997)^{12}$ correct to three decimal places

Now $(1 - x)^{12} = 1 + {}^{12}C_1(-x) + {}^{12}C_2(-x)^2 + {}^{12}C_3(-x)^3 + \cdots$

$$= 1 - 12x + \frac{12.11}{2.1}x^2 - \frac{12.11.10}{3.2.1}x^3 + \cdots$$

$$= 1 - 12x + 66x^2 - 220x^3 + \cdots$$

Now if $x = 0.003$

$$(1 - x)^{12} = (1 - 0.003)^{12} = 0.997^{12}$$

$$= 1 - 12(0.003) + 66(0.003)^2 - 220(0.003)^3 + \cdots$$

$$= 1 - 0.036 + 0.000594 - 0.00000594$$

$$= 0.964588$$

$$= 0.965 \text{ to three decimal places}$$

The binomial theorem for any index

We have seen that if the index is a positive integer the series terminates after $n + 1$ terms. This is not the case if n is a positive or negative fraction or negative integer. In addition, the ^nC_r notation has no meaning and so the series is given by

$$(1 + x)^n = 1 + nx + \frac{n(n-1)}{2!}x^2 + \frac{n(n-1)(n-2)}{3!}x^3$$
$$+ \cdots \frac{n(n-1)(n-2)\cdots(n-r+1)}{r!}x^r + \cdots$$

This form of the series does not produce a sum to infinity for all values of x, only if $-1 < x < 1$.

Note that if $n = -1$ we produce a G.P. and its sum to infinity

$$1 - x + x^2 - x^3 + \cdots = \frac{1}{1+x}$$

Compare with the sum to infinity of a G.P. whose common ratio is $-x$.

If we replace x with $-x$,

$$1 + x + x^2 + x^3 + \cdots = \frac{1}{1-x}$$

This is a G.P. whose common ratio is x. Both of these series converge if $-1 < x < 1$ as we found in the case of a G.P. By putting $n = -2$ we produce an arithmetico–geometrical series

$$1 + 2x + 3x^2 + 4x^3 + \cdots = \frac{1}{(1-x)^2}$$

Example 22 Expand $(1 - 2x)^{-1/2}$ in ascending powers of x as far as the term in x^3.

By using the binomial theorem

$$(1 - 2x)^{-1/2} = 1 + (-\tfrac{1}{2})(-2x) + \frac{(-\tfrac{1}{2})(-\tfrac{3}{2})}{2!}(-2x)^2$$
$$+ \frac{(-\tfrac{1}{2})(-\tfrac{3}{2})(-\tfrac{5}{2})}{3!}(-2x)^3 + \cdots$$
$$= 1 + x + \tfrac{3}{2}x^2 + \tfrac{5}{2}x^3 + \cdots$$

Example 23 Show that $\sqrt{\dfrac{1-x}{1+x}} = 1 - x + \dfrac{x^2}{2}$ if x is sufficiently small so that the cube and higher powers may be neglected.

Now $\sqrt{\dfrac{1-x}{1+x}} = (1-x)^{1/2}(1+x)^{-1/2} = (1-x)(1-x^2)^{-1/2}$

$\therefore \quad (1-x)(1-x^2)^{-1/2} = (1-x)\left[1 + \frac{1}{2}x^2 + \frac{(-\frac{1}{2})(-\frac{3}{2})}{2!}(-x^2)^2 + \cdots\right]$

$$= (1-x)(1 + \frac{1}{2}x^2 + \frac{3}{8}x^4 + \cdots)$$

If x^3 and higher powers are neglected

$$\sqrt{\frac{1-x}{1+x}} = 1 + \frac{1}{2}x^2 - x = 1 - x + \frac{1}{2}x^2$$

Example 24 Find the sum to infinity of the series, stating the range of validity of x.

$$1 - x + \frac{1.3}{1.2}x^2 - \frac{1.3.5}{1.2.3}x^3 + \frac{1.3.5.7}{1.2.3.4}x^4 - \cdots$$

By rearranging the coefficients in each term the series can be written

$$1 + (-\tfrac{1}{2})(2x) + \frac{(-\frac{1}{2})(-\frac{3}{2})}{2!}(2x^2) + \frac{(-\frac{1}{2})(-\frac{3}{2})(-\frac{5}{2})}{3!}(2x^3)$$
$$+ \frac{(-\frac{1}{2})(-\frac{3}{2})(-\frac{5}{2})(-\frac{7}{2})}{4!}(2x^4)$$

which is clearly the expansion of $(1+2x)^{-1/2}$.

Hence the sum to infinity is given by $(1+2x)^{-1/2}$.
The series will possess this sum if $-1 < 2x < 1$, i.e. $|x| < \frac{1}{2}$.

Example 25 Expand the function $1/(1+x+x^2)$ as far as the term in x^3.

Consider $x + x^2 = u$

$\therefore \quad (1+x+x^2)^{-1} = (1+u)^{-1}$

$$= 1 - u + \frac{(-1)(-2)}{2!}u^2 + \frac{(-1)(-2)(-3)}{3!}u^3 + \cdots$$

$$= 1 - u + u^2 - u^3 + \cdots$$

$\therefore \quad (1+x+x^2)^{-1} = 1 - (x+x^2) + (x+x^2)^2 - (x+x^2)^3 + \cdots$

$$= 1 - x - x^2 + x^2 + 2x^3 - x^3$$

including terms as far as x^3

$\therefore \quad (1+x+x^2)^{-1} = 1 - x + x^3 + \cdots$

Infinite series and convergence

We have already considered two infinite series in the G.P. and the binomial expansion for a rational or negative index. These, we noted, converged for a fixed range of values of x only. Generally we say that if the first n terms of an infinite series $\sum_{r=1}^{\infty} u_r$ tend to a limit as n tends to infinity then the series is **convergent**.

Consider the series $\dfrac{1}{1.2.3} + \dfrac{1}{2.3.4} + \dfrac{1}{3.4.5} + \cdots$

We can find the sum of the first n terms by the difference method, which gives

$$S_n = \sum_{r=1}^{n} \frac{1}{r(r+1)(r+2)} = \frac{1}{4} - \frac{1}{2(n+1)(n+2)}$$

As $n \to \infty$, $S_n \to \frac{1}{4}$ and the series is convergent.
Conversely, if $S_n = u_1 + u_2 + u_3 + \cdots + u_n$ and S_n does not tend to a limit as $n \to \infty$ the series is said to be divergent.

Tests for convergence

There are many methods by which the convergence or divergence of a series may be discovered. Some are applicable only to special types of series, e.g., where all the terms are positive.

A mention is included here of the more common techniques and results.

1. If $u_1 + u_2 + u_3 + \cdots$ is convergent and possesses a sum to infinity S, then $u_{m+1} + u_{m+2} + \cdots$ is convergent with a sum to infinity of $S - (u_1 + u_2 + \cdots + u_m)$.

$$\lim_{n \to \infty} (u_{m+1} + u_{m+2} + \cdots + u_{m+n})$$

$$= \lim_{n \to \infty} [(u_1 + u_2 + \cdots u_{m+n}) - (u_1 + u_2 + \cdots + u_m)]$$

$$= S - (u_1 + u_2 + \cdots + u_m)$$

Hence a finite number of terms at the beginning of a series does not affect its convergence.

2. If $u_1 + u_2 + u_3 + \cdots$ is convergent with a sum to infinity S then $ku_1 + ku_2 + ku_3 + \cdots$ where k is any constant, is also convergent with a sum to infinity kS.

Clearly $\lim_{n \to \infty} (ku_1 + ku_2 + ku_3 + \cdots + ku_n)$

$$= k \lim_{n \to \infty} (u_1 + u_2 + \cdots + u_n) = kS$$

3. If two series $u_1 + u_2 + u_3 + \cdots$ and $v_1 + v_2 + v_3 + \cdots$ are both convergent with sums to infinity S_1 and S_2 respectively then the sum of any multiples of these series, say $(\lambda u_1 + \mu v_1) + (\lambda u_2 + \mu v_2) + \cdots$, also converges to a sum $\lambda S_1 + \mu S_2$.

$$\lim_{n \to \infty} [(\lambda u_1 + \mu v_1) + (\lambda u_2 + \mu v_2) + \cdots (\lambda u_n + \mu v_n)]$$

$$= \lambda \lim_{n \to \infty} (u_1 + u_2 + \cdots u_n) + \mu \lim_{n \to \infty} (v_1 + v_2 + v_3 + \cdots v_n)$$

$$= \lambda S_1 + \mu S_2$$

4. If $u_1 + u_2 + u_3 + \cdots$ is convergent, $\lim\limits_{n \to \infty} u_n = 0$.

$$\lim_{n \to \infty} u_n = \lim_{n \to \infty} [(u_1 + u_2 + u_3 + \cdots + u_n)$$
$$- (u_1 + u_2 + u_3 + \cdots + u_{n-1})]$$
$$= S - S = 0$$

The converse is not true, i.e. if $\lim\limits_{n \to \infty} u_n = 0$ it does not necessarily mean that the series converges. We say that this is a necessary but not sufficient condition for convergence.
The usual example to illustrate this is

$$1 + \frac{1}{2} + \frac{1}{3} + \frac{1}{4} + \cdots \frac{1}{n} + \cdots$$

By grouping $\frac{1}{3} + \frac{1}{4} > \frac{1}{2}$

$$\frac{1}{5} + \frac{1}{6} + \frac{1}{7} + \frac{1}{8} > \frac{4}{8} = \frac{1}{2}$$

$$\frac{1}{9} + \frac{1}{10} + \frac{1}{11} + \frac{1}{12} + \frac{1}{13} + \frac{1}{14} + \frac{1}{15} + \frac{1}{16} > \frac{8}{16} = \frac{1}{2}$$

Hence $S_4 > 1 + \frac{1}{2} + \frac{1}{2}$, $S_8 > 1 + \frac{1}{2} + \frac{1}{2} + \frac{1}{2}$, $S_{16} > 1 + \frac{1}{2} + \frac{1}{2} + \frac{1}{2} + \frac{1}{2}$.

$\therefore \ S_{2n} > 1 + \frac{1}{2}n$ and so $S_n \to \infty$ as $n \to \infty$

$\therefore \ 1 + \frac{1}{2} + \frac{1}{3} + \frac{1}{4} + \frac{1}{5} + \cdots$ is divergent although $\lim\limits_{n \to \infty} \left(\frac{1}{n}\right) = 0$

5. **Comparison test** If $u_1 + u_2 + u_3 + \cdots$ and $v_1 + v_2 + v_3 + \cdots$ are two series of positive terms then if $\Sigma\, v_n$ converges, $\Sigma\, u_n$ will converge if either

(i) $u_n \leqslant kv_n$ for all $n \geqslant$ some value N, or

(ii) $\lim\limits_{n \to \infty} \dfrac{u_n}{v_n} = S$ where k and S are positive constants.

(i) If V is the sum to infinity of $v_1 + v_2 + v_3 + \cdots$ then $v_1 + v_2 + v_3 + \cdots + v_n < V$ for all n

$$\therefore \quad u_{p+1} + u_{p+2} + \cdots + u_n \leqslant k(v_{p+1} + v_{p+2} + \cdots + v_n) < kV \quad \text{for}$$
all $n > p$.

But since p is fixed $u_1 + u_2 + u_3 + \cdots$ is convergent.

(ii) Since $\lim\limits_{n \to \infty} \left(\dfrac{u_n}{v_n} \right) = S$ then $\left| \dfrac{u_n}{v_n} - S \right| < \epsilon$ for $n \geqslant N$ where ϵ is some positive quantity. If $0 < \epsilon < S$ then

$$0 < S - \epsilon < \frac{u_n}{v_n} < S + \epsilon$$

i.e. $0 < (S - \epsilon)v_n < u_n < (S + \epsilon)v_n$

$\therefore \quad u_n < (S + \epsilon)v_n$ for all $n \geqslant N$, and from part (i) it follows that if $\Sigma \, v_r$ is convergent so is $\Sigma \, u_r$.

Also since $u_n > (S - \epsilon)v_n$ for all $n \geqslant N$ it follows that if $\Sigma \, v_r$ is divergent so is $\Sigma \, u_r$.

Example 26 Does the series $\displaystyle\sum_{r=1}^{\infty} u_r$ where $u_r = \dfrac{4r+1}{3r^2-1}$ converge or diverge?

$\dfrac{4r+1}{3r^2-1} > \dfrac{4r}{3r^2} = \dfrac{4}{3} \cdot \dfrac{1}{r}$ and $\displaystyle\sum \dfrac{1}{r}$ is divergent.

Hence by comparison $\displaystyle\sum_{r=1}^{\infty} \dfrac{4r+1}{3r^2-1}$ diverges.

Note that in example 26 we have used the corresponding result for divergence, namely, that a series of positive terms diverges if its terms are greater than the corresponding terms of a divergent series.

The G.P. and the harmonic series $1 + \frac{1}{2} + \frac{1}{3} + \frac{1}{4} + \frac{1}{5} + \cdots$ are useful series for comparison. A more general form of the second one is also useful i.e. the series $\displaystyle\sum_{r=1}^{\infty} \left(\dfrac{1}{r^p} \right)$.

Consider the series

$$\frac{1}{1^p} + \frac{1}{2^p} + \frac{1}{3^p} + \frac{1}{4^p} + \cdots + \frac{1}{n^p} + \cdots$$

If $p > 1$, let $S_n = \dfrac{1}{1^p} + \dfrac{1}{2^p} + \dfrac{1}{3^p} + \dfrac{1}{4^p} + \cdots + \dfrac{1}{n^p} + \cdots$

Now if we group terms as follows,

$$S_n = \left(\frac{1}{1^p} \right) + \left(\frac{1}{2^p} + \frac{1}{3^p} \right) + \left(\frac{1}{4^p} + \cdots + \frac{1}{7^p} \right) + \cdots \text{ to } n \text{ brackets}$$

215

$$\therefore \quad S_n < \frac{1}{1^p} + \frac{2}{2^p} + \frac{4}{4^p} + \cdots \text{ to } n \text{ terms}$$

$$\therefore \quad S_n < 1 + \frac{1}{2^{p-1}} + \frac{1}{4^{p-1}} + \cdots \text{ to } n \text{ terms}$$

This is a G.P. with common ratio $(\frac{1}{2})^{p-1}$. Since $(\frac{1}{2})^{p-1} < 1$ if $p > 1$ it possesses a sum to infinity

$$S_n < \frac{1}{1 - (\frac{1}{2})^{p-1}} \text{ for all } n.$$

Hence if $p > 1$, the series converges.

If $p = 1$ the series becomes $1 + \frac{1}{2} + \frac{1}{3} + \frac{1}{4} + \frac{1}{5}$ which has already been shown to be divergent.

If $p < 1$, $1/r^p > 1/r$ for $r > 1$.

Hence each term of $\Sigma (1/r^p)$ after the first is greater than the corresponding term of the series $\Sigma \dfrac{1}{r}$ which is divergent.

$\therefore$ if $p < 1$ the series diverges.

Hence $\dfrac{1}{1^p} + \dfrac{1}{2^p} + \dfrac{1}{3^p} + \cdots \dfrac{1}{n^p} + \cdots$ converges if $p > 1$ but diverges if $p \le 1$.

6. **The ratio test** If Σu_r is a series of positive terms and $\lim\limits_{n \to \infty} \left(\dfrac{u_n}{u_{n+1}} \right) = l$, then the series converges if $l > 1$ and diverges if $l < 1$.

Now since $\lim\limits_{n \to \infty} \left(\dfrac{u_n}{u_{n+1}} \right) = l > 1$, $\dfrac{u_n}{u_{n+1}}$ can be made as near l as we please for large values of n. Therefore if c is a value such that $l > c > 1$ then $\dfrac{u_n}{u_{n+1}} > c > 1$ for all sufficiently large values of n, say $\ge m$.

$\therefore \quad u_m > cu_{m+1}, \qquad u_{m+1} > cu_{m+2},$ and so on

$\therefore \quad u_{m+1} < \dfrac{u_m}{c}, \qquad u_{m+2} < \dfrac{u_{m+1}}{c} < \dfrac{u_m}{c^2} \cdots$

Hence the series $u_{m+1} + u_{m+2} + u_{m+3} + \cdots$ has each term less than the corresponding term of the G.P. $\dfrac{u_m}{c} + \dfrac{u_m}{c^2} + \dfrac{u_m}{c^3} + \cdots$

$\therefore$ Since m is fixed, Σu_r is convergent, since the G.P. is convergent.

216

A similar argument holds for the divergent property.

Example 27 Show by the ratio test that the series $\sum \dfrac{1}{n(n+2)3^n}$ converges.

Now $\lim\limits_{n \to \infty} \left(\dfrac{u_n}{u_{n+1}}\right) = \lim\limits_{n \to \infty} \left[\dfrac{(n+1)(n+3)3^{n+1}}{n(n+2)3^n}\right]$

$$= \lim\limits_{n \to \infty} \left[3\left(1 + \frac{1}{n}\right)\left(1 + \frac{1}{n+2}\right)\right] = 3$$

Hence the series converges.

Note that if $l = 1$ no definite conclusion can be reached about the convergence of the series.

Rolle's theorem If $f(x)$ is a single valued continuous function for $a \leqslant x \leqslant b$ and $f'(x)$ exists for $a < x < b$ and if $f(a) = f(b)$, then $f'(x)$ vanishes for at least one value of x between a and b.
If $f(a) = f(b) = k$, consider $g(x) = f(x) - k$.
Clearly $f'(x) = g'(x)$ and $g(a) = g(b) = 0$.
Assuming $g(x) \neq 0$ it must have a greatest value U and a least value L where $U \neq L$. Let $U = g(c)$ where $a < c < b$. Assume that $g(x)$ is sometimes positive in the range. If $g'(c) > 0$ we can take x sufficiently near c ($x > c$) such that $g(x) > g(c)$. But this is impossible since $g(c) = U$ (the upper bound).
Similarly if $g'(c) < 0$ we can take a value of x near c ($x > c$) such that $g(x) > g(c)$ which is again impossible. But $f'(c)$ exists $\Rightarrow g'(c)$ exists but $g'(c)$ is not positive or negative.

$\therefore \quad g'(c) = 0$, i.e. $f'(c) = 0$ for $a < c < b$.

A similar argument holds if $g(x)$ is negative in the range.

First mean value theorem
If $f(x)$ is continuous in the interval $a \leqslant x \leqslant b$ and $f'(x)$ exists in the interval $a < x < b$, there is a value c of x in $a < x < b$ such that $f(b) - f(a) = (b - a)f'(c)$.

Clearly the function $f(b) - f(x) - \left(\dfrac{b-x}{b-a}\right)\{f(b) - f(a)\}$ vanishes when $x = a$ and $x = b$ and hence by Rolle's theorem there exists a value c of x in $a < x < b$ for which its derivative vanishes.

Hence $f'(c) + \dfrac{1}{b-a}\{f(b) - f(a)\} = 0$, i.e.

$$f(b) - f(a) = (b - a)f'(c)$$

An alternative form is obtained by writing $b = a + h$ and $c = a + \theta h$ where $0 < \theta < 1$. Note that this gives c between a and $a + h$.

$$\therefore \quad f(a + h) = f(a) + hf'(a + \theta h)$$

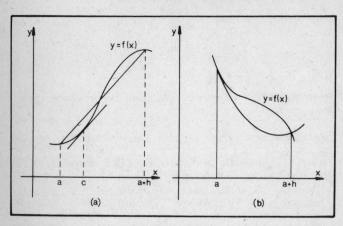

Figure 96

This means that we have approximated $f(x)$ by a linear function (see fig. 96) but it is possible to improve this approximation by using polynomials of higher degree.

By using a quadratic function (fig. 96) we can show that

$$f(a + h) = f(a) + hf'(a) + \tfrac{1}{2}h^2 f''(a + \theta h)$$

Taylor's theorem

A general result known as **Taylor's theorem** can be formed by extending the above process.

If $f(x)$ and its first $(n - 1)$ derivatives are continuous in $a \le x \le b$ and its nth derivative exists in $a < x < b$ then

$$f(a + h) = f(a) + hf'(a) + \frac{h^2}{2!} f''(a) + \cdots + \frac{h^{n-1}}{(n-1)!} f^{(n-1)}(a) + R_n$$

$$\text{where } R_n = \frac{h^n}{n!} f^{(n)}(a + \theta h), \ 0 < \theta < 1$$

The last term is usually called the remainder R_n and this form is known as the **Lagrange Remainder.** We have expanded $f(a + h)$ as a polynomial in h of degree n. To find R_n precisely,

218

a value for θ has to be found, which is not possible in general. We can, however, find an approximate value of the expansion by neglecting R_n when it tends to zero, giving a **Taylor Series,**

i.e. $f(a + h) = f(a) + hf'(a) + \dfrac{h^2}{2!} f''(a) + \cdots$

As a particular case we take $h = x$ and $a = 0$ which gives the **Maclaurin series**

i.e. $f(x) = f(0) + xf'(0) + \dfrac{x^2}{2!} f''(0) + \cdots \dfrac{x^n}{n!} f^{(n)}(0) + \cdots$

Series for sin x

$\quad f(x) = \sin x \Rightarrow f^{(n)}(x) = \sin (x + \tfrac{1}{2}n\pi)$ (Leibnitz's theorem)

$\qquad\qquad f^{(n)}(0) = \sin \tfrac{1}{2}n\pi$

Hence $f^{(n)}(0)$ takes successively values of $0, 1, 0, -1, \ldots$ and the series is given by

$$\sin x = x - \frac{x^3}{3!} + \frac{x^5}{5!} - \frac{x^7}{7!} + \cdots + \frac{(-1)^r x^{2r+1}}{(2r+1)!} + \cdots$$

Note that for this series, since $f^{(n)}x = \sin (x + \tfrac{1}{2}n\pi)$ the Lagrange form of the remainder $R_n = \sin (\theta x + \tfrac{1}{2}n\pi) \dfrac{x^n}{n!} (a = 0)$

Now since $|\sin (\theta x + \tfrac{1}{2}n\pi)| \leqslant 1$ and for all large n, $\left(\dfrac{x^n}{n!}\right) \to 0$, $R_n \to 0$.

Series for cos x

Now $f(x) = \cos x \Rightarrow f'(x) = \cos (x + \tfrac{1}{2}\pi)$

$\therefore \qquad\qquad f^{(n)}(x) = \cos (x + \tfrac{1}{2}n\pi)$

$\therefore \qquad\qquad f^n(0) = \cos \tfrac{1}{2}n\pi$

Hence $f^{(n)}(0)$ takes successively values of $1, 0, -1, 0, \ldots$ and the series is given by

$$\cos x = 1 - \frac{x^2}{2!} + \frac{x^4}{4!} - \frac{x^6}{6!} + \cdots + \frac{(-1)^r x^{2r}}{(2r)!} + \cdots$$

Both of the series above for $\sin x$ and $\cos x$ converge for all values of x and hence the expansions are valid for all the values of x.

The exponential series

If $f(x) = e^x$, then $f^{(n)}(x) = e^x$.

Thus $f^{(n)}(0) = 1$.

Hence the series can be written using Maclaurin's theorem

$$e^x = 1 + x + \frac{x^2}{2!} + \frac{x^3}{3!} + \frac{x^4}{4!} + \cdots + \frac{x^r}{r!} + \cdots$$

This series is convergent for all values of x and is sometimes written $\exp(x)$.

The ratio test gives $\lim_{n \to \infty} \left| \frac{u_n}{u_{n+1}} \right| = \frac{n}{|x|} > 1$ for any finite value of x

and hence the series is convergent.

This can be derived by assuming that the function can be expanded as a power series and that the series can be differentiated term by term. This holds for a convergent series.

Let $e^x = a_0 + a_1 x + a_2 x^2 + a_3 x^3 + \cdots + a_n x^n + \cdots$

Differentiating successively

$$e^x = a_1 + 2a_2 x + 3a_3 x^2 + \cdots + na_n x^{n-1} + \cdots$$

$$e^x = \quad\quad 2a_2 + 3.2a_3 x + \cdots + n(n-1)a_n x^{n-2} + \cdots$$

$$e^x = \quad\quad\quad\quad 3.2.1a_3 + \cdots + n(n-1)(n-2)a_n x^{n-3} + \cdots$$

and after differentiating n times

$$e^x = n!a_n + \cdots$$

When $x = 0$ these results yield $a_0 = 1$, $a_1 = 1$, $a_2 = \frac{1}{2}$, $a_3 = 1/(3.2)$
and in general $a_n = \frac{1}{n!}$.

Using the original power series and substituting for a_r

$$e^x = 1 + x + \frac{x^2}{2!} + \frac{x^3}{3!} + \frac{x^4}{4!} + \cdots + \frac{x^r}{r!} + \cdots$$

By putting $x = 1$ we can find the value of e.

i.e. $e = 1 + 1 + \frac{1}{2!} + \frac{1}{3!} + \frac{1}{4!} + \frac{1}{5!} + \frac{1}{6!} + \frac{1}{7!} + \cdots$

$\quad\quad = 2 + 0 \cdot 5 + 0 \cdot 16667 + 0 \cdot 04167 + 0 \cdot 00833 + 0 \cdot 00139$

$\quad\quad\quad + 0 \cdot 00020 + 0 \cdot 00002 + 0 \cdot 00000$

$\therefore \quad e \simeq 2 \cdot 71828$

Example 28 Find the first three terms of the expansion $(1+2x)e^{-2x}$.

$$(1+2x)e^{-2x} = (1+2x)\left\{1 + (-2x) + \frac{(-2x)^2}{2!} + \frac{(-2x)^3}{3!} + \frac{(-2x)^4}{4!} + \cdots\right\}$$

$$= (1+2x)\left(1 - 2x + 2x^2 - \frac{4}{3}x^3 + \frac{2}{3}x^4 - \cdots\right)$$

Multiplying out and retaining only terms up to and including x^3

$$(1+2x)e^{-2x} = 1 - 2x + 2x^2 - \frac{4}{3}x^3 + 2x - 4x^2 + 4x^3 + \cdots$$

$$= 1 - 2x^2 + \frac{8}{3}x^3 + \cdots$$

The logarithmic series

If $f(x) = \ln(1+x)$, $f'(x) = \dfrac{1}{1+x} = (1+x)^{-1}$

Hence $f''(x) = -(1+x)^{-2}$, $f'''(x) = (-1)(-2)(1+x)^{-3}$

and
$$f^{(n)}(x) = \frac{(-1)^{n-1}(n-1)!}{(1+x)^n}$$

When $x = 0$, $f(0) = 0$, $f'(0) = 1$, $f''(0) = -1$, $f'''(0) = 2$ and in general $f^{(n)}(0) = (-1)^{n-1}(n-1)!$

Using Maclaurin's theorem

$$\boldsymbol{\ln(1+x) = x - \frac{x^2}{2} + \frac{x^3}{3} - \frac{x^4}{4} + \cdots + \frac{(-1)^{r-1}x^r}{r} + \cdots}$$

Considering $\displaystyle\lim_{n \to \infty}\left|\frac{u_n}{u_{n+1}}\right| = \lim_{n \to \infty}\left(\frac{n+1}{n}\right)\frac{1}{|x|} = \lim_{n \to \infty}\left(1 + \frac{1}{n}\right)\frac{1}{|x|} = \frac{1}{|x|}$

Hence the series converges for $|x| < 1$ but the special case when $x = 1$ also gives a convergent series $1 - \frac{1}{2} + \frac{1}{3} - \frac{1}{4} + \cdots$.

$\therefore$ The range of convergence is $-1 < x \leqslant 1$.

Note that we have given a series for $\ln(1+x)$ and not $\ln x$. This is because $\ln x$ and its derivatives do not exist at $x = 0$, and hence $\ln x$ cannot be represented as a power series.

By replacing x with $-x$ we can form the alternative series

$$\boldsymbol{\ln(1-x) = -x - \frac{x^2}{2} - \frac{x^3}{3} - \frac{x^4}{4} - \cdots - \frac{x^r}{r} - \cdots}$$

provided $-1 \leqslant x < 1$.

By combining these two results we form yet another two logarithmic expansions.

$$\ln(1+x) + \ln(1-x) = \ln(1-x^2)$$

$$\ln(1-x^2) = 2\left\{-\frac{x^2}{2} - \frac{x^4}{4} - \frac{x^6}{6} - \cdots - \frac{x^{2r}}{2r} - \cdots\right\}$$

Also

$$\ln\left(\frac{1+x}{1-x}\right) = 2\left\{x + \frac{x^3}{3} + \frac{x^5}{5} + \cdots + \frac{x^{2r-1}}{2r-1} + \cdots\right\}$$

These expansions are valid if $-1 \leqslant x < 1$ and $-1 < x \leqslant 1$ and so for both to be valid $-1 < x < 1$.

Example 29 Expand $\ln(2-5x)$ as far as the term in x^3 and give the general term.

$$\ln(2-5x) = \ln 2\left(1 - \frac{5}{2}x\right) = \ln 2 + \ln\left(1 - \frac{5}{2}x\right)$$

$$\ln(2-5x) = \ln 2 + \left[\left(-\frac{5}{2}x\right) - \frac{1}{2}\left(-\frac{5}{2}x\right)^2 + \frac{1}{3}\left(-\frac{5}{2}x\right)^3 + \cdots\right.$$
$$\left. + \frac{1}{r}\left(-\frac{5}{2}x\right)^r + \cdots\right]$$

$$= \ln 2 - \frac{5}{2}x - \frac{25}{8}x^2 - \frac{125}{24}x^3 + \cdots - \frac{5^r}{2^r \cdot r}x^r - \cdots$$

This expansion is valid when $-1 \leqslant \frac{5}{2}x < 1$, i.e. $-\frac{2}{5} \leqslant x < \frac{2}{5}$

Series for hyperbolic functions (defined in chapter 5)

Since $\cosh x = \frac{1}{2}(e^x + e^{-x})$ and $\sinh x = \frac{1}{2}(e^x - e^{-x})$ we can form the series from the exponential series.

$$e^x = 1 + x + \frac{x^2}{2!} + \frac{x^3}{3!} + \cdots + \frac{x^r}{r!} + \cdots$$

$$e^{-x} = 1 - x + \frac{x^2}{2!} - \frac{x^3}{3!} + \cdots + \frac{(-1)^r x^r}{r!} + \cdots$$

$$\therefore \quad \cosh x = \frac{1}{2}\left(2 + \frac{2x^2}{2!} + \frac{2x^4}{4!} + \cdots + \frac{2x^{2r}}{(2r)!} + \cdots\right)$$

$$\cosh x = 1 + \frac{x^2}{2!} + \frac{x^4}{4!} + \cdots + \frac{x^{2r}}{(2r)!} + \cdots$$

Similarly, $\mathbf{sinh}\ x = x + \dfrac{x^3}{3!} + \dfrac{x^5}{5!} + \cdots + \dfrac{x^{2r-1}}{(2r-1)!} + \cdots$

These series are valid for all values of x.

Since $y = \tanh^{-1} x$, $x = \tanh y = \dfrac{e^y - e^{-y}}{e^y + e^{-y}} = \dfrac{e^{2y} - 1}{e^{2y} + 1}$ see page 125.

$\therefore\ e^{2y} = \dfrac{1+x}{1-x}$ and thus $y = \tanh^{-1} x = \dfrac{1}{2} \ln\left(\dfrac{1+x}{1-x}\right)$

$\therefore$ the series for $\tanh^{-1} x$ is given by

$$\tanh^{-1} x = \frac{1}{2} \ln\left(\frac{1+x}{1-x}\right) = x + \frac{x^3}{3} + \frac{x^5}{5} + \cdots + \frac{x^{2r-1}}{(2r-1)} + \cdots$$

Series for $\tan^{-1} x$

We can obtain the series for this function by integration. If an expansion of $f(x)$ is required and the expansion for $f'(x)$ is obtainable then we can obtain the expansion of $f(x)$ by integration.

Considered graphically this is equivalent to saying that if two curves approximate over a range of values of x then the areas under these curves will be approximately equal over that range.

$$\frac{1}{1+x^2} = (1+x^2)^{-1} = 1 - x^2 + x^4 - x^6 + \cdots$$

Hence $\displaystyle\int \frac{dx}{1+x^2} = \int 1\,dx - \int x^2\,dx + \int x^4\,dx - \cdots$

$$\tan^{-1} x = x - \frac{x^3}{3} + \frac{x^5}{5} - \cdots$$

We can obtain the expansion of $\ln(1+x)$ in this way.

$$\left[\ln(1+u)\right]_0^x = \int_0^x \frac{1}{1+u}\,du$$

$$= \int_0^x (1 - u + u^2 - u^3 + \cdots)\,du \text{ for } -1 < u < 1$$

$$= \left[u - \tfrac{1}{2}u^2 + \tfrac{1}{3}u^3 - \tfrac{1}{4}u^4 + \cdots\right]_0^x$$

$$\ln(1+x) = x - \frac{x^2}{2} + \frac{x^3}{3} - \frac{x^4}{4} + \cdots$$

The expansion for $\tan^{-1} x$ is known as **Gregory's series** and can be used to find a value for π.

Let $x = 1 \Rightarrow \tan^{-1}(1) = \dfrac{\pi}{4} = 1 - \dfrac{1}{3} + \dfrac{1}{5} - \dfrac{1}{7} + \dfrac{1}{9} - \cdots$

It is clear that this series converges very slowly and a better relation is given by $\tan^{-1}(\frac{1}{2}) + \tan^{-1}(\frac{1}{3}) = \dfrac{\pi}{4}$ or an even better one is $\dfrac{\pi}{4} = 4 \tan^{-1}\left(\dfrac{1}{5}\right) - \tan^{-1}\left(\dfrac{1}{239}\right)$ which converges far more rapidly.

Example 30 Use the series for $\cos x$ and $\sin x$ to deduce the expansion of $\tan x$ as far as the term in x^5.

Since $\tan x$ is an odd function we can assume that the power series will be of the form $a_1 x + a_3 x^3 + a_5 x^5 + \cdots$

Hence as $\tan x = \dfrac{\sin x}{\cos x} = a_1 x + a_3 x^3 + a_5 x^5 + \cdots$

$$x - \dfrac{x^3}{3!} + \dfrac{x^5}{5!} - \cdots = \left(1 - \dfrac{x^2}{2!} + \dfrac{x^4}{4!} - \cdots\right)(a_1 x + a_3 x^3 + a_5 x^5 + \cdots)$$

$$= a_1 x + \left(a_3 - \dfrac{a_1}{2!}\right)x^3 + \left(a_5 - \dfrac{a_3}{2!} + \dfrac{a_1}{4!}\right)x^5 + \cdots$$

Equating coefficients

$$a_1 = 1, \qquad a_3 - \dfrac{a_1}{2} = \dfrac{-1}{3!} \Rightarrow a_3 = \dfrac{1}{3}$$

$$a_5 - \dfrac{a_3}{2} + \dfrac{a_1}{24} = \dfrac{1}{120} \Rightarrow a_5 = \dfrac{2}{15}$$

$$\tan x = x + \dfrac{1}{3}x^3 + \dfrac{2}{15}x^5 + \cdots$$

Key terms

The **arithmetical progression** (A.P.)

$$a + (a + d) + (a + 2d) + \cdots + [a + (n - 1)d] + \cdots$$

The sum to n terms is $\dfrac{n}{2}[2a + (n - 1)d]$

The **geometrical progression** (G.P.)

$$a + ar + ar^2 + ar^3 + \cdots + ar^{n-1} + \cdots$$

The sum to n terms is $\dfrac{a(1-r^n)}{1-r}$ and the sum to infinity is $\dfrac{a}{1-r}$.

The series of natural numbers

(i) $\displaystyle\sum_{r=1}^{n} r = \tfrac{1}{2}n(n+1)$ (ii) $\displaystyle\sum_{r=1}^{n} r^2 = \tfrac{1}{6}n(n+1)(2n+1)$

(iii) $\displaystyle\sum_{r=1}^{n} r^3 = \tfrac{1}{4}n^2(n+1)^2$

A **permutation** is an arrangement of items in a particular order. We choose r items from n different items in

$$^nP_r = \frac{n!}{(n-r)!} \text{ ways.}$$

A **combination** is a selection of objects where the order is not important.

We can form groups of r objects from n in $^nC_r = \dfrac{n!}{(n-r)!\,r!}$

ways.

The binomial theorem for any rational index gives

$$(1+x)^n = 1 + nx + \frac{n(n-1)}{2!}x^2 + \frac{n(n-1)(n-2)}{3!}x^3 + \cdots$$

If n is a positive integer the series terminates after $(n+1)$ terms.

The Taylor series

$$f(a+h) = f(a) + hf'(a) + \frac{h^2}{2!}f''(a) + \cdots + \frac{h^n}{n!}f^{(n)}(a) + \cdots$$

A special case is the **Maclaurin series**

$$f(x) = f(0) + xf'(0) + \frac{x^2}{2!}f''(0) + \cdots + \frac{x^n}{n!}f^{(n)}(0) + \cdots$$

$$\sin x = x - \frac{x^3}{3!} + \frac{x^5}{5!} - \frac{x^7}{7!} + \cdots + (-1)^r \frac{x^{2r+1}}{(2r+1)!} + \cdots$$

$$\cos x = 1 - \frac{x^2}{2!} + \frac{x^4}{4!} - \frac{x^6}{6!} + \cdots + (-1)^r \frac{x^{2r}}{(2r)!} + \cdots$$

$$e^x = 1 + x + \frac{x^2}{2!} + \frac{x^3}{3!} + \frac{x^4}{4!} + \cdots + \frac{x^r}{r!} + \cdots$$

$$\ln(1+x) = x - \frac{x^2}{2} + \frac{x^3}{3} - \frac{x^4}{4} + \frac{x^5}{5} - \cdots + (-1)^{r-1}\frac{x^r}{r} + \cdots$$

Chapter 9
Complex Numbers

In the algebraic work on polynomials we have seen methods by which they can be factorized. Clearly, by inspection, the function $P(x) = x^2 - x - 6$ can be written as $(x-3)(x+2)$. For a cubic or polynomial of higher degree it may be necessary to use the factor theorem (see page 12). If $F(x) = x^3 - 3x^2 - 6x + 8$ we see that $F(1) = 0$ and hence $x - 1$ is a factor. The second factor can be obtained by division and we find

$$F(x) = (x-1)(x^2 - 2x - 8)$$

The quadratic factor can also be factorized, so that $F(x)$ can be completely factorized into three linear factors.

i.e. $F(x) = x^3 - 3x^2 - 6x + 8 = (x-1)(x+2)(x-4)$

It is not always possible to go as far as this and, for example,

$$g(x) = x^3 + 4x^2 + 4x + 3 = (x+3)(x^2 + x + 1)$$

The quadratic factor $x^2 + x + 1$ cannot be factorized further over the real numbers.

If we consider the equations $P(x) = 0$, $F(x) = 0$ and $g(x) = 0$,

$P(x) = x^2 - x - 6 = (x-3)(x+2) = 0 \Rightarrow x = 3$ or -2

$F(x) = x^3 - 3x^2 - 6x + 8$

$\qquad = (x-1)(x+2)(x-4) = 0 \Rightarrow x = 1, -2$ or 4

$g(x) = x^3 + 4x^2 + 4x + 3 = (x+3)(x^2 + x + 1) = 0 \Rightarrow x = -3$

only over the real field.

These results can be illustrated graphically for the solutions give the cutting points on the x axis. For $P(x)$ and $F(x)$ there are two and three distinct points respectively, whereas for $g(x)$ there is only one real point. See fig. 97.

Thus we see that a quadratic may have 0 or 2 real roots and a cubic may have 1 or 3 real roots (it cannot have two real roots since this would necessitate there being a third). It would be much more convenient if we could simply say that every quadratic has two roots, every cubic has three roots and, generally, every equation of degree n has n roots. To do this we will have to consider equations such as $x^2 + 4 = 0$ and $x^2 + x + 1 = 0$ which have no real solutions.

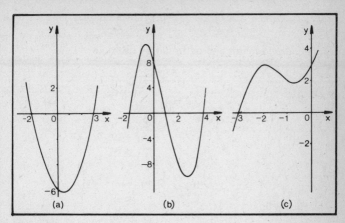

Figure 97

We define a new number, denoted by j where $j^2 = -1$, i.e. $j = \sqrt{-1}$. If this new number is combined with the real numbers we form a more general number in the form $a + bj$ where a and b are real numbers.

Such a number is called a **complex number**.

Remember that a and b may be negative or fractional but must be real.

Thus $3 + 2j$, $4 - 7j$, $-\frac{1}{2} + \frac{1}{3}j$ are examples of complex numbers.

To return to the equation $x^2 + 4 = 0$. This can be written as

$$x^2 = -4 \text{ or } x^2 = (-1)(4) \Rightarrow x = \sqrt{-1} \times \sqrt{4} = \pm 2j$$

Similarly we can now find solutions for the equation $x^2 + x + 1 = 0$ but we need the formula solution.

The solution of $ax^2 + bx + c = 0$ is $x = \dfrac{-b \pm \sqrt{b^2 - 4ac}}{2a}$

and hence, in this case $a = 1$, $b = 1$, $c = 1$

$$\therefore \quad x = \frac{-1 \pm \sqrt{1 - 4(1)(1)}}{2} \Rightarrow x = \tfrac{1}{2}(-1 \pm \sqrt{-3})$$

Hence the solutions of the equation $x^2 + x + 1 = 0$ are $x = \tfrac{1}{2}(-1 \pm \sqrt{3}j)$.

227

Example 1 Solve the equation $x^3 + 2x^2 - 3x - 10 = 0$.

Let $f(x) = x^3 + 2x^2 - 3x - 10$.

$f(1) = 1 + 2 - 3 - 10 \neq 0$ $\therefore$ $x - 1$ is not a factor.
$f(2) = 8 + 8 - 6 - 10 = 0$ $\therefore$ $x - 2$ is a factor.

The quadratic factor can be found by division.

$$
\begin{array}{r}
x^2 + 4x + 5 \\
x - 2 \overline{)\, x^3 + 2x^2 - 3x - 10} \\
\underline{x^3 - 2x^2} \\
4x^2 - 3x \\
\underline{4x^2 - 8x} \\
5x - 10 \\
\underline{5x - 10} \\
\cdots
\end{array}
$$

Hence $x^3 + 2x^2 - 3x - 10 = (x - 2)(x^2 + 4x + 5) = 0$

$\therefore$ $x - 2 = 0$ or $x^2 + 4x + 5 = 0$

$\therefore$ $x = 2$ or by the formula $x = \dfrac{-4 \pm \sqrt{16 - 20}}{2}$

$$= \tfrac{1}{2}[-4 \pm 2j] = -2 \pm j$$

Hence the roots of $x^3 + 2x^2 - 3x - 10 = 0$ are $x = 2, -2 \pm j$.

The algebra of complex numbers

As with all algebras we need to define our rules for combining elements (cf. the algebra of vectors, page 269, and the algebra of matrices, page 170).

We adopt the same operations for complex numbers as for real numbers, i.e. addition, subtraction, multiplication and division. We apply the same rules of algebra to give

$$(a + bj) + (c + dj) = (a + c) + (b + d)j$$

$$(a + bj) - (c + dj) = (a - c) + (b - d)j$$

$$(a + bj)(c + dj) = ac + adj + bcj + bdj^2$$
$$= (ac - bd) + (ad + bc)j \text{ since } j^2 = -1$$

Before considering division we need to note a useful result.

Conjugate complex numbers

If $z = a + bj$ is a given complex number, then the number $z = a - bj$ is called its **conjugate**.

Note that the conjugate is obtained by changing the sign of the imaginary part of the complex number.

The product of two complex conjugates is always real.

$$(p + qj)(p - qj) = p^2 - q^2j^2 = p^2 + q^2$$

We can use this to define the operation of division.

Consider $\dfrac{a + bj}{c + dj}$ and multiply the numerator and denominator by $(c - dj)$ (the conjugate of the denominator) to give

$$\frac{a + bj}{c + dj} = \frac{(a + bj)(c - dj)}{(c + dj)(c - dj)} = \frac{(ac + bd) + j(bc - ad)}{c^2 + d^2}$$

$$= \left(\frac{ac + bd}{c^2 + d^2}\right) + \left(\frac{bc - ad}{c^2 + d^2}\right)j$$

This is clearly another complex number in the form $x + yj$.

Hence each of the standard rules of algebra can be applied to combine complex numbers, each giving a new complex number in the form $a + bj$.

Example 2 Evaluate the following

(i) $(3 + 2j) + (4 - 3j) = 7 - j$

(ii) $(2 - 3j) - (5 - 6j) = -3 + 3j$

(iii) $(4 + 3j)(2 - 3j) = 8 - 12j + 6j - 9j^2 = 17 - 6j$

(iv) $\dfrac{(2 + 3j)}{(1 - j)} = \dfrac{(2 + 3j)(1 + j)}{(1 - j)(1 + j)} = \dfrac{2 + 5j + 3j^2}{2} = -\dfrac{1}{2} + \dfrac{5j}{2}$

When a complex number is written in the form $x + yj$, x and y are known as the **real** and **imaginary parts** respectively. Two complex numbers $a + bj$ and $c + dj$ are equal if and only if, $a = c$ and $b = d$, i.e. if the real and imaginary parts are equal.

Remember that $j^3 = j^2 \cdot j = -j$

$$j^4 = j^2 \cdot j^2 = (-1)(-1) = 1$$

$$j^5 = j^4 \cdot j = 1 \cdot j = j \text{ and so on}$$

Example 3 Evaluate $(2 + 3j)^4$.

We can use the binomial expansion (see page 208) to give

$$(2 + 3j)^4 = 2^4 + 4 \times 2^3(3j) + 6 \times 2^2(3j)^2 + 4 \times 2(3j)^3 + (3j)^4$$

$$= 16 + 96j + 216j^2 + 216j^3 + 81j^4$$

$$= 16 + 96j - 216 - 216j + 81 = -119 - 120j$$

The Argand diagram

Since any complex number $a + bj$ is determined by the values of a and b we can uniquely associate an ordered pair (a, b) with the complex number $a + bj$. The order is important since $a + bj \neq b + aj$. Thus, we can represent a complex number $a + bj$ by a point in a cartesian plane whose coordinates are (a, b). This plane is called an **Argand diagram** (after J.R. Argand) or a **complex plane.**

Notice the similarity between the correspondence here of a complex number and a point in a plane and the one-to-one correspondence between the real numbers and a point on a straight line.

Figure 98 shows an Argand diagram with the points A, B and C representing the complex numbers $3 + 2j$, $-4 + 3j$ and $2 - 2j$ respectively.

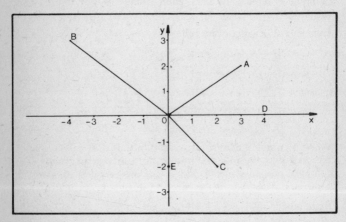

Figure 98

On the diagram point D represents the complex number $4 + 0j$ (i.e. the real number 4) and point E represents the complex number $0 - 2j$ (i.e. $-2j$). Thus real numbers are represented by points on the x axis and complex numbers in which the real part is zero are represented by points on the y axis.

Notice that each point in the plane has an associated radius vector. The radius vectors in fig. 98 above are OA, OB, OC,

OD and *OE*.

Consider fig. 99 in which a complex number $x + yj$ is represented by point P. Its radius vector is OP.

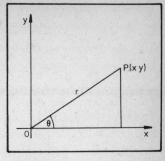

The magnitude of OP (usually denoted by r) is called the **modulus** of the complex number $x + yj$.

Thus $r = \sqrt{x^2 + y^2}$. A special notation is used for the modulus, namely $|x + yj|$. Hence $|x + yj| = r = \sqrt{x^2 + y^2}$.

Figure 99

The angle that OP makes with the real axis is called the **argument** of the complex number $x + yj$. This is abbreviated to $\arg(x + yj)$. Since this is a many valued function we take, by convention, the principal value of the argument so that

$$-\pi < \arg(x + yj) \leqslant \pi$$

In fig. 100 several complex numbers have been represented by points in the Argand diagram. Their associated radius vectors are drawn from the origin.

In each case the modulus and argument has been calculated. We can adopt the technique used in vectors by using a single letter to describe a complex number,

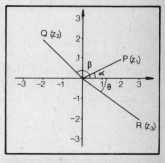

Figure 100

i.e. $z = 3 + 4j$.

(i) $z_1 = 2 + j$ $\quad \therefore \quad |z_1| = \sqrt{2^2 + 1^2} = \sqrt{5}$

$\arg z_1 = \alpha = \tan^{-1}\left(\tfrac{1}{2}\right) = 26°34'$

(ii) $z_2 = -2 + 2j$ $\therefore$ $|z_2| = \sqrt{(-2)^2 + 2^2} = 2\sqrt{2}$

$\arg z_2 = \beta = \pi - \tan^{-1}(1) = 135°$

(iii) $z_3 = 3 - 2j$ $\quad \therefore \quad |z_3| = \sqrt{3^2 + (-2)^2} = \sqrt{13}$

$\arg z_3 = -\theta = -\tan^{-1}\left(\tfrac{2}{3}\right) = -33°41'$

Since any complex number $z = x + yj$ has a modulus, r, given by $\sqrt{x^2 + y^2}$ and an argument θ given by $\tan^{-1}\left(\dfrac{y}{x}\right)$ it follows that z can be expressed in terms of r and θ,

i.e. $z = x + yj = r(\cos\theta + \sin\theta\, j)$

where $x = r\cos\theta$ and $y = r\sin\theta$. (See fig. 99 on page 231.) This is known as the **modulus-argument** or **polar form** of a complex number.

Example 4 Express (i) $z = 1 + j$, (ii) $z = 1 - j$, in polar form.

(i) Since $z = 1 + j$, $|z| = \sqrt{2}$ and $\arg z = \theta = \tan^{-1}(1) = \pi/4$

$\therefore$ $1 + j$ can be written in the form $\sqrt{2}\left(\cos\dfrac{\pi}{4} + j\sin\dfrac{\pi}{4}\right)$

(ii) For $z = 1 - j$, $|z| = \sqrt{2}$ and $\arg z = \alpha = \tan^{-1}(-1) = -\pi/4$ $\therefore$

$1 - j$ can be written in the form $\sqrt{2}\left\{\cos\left(\dfrac{-\pi}{4}\right) + j\sin\left(\dfrac{-\pi}{4}\right)\right\}$

But $\cos\left(\dfrac{-\pi}{4}\right) = \cos\left(\dfrac{\pi}{4}\right)$ and $\sin\left(\dfrac{-\pi}{4}\right) = -\sin\left(\dfrac{\pi}{4}\right)$

$\therefore$ $1 - j = \sqrt{2}\left(\cos\dfrac{\pi}{4} - \sin\dfrac{\pi}{4}\right)$

Geometrical representation for the addition and subtraction of two complex numbers

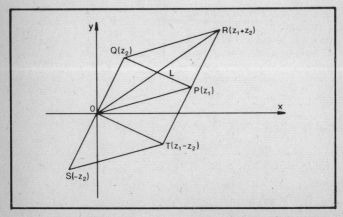

Figure 101

Consider fig. 101 which shows two complex numbers z_1 and z_2 with their associated radius vectors OP and OQ, respectively.

Let R be the fourth vertex of the completed parallelogram $OPRQ$. If $z_1 = x_1 + y_1j$ and $z_2 = x_2 + y_2j$ then P is the point (x_1, y_1) and Q the point (x_2, y_2). Since the diagonals bisect each other, L has coordinates $(\frac{1}{2}(x_1 + x_2), \frac{1}{2}(y_1 + y_2))$.

$\therefore$ The coordinates of R are $((x_1 + x_2), (y_1 + y_2))$ and hence the point R represents the complex number $(x_1 + x_2) + (y_1 + y_2)j$, i.e. $z_1 + z_2$.

Note that the length of $OR = |z_1 + z_2|$.

In order to subtract two complex numbers, say $z_1 - z_2$, we use the fact that $z_1 - z_2 = z_1 + (-z_2)$.

Now $(-z_2)$ is represented by the point S in figure 101 and hence if this is added to z_1 we complete the parallelogram $OSTP$ and point T represents $z_1 - z_2$.

Note that the length of $OT = |z_1 - z_2|$.

Since in fig. 101 $OT = QP$ (opposite sides of parallelogram $OTPQ$) we have $QP = |z_1 - z_2|$.

i.e. in the original parallelogram $OPRQ$ the longer diagonal, OR, represents $|z_1 + z_2|$ and the shorter diagonal, PQ, represents $|z_1 - z_2|$.

Two inequalities

(i) $|z_1 + z_2| \leqslant |z_1| + |z_2|$

Since in fig. 101 OPR is a triangle, $OR \leqslant OP + PR$.

Now $OR = |z_1 + z_2|$, $OP = |z_1|$ and $PR = OQ = |z_2|$.

Hence $|z_1 + z_2| \leqslant |z_1| + |z_2|$.

Thus the modulus of the sum of two complex numbers is not greater than the sum of the moduli of the separate numbers. The equality only occurs when the arguments of the two numbers are the same, i.e. $\triangle OPR$ reduces to a straight line.

(ii) $|z_1 - z_2| \geqslant ||z_1| - |z_2||$

Clearly in $\triangle OPQ$, $OP + PQ \geqslant OQ$

$\therefore$ $PQ \geqslant OQ - OP \Rightarrow |z_1 - z_2| \geqslant |z_2| - |z_1|$

But also $OQ + QP \geqslant OP$

$\therefore$ $QP \geqslant OP - OQ \Rightarrow |z_1 - z_2| \geqslant |z_1| - |z_2|$.

Hence $|z_1 - z_2| \geqslant ||z_1| - |z_2||$.

It is worth noting that the complex numbers under addition are isomorphic to the two dimensional vectors under addition.

For if we use the correspondence $\begin{pmatrix} a \\ b \end{pmatrix} \to a + bj$, $\begin{pmatrix} c \\ d \end{pmatrix} \to c + dj$

then under addition $\begin{pmatrix} a + c \\ b + d \end{pmatrix} \to (a + c) + (b + d)j$

We therefore conclude that the discussion above concerning complex numbers in the complex plane is equivalent to the corresponding work on two dimensional vectors. Indeed, for addition in fig. 101 we have the law of vector addition.

Multiplication and division in polar form

Consider two complex numbers written in modulus-argument form

$$z_1 = r_1(\cos \theta + j \sin \theta) \text{ and } z_2 = r_2(\cos \phi + j \sin \phi)$$

Now $z_1 z_2 = r_1 r_2 (\cos \theta + j \sin \theta)(\cos \phi + j \sin \phi)$

$\qquad = r_1 r_2 \{\cos \theta \cos \phi - \sin \theta \sin \phi$

$\qquad\qquad + j(\cos \theta \sin \phi + \sin \theta \cos \phi)\}$

$\therefore \quad z_1 z_2 = r_1 r_2 \{\cos (\theta + \phi) + j \sin (\theta + \phi)\}$

This product $z_1 z_2$ is in the standard polar form with a modulus $r_1 r_2$ and an argument $(\theta + \phi)$.

Hence to multiply two complex numbers in polar form we multiply the moduli and add the arguments.
Note that it can be extended to any number of terms.

Example 5 Evaluate $z_1 z_2$ when $z_1 = 2\left(\cos \dfrac{\pi}{3} + j \sin \dfrac{\pi}{3}\right)$ and $z_2 = 4\left(\cos \dfrac{\pi}{6} + j \sin \dfrac{\pi}{6}\right)$.

Hence $z_1 z_2 = 8 \cos \left(\dfrac{\pi}{3} + \dfrac{\pi}{6}\right) + j \sin \left(\dfrac{\pi}{3} + \dfrac{\pi}{6}\right)$

$\qquad = 8\left(\cos \dfrac{\pi}{2} + j \sin \dfrac{\pi}{2}\right) = 8j$

This can be checked by converting z_1 and z_2 to an algebraic form first.

$$z_1 = 1 + \sqrt{3}j \text{ and } z_2 = 2(\sqrt{3} + j)$$

$\therefore \quad z_1 z_2 = 2(1 + \sqrt{3}j)(\sqrt{3} + j) = 2\{(\sqrt{3} - \sqrt{3}) + j(3 + 1)\} = 8j$

For division of two complex numbers in polar form we make use of the complex conjugate.

If $z_1 = r_1(\cos \theta + j \sin \theta)$ and $z_2 = r_2(\cos \phi + j \sin \phi)$ we have

$$\frac{z_1}{z_2} = \frac{r_1}{r_2} \cdot \frac{(\cos \theta + j \sin \theta)}{(\cos \phi + j \sin \phi)}$$

Multiplying the numerator and denominator by the complex conjugate of the denominator we obtain

$$\frac{z_1}{z_2} = \frac{r_1}{r_2} \frac{(\cos \theta + j \sin \theta)(\cos \phi - j \sin \phi)}{(\cos \phi + j \sin \phi)(\cos \phi - j \sin \phi)}$$

$$= \frac{r_1}{r_2} \frac{[\cos \theta \cos \phi + \sin \theta \sin \phi + j(\sin \theta \cos \phi - \cos \theta \sin \phi)]}{\cos^2 \phi + \sin^2 \phi}$$

$$\therefore \quad \frac{z_1}{z_2} = \frac{r_1}{r_2} \{\cos (\theta - \phi) + j \sin (\theta - \phi)\}$$

Hence to divide two complex numbers in polar form we divide the moduli and subtract the arguments. Remember that the order is important in division.

Example 6 Find the quotient z_1/z_2 where $z_1 = 1 - 2j$ and $z_2 = 2 + j$ by converting to polar form.

Since $z_1 = 1 - 2j$, $|z_1| = \sqrt{5}$ and $\arg(z_1) = -63°26'$

also $z_2 = 2 + j$, $|z_2| = \sqrt{5}$ and $\arg(z_2) = 26°34'$

Hence $z_1 = \sqrt{5}\{\cos(-63°26') + j \sin(-63°26')\}$

and $\quad z_2 = \sqrt{5}\{\cos(26°34') + j \sin(26°34')\}$

$$\therefore \quad \frac{z_1}{z_2} = 1\{\cos(-90°) + j \sin(-90°)\} = -j$$

Geometrical representation for multiplication and division of two complex numbers

In order to find a point R representing the product of two complex numbers z_1 and z_2 where $z_1 = r_1(\cos \theta + j \sin \theta)$ and $z_2 = r_2(\cos \phi + j \sin \phi)$ we need a point whose polar coordinates are $(r_1 r_2, \theta + \phi)$.

Figure 102 shows z_1 and z_2 with their associated radius vectors OP and OQ and the construction necessary to find R.

Take a point $A(1, 0)$ on the real axis and complete the triangle OAP. Now construct triangle OQR similar to triangle OAP.

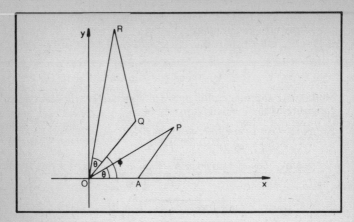

Figure 102

The vertex, R, of this triangle represents the complex number z_1z_2, since its argument is clearly $(\theta + \phi)$ and its modulus is given by OR.

But the triangles are similar $\Rightarrow \dfrac{OR}{OQ} = \dfrac{OP}{OA} \Rightarrow OR = \dfrac{|z_1| \times |z_2|}{1}$.

Hence, since $OP = |z_1| = r_1$ and $OQ = |z_2| = r_2$, $OR = r_1r_2$.

For division, we need to find a point whose polar coordinates are $(r_1/r_2, \theta - \phi)$. Two complex numbers z_1 and z_2 are shown in fig. 103 with their associated radius vectors OP and OQ.

The construction can be completed by taking a point $A(1, 0)$ as before and forming triangle OAS similar to triangle OQP.

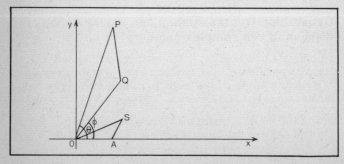

Figure 103

236

The vertex, S, of this triangle represents the complex number z_1/z_2, since its argument is clearly $(\theta - \phi)$ and its modulus is given by OS.

But the triangles are similar $\Rightarrow \dfrac{OS}{OA} = \dfrac{OP}{OQ} \Rightarrow OS = \dfrac{|z_1|}{|z_2|} \times 1$.

Hence, since $OP = |z_1| = r_1$ and $OQ = |z_2| = r_2$, $OS = r_1/r_2$.

Note that fig. 102 is, in fact, illustrating the product of

$$z_1 = 2(\cos 30° + j \sin 30°) \qquad \text{and} \qquad z_2 = \frac{3}{2}(\cos 50° + j \sin 50°)$$

Similarly, fig. 103 shows the quotient of

$$z_1 = 3 \cos (70° + j \sin 70°) \quad \text{and} \quad z_2 = 2(\cos 50° + j \sin 50°).$$

De Moivre's theorem

This states that for any integral or rational value of n

$$(\cos \theta + j \sin \theta)^n = \cos n\theta + j \sin n\theta.$$

The proof is in three stages.

(i) Use induction for the case when n is a positive integer. Assume that the result is true for a value $n = k$,

i.e. $(\cos \theta + j \sin \theta)^k = \cos k\theta + j \sin k\theta$

$\therefore (\cos \theta + j \sin \theta)^{k+1} = (\cos k\theta + j \sin k\theta)(\cos \theta + j \sin \theta)$

$$= (\cos k\theta \cos \theta - \sin k\theta \sin \theta)$$
$$+ j(\sin k\theta \cos \theta + \cos k\theta \sin \theta)$$
$$= \cos (k + 1)\theta + j \sin (k + 1)\theta$$

Thus if the result is true for $n = k$ it also holds for $n = k + 1$. But the result is clearly true for $n = 1$ and therefore for $n = 2$ and for any positive integral value of n.

(ii) If n is a negative integer, let $n = -m$ where m is a positive integer.

Thus

$$(\cos \theta + j \sin \theta)^n = (\cos \theta + j \sin \theta)^{-m} = \frac{1}{(\cos \theta + j \sin \theta)^m}$$

$$= \frac{1}{(\cos m\theta + j \sin m\theta)}$$

since m is a positive integer.

Multiply numerator and denominator by $\cos m\theta - j \sin m\theta$.

$$\therefore \quad (\cos\theta + j\sin\theta)^n = \frac{\cos m\theta - j\sin m\theta}{1}$$

$$= \cos(-m\theta) + j\sin(-m\theta)$$

$$= \cos n\theta + j\sin n\theta$$

Thus the result holds if n is a negative integer.

(iii) If n is a rational value, let $n = p/q$ where p and q are integral and co-prime. Assume that q is positive.

Now as $q > 0$ $\quad \left(\cos\left(\frac{\theta}{q}\right) + j\sin\left(\frac{\theta}{q}\right)\right)^q = \cos\theta + j\sin\theta$

i.e. $\cos\left(\frac{\theta}{q}\right) + j\sin\left(\frac{\theta}{q}\right)$ is one qth root of $\cos\theta + j\sin\theta$

i.e. one value of $(\cos\theta + j\sin\theta)^{1/q}$

Taking the pth power of each side

$\cos\left(\frac{p\theta}{q}\right) + j\sin\left(\frac{p\theta}{q}\right)$ is one value of $(\cos\theta + j\sin\theta)^{p/q}$

Hence we see that De Moivre's theorem holds for any positive or negative rational value of n.

The nth roots of unity

To find the solutions of the equations $z^n - 1 = 0$ we write $z = 1^{1/n}$.

Now 1 can be expressed as a complex number in polar form as $\cos 2r\pi + j\sin 2r\pi$ where $r = 0, 1, 2, \ldots$
Thus $z = (\cos 2r\pi + j\sin 2r\pi)^{1/n}$

By applying De Moivre's theorem we have

$$z = \cos\left(\frac{2r\pi}{n}\right) + j\sin\left(\frac{2r\pi}{n}\right) \quad r = 0, 1, 2, \ldots (n-1)$$

Hence as r takes successively the values $0, 1, 2, 3, \ldots (n-1)$ we obtain n different values for z which are the n roots of the equation $z^n - 1 = 0$.

Example 7 Find the roots of the equation $z^5 - 1 = 0$.

The roots of the equation are given by

$$z = (\cos 2r\pi + j\sin 2r\pi)^{1/5} \text{ for } r = 0, 1, 2, 3, 4.$$

i.e. the roots are given by $z = \cos\left(\frac{2r\pi}{5}\right) + j\sin\left(\frac{2r\pi}{5}\right)$

when $r = 0$, $z = \cos 0 + j \sin 0 = 1$.

$$r = 1 \quad z = \cos\left(\frac{2\pi}{5}\right) + j \sin\left(\frac{2\pi}{5}\right)$$

$$r = 2 \quad z = \cos\left(\frac{4\pi}{5}\right) + j \sin\left(\frac{4\pi}{5}\right)$$

$$r = 3 \quad z = \cos\left(\frac{6\pi}{5}\right) + j \sin\left(\frac{6\pi}{5}\right)$$

$$r = 4 \quad z = \cos\left(\frac{8\pi}{5}\right) + j \sin\left(\frac{8\pi}{5}\right)$$

We note from the results that
  (i) No more solutions may be obtained by substituting further values of r since $r = 5$ repeats the value given when $r = 0$, and so on.
 (ii) All the roots have unit modulus and are equally spread around the circumference of a circle of unit radius in the Argand diagram.
(iii) The n roots will be symmetrical about the real axis on the Argand diagram.
(iv) The nth roots of any real number a can be written as $\sqrt[n]{a}\left(\cos\frac{2r\pi}{n} + j \sin\frac{2r\pi}{n}\right)$ where $\sqrt[n]{a}$ denotes the ordinary positive nth root of a.

Thus $16^{1/4} = 2\left(\cos\frac{2r\pi}{4} + j \sin\frac{2r\pi}{4}\right)$ for $r = 0, 1, 2, 3$.

$\therefore$ the roots are ± 2, $\pm 2j$.

The cube roots of ± 1

The cube roots of unity are a special case.
The solution of $z^3 - 1 = 0$ over the complex numbers gives three roots,

i.e. $z = 1$, $z = \cos\left(\frac{2\pi}{3}\right) + j \sin\left(\frac{2\pi}{3}\right)$, $z = \cos\left(\frac{4\pi}{3}\right) + j \sin\left(\frac{4\pi}{3}\right)$

or $z = 1$, $-\frac{1}{2} + \frac{\sqrt{3}}{2}j$ and $-\frac{1}{2} - \frac{\sqrt{3}}{2}j$

The cube roots of unity are $z = 1$, $-\frac{1}{2}(1 + \sqrt{3}j)$, $-\frac{1}{2}(1 - \sqrt{3}j)$, and are usually denoted by $1, \omega_1, \omega_2$.

We can also derive these results in this special case by solving the algebraic equation $z^3 - 1 = (z - 1)(z^2 + z + 1) = 0$.

$$\therefore \quad z = 1 \text{ or } z^2 + z + 1 = 0 \Rightarrow z = \frac{-1 \pm \sqrt{1-4}}{2}$$

$$= -\tfrac{1}{2}(1 \pm \sqrt{3}j)$$

i.e. 1, ω_1, ω_2 as before.

Note that
$\omega_1{}^2 = [-\tfrac{1}{2}(1 + \sqrt{3}j)]^2 = \tfrac{1}{4}(1 - 3 + 2\sqrt{3}j) = -\tfrac{1}{2}(1 - \sqrt{3}j) = \omega_2.$
Likewise $\omega_2{}^2 = \omega_1$ and so the cube roots of unity are usually given as 1, ω and ω^2.

By considering the roots of a cubic equation we observe that
 (i) $\omega^3 = 1$, since the product of the roots is 1.
 (ii) $1 + \omega + \omega^2 = 0$, since the sum of the roots $= 0$ (the coefficient of the term z^2).

Similarly the cube roots of -1 can be found by solving the equation $z^3 + 1 = (z + 1)(z^2 - z + 1) = 0$.

$$\Rightarrow z = -1 \text{ or } z^2 - z + 1 = 0 \Rightarrow z = \tfrac{1}{2}(1 \pm \sqrt{3}j)$$

Hence the roots are -1, $\tfrac{1}{2}(1 \pm \sqrt{3}j)$.

Trigonometrical identities

We can use De Moivre's to prove some trigonometrical identities by making use of the fact that if two complex numbers are equal the real and imaginary parts of each can be equated, i.e. $a + bj = c + dj \Rightarrow a = c$ and $b = d$.

Example 8 Prove that (i) $\cos 3\theta = 4\cos^3 \theta - 3\cos \theta$ and (ii) $\sin 3\theta = 3 \sin \theta - 4 \sin^3 \theta$.

Now by De Moivre's theorem

$\cos 3\theta + j \sin 3\theta$
$\qquad = (\cos \theta + j \sin \theta)^3$
$\qquad = \cos^3 \theta + 3j \cos^2 \theta \sin \theta + 3j^2 \cos \theta \sin^2 \theta + j^3 \sin^3 \theta$
$\qquad = \cos^3 \theta - 3 \cos \theta \sin^2 \theta + j(3 \cos^2 \theta \sin \theta - \sin^3 \theta)$

$\therefore$ Equating real and imaginary parts

$$\cos 3\theta = \cos^3 \theta - 3 \cos \theta \sin^2 \theta \qquad (1)$$

and $\qquad\qquad \sin 3\theta = 3 \cos^2 \theta \sin \theta - \sin^3 \theta \qquad (2)$

Using the identity $\cos^2 \theta + \sin^2 \theta = 1$ we have

$$\cos 3\theta = 4\cos^3 \theta - 3 \cos \theta \text{ and } \sin 3\theta = 3 \sin \theta - 4 \sin^3 \theta$$

Notice that an expression for $\tan 3\theta$ can be obtained by dividing the two results (1) and (2).

i.e. $\tan 3\theta = \dfrac{3\cos^2\theta \sin\theta - \sin^3\theta}{\cos^3\theta - 3\cos\theta \sin^2\theta} = \dfrac{3\tan\theta - \tan^3\theta}{1 - 3\tan^2\theta}$

after dividing by $\cos^3\theta$.

Powers of $\sin\theta$ and $\cos\theta$ in terms of multiple angles

The above process is best when expressing the sine or cosine of a multiple angle in terms of powers of sine or cosine, but an alternative method is suitable when expressing powers in terms of multiple angles.

By De Moivre's theorem,

$$z = \cos\theta + j\sin\theta \Rightarrow z^n = \cos n\theta + j\sin n\theta$$

and $z^{-1} = \dfrac{1}{z} = \cos\theta - j\sin\theta \Rightarrow z^{-n} = \dfrac{1}{z^n} = \cos n\theta - j\sin n\theta$

Hence $\quad z + \dfrac{1}{z} = 2\cos\theta \quad$ (3) $\qquad z^n + \dfrac{1}{z^n} = 2\cos n\theta \quad$ (5)

$$z - \dfrac{1}{z} = 2j\sin\theta \quad \text{(4)} \qquad z^n - \dfrac{1}{z^n} = 2j\sin n\theta \quad \text{(6)}$$

Example 9 Prove that

$$\cos^6\theta = \frac{1}{32}(\cos 6\theta + 6\cos 4\theta + 15\cos 2\theta + 10).$$

Using relation (3) $\qquad (2\cos\theta)^6 = \left(z + \dfrac{1}{z}\right)^6$

$\therefore \quad 64\cos^6\theta = z^6 + 6z^4 + 15z^2 + 20 + \dfrac{15}{z^2} + \dfrac{6}{z^4} + \dfrac{1}{z^6}$

$$= \left(z^6 + \frac{1}{z^6}\right) + 6\left(z^4 + \frac{1}{z^4}\right) + 15\left(z^2 + \frac{1}{z^2}\right) + 20$$

using (5) $\qquad = 2\cos 6\theta + 6(2\cos 4\theta) + 15(2\cos 2\theta) + 20$

$\therefore \quad \cos^6\theta = \dfrac{1}{32}(\cos 6\theta + 6\cos 4\theta + 15\cos 2\theta + 10)$

Series of Complex terms

The exponential series for e^x, when x is real, is given by

$$e^x = 1 + x + \frac{x^2}{2!} + \frac{x^3}{3!} + \cdots$$

which is valid for all values of x.

If z is complex, we define e^z as the sum of the series $1 + z + \dfrac{z^2}{2!} + \dfrac{z^3}{3!} + \cdots$

i.e. $e^z = 1 + z + \dfrac{z^2}{2!} + \dfrac{z^3}{3!} + \dfrac{z^4}{4!} + \cdots + \dfrac{z^r}{r!} + \cdots$

In particular, if $z = j\theta$, we have

$$e^{j\theta} = 1 + j\theta + \frac{(j\theta)^2}{2!} + \frac{(j\theta)^3}{3!} + \frac{(j\theta)^4}{4!} + \cdots$$

$$= \left(1 - \frac{\theta^2}{2!} + \frac{\theta^4}{4!} - \cdots\right) + j\left(\theta - \frac{\theta^3}{3!} + \frac{\theta^5}{5!} - \cdots\right)$$

$$\therefore \quad e^{j\theta} = \cos\theta + j\sin\theta \tag{1}$$

using the series for $\cos\theta$ and $\sin\theta$ (see page 219).

Thus the complex number $z = r(\cos\theta + j\sin\theta)$ can be written as $z = re^{j\theta}$. This is known as the **exponential form** of a complex number.

By writing $-\theta$ for θ in equation (1) we have

$$e^{-j\theta} = \cos\theta - j\sin\theta \tag{2}$$

Combining (1) and (2) we obtain

$$\cos\theta = \frac{1}{2}(e^{j\theta} + e^{-j\theta}) \quad \text{and} \quad \sin\theta = \frac{1}{2j}(e^{j\theta} - e^{-j\theta}) \tag{3}$$

Since the hyperbolic functions are defined in terms of e^θ and $e^{-\theta}$ we can find a relation between the circular and hyperbolic functions.

$$\cos j\theta = \tfrac{1}{2}(e^{-\theta} + e^\theta) = \cosh\theta$$

$$\sin j\theta = \frac{1}{2j}(e^{-\theta} - e^\theta) = j\sinh\theta$$

Dividing we have, $\tan j\theta = j\tanh\theta$.
Alternatively these could be written as

$$j\sin\theta = \sinh j\theta, \qquad \cos\theta = \cosh j\theta, \qquad j\tan\theta = \tanh j\theta$$

Applications to differentiation and integration

We can use the exponential form of a complex number to help in differentiating or integrating certain functions.

Example 10 Differentiate $e^{2x}\cos 3x$ and $e^{2x}\sin 3x$ with respect to x.
Let $C = e^{2x}\cos 3x$ and $S = e^{2x}\sin 3x$

$$\therefore \quad C + jS = e^{2x}(\cos 3x + j\sin 3x) = e^{2x} \times e^{3jx} = e^{(2+3j)x}$$

Let $z = C + jS \Rightarrow z = e^{(2+3j)x}$

$\therefore \dfrac{dz}{dx} = (2 + 3j)e^{(2+3j)x}$

$\therefore \dfrac{d}{dx}\{e^{2x}(\cos 3x + j\sin 3x)\} = (2 + 3j)e^{(2+3j)x}$

$$= (2 + 3j)\{\cos 3x + j\sin 3x\}e^{2x}$$

Equating real and imaginary parts gives

$$\frac{d}{dx}(e^{2x}\cos 3x) = (2\cos 3x - 3\sin 3x)e^{2x}$$

and $\qquad \dfrac{d}{dx}(e^{2x}\sin 3x) = (3\cos 3x + 2\sin 3x)e^{2x}.$

Loci

Certain well known curves and loci that we study using cartesian or polar coordinates have simple representations using complex numbers, and transformations using complex numbers can be very illuminating.

Example 11 Describe the locus of points $z = a + jb$ where $|z| = 2$.

$|z|$ is the modulus of z or the distance of the point (a, b) from the origin. $|z| = 2$ represents all points whose distance from the origin is 2, i.e. the circle centre $(0, 0)$, radius 2.
Or, $z = x + jy \Rightarrow |z| = \sqrt{x^2 + y^2}$ so $|z| = 2 \Rightarrow x^2 + y^2 = 4$. This is the cartesian equation of a circle centre $(0, 0)$ radius 2.

Example 12 $|z - (2 + 3j)| = 3$.
For a general point P representing $z = x + jy$ where A represents $2 + 3j$.

$\mathbf{AP} = \mathbf{OB} = \mathbf{OP} + \mathbf{PB}$

$\qquad = \mathbf{OP} - \mathbf{OA} = \begin{pmatrix} x \\ y \end{pmatrix} - \begin{pmatrix} 2 \\ 3 \end{pmatrix}$

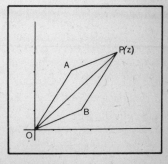

$|z - (2 + 3j)|$ is the distance of P from A so all points which are distant 3 from A satisfy the locus $|z - (2 + 3j)| = 3$.
So $|z - (2 + 3j)| = 3$ represents a circle radius 3, centre $A(2, 3)$.

Figure 104

Alternatively if $z = x + jy$

$$|z - (2 + 3j)| = |x + jy - (2 + 3j)| = |x - 2 + j(y - 3)|$$
$$= \sqrt{(x-2)^2 + (y-3)^2} = 3$$
$$|z - (2 + 3j)| = 3 \Rightarrow (x-2)^2 + (y-3)^2 = 9$$

which represents a circle centre $(2, 3)$ and radius 3.

Example 13 $|z - 2| = |z - 4j|$.

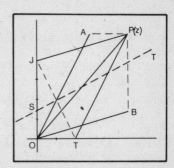

If OP represents $z = x + jy$
$AO = PT$ represents $|z - 2|$.
$OB = JP$ represents $|z - 4j|$
and $|z - 2| = |z - 4j|$

$$\Rightarrow JP = PT$$

The locus is the set of points which are equidistant from J and T, i.e. the mediator ST (perpendicular bisector) of JT. Alternatively

Figure 105

$$|z - 2| = |z - 4j| \Rightarrow |x - 2 + jy|$$
$$= |x + j(y - 4)| \Rightarrow (x-2)^2 + y^2 = x^2 + (y-4)^2$$
$$\Rightarrow -4x + 4 = -8y + 16 \Rightarrow 8y = 4x + 12 \Rightarrow y = \tfrac{1}{2}x + \tfrac{3}{2}$$

which is the mediator of JT (shown dotted in fig. 105).
In general $|z - (a + jb)| = |z - (c + jd)|$ represents the mediator of AC where A is (a, b) and C is (c, d).

Example 14 $|z + 2| = 2|z - 1|$.

$|z + 2|$ represents the distance of a general point z from $T(-2, 0)$, and $|z - 1|$ represents the distance of $P(x + jy)$ from $W(1, 0)$ (fig. 106). The locus of point P satisfies

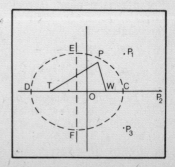

$$PT = 2 \times PW$$

and $O(0, 0)$, $P_1(2, 2)$, $P_2(4, 0)$, and $P_3(2, -2)$ satisfy this relation. In fact the locus is a circle (circle of Apollonius) centre $C(2, 0)$, radius 2 but

Figure 106

244

this can more easily be seen from the cartesian form of the equation.

$$|z+2| = 2|z-1| \Rightarrow |(x+2)+jy| = 2|(x-1)+jy|$$
$$\Rightarrow (x+2)^2 + y^2 = 4[(x-1)^2 + y^2]$$
$$\Rightarrow x^2 + 4x + 4 + y^2 = 4x^2 - 8x + 4 + 4y^2$$
$$\Rightarrow 3x^2 + 3y^2 - 12x = 0$$
$$\Rightarrow x^2 + y^2 - 4x = 0$$
$$\Rightarrow (x-2)^2 + y^2 = 4$$

which represents a circle centre $(2, 0)$, radius 2.

Example 15 $|z+2| + |z-1| = 5$.

From fig. 106 this locus is satisfied by points P such that $PT + PW = 5$. Two points satisfying this locus are $C(2, 0)$ and $D(-3, 0)$. The locus is an ellipse (you may know this property of the ellipse) with T and W as foci and one axis CD. The other axis is on $x = -\frac{1}{2}$ (by symmetry and this axis is, in fact, EF where E is $(-\frac{1}{2}, 2)$ and D is $(-\frac{1}{2}, -2)$. The equation of this ellipse is given by

$$|x+2+jy| + |x-1+jy| = 5$$
$$\Rightarrow \sqrt{(x+2)^2 + y^2} + \sqrt{(x-1)^2 + y^2} = 5$$
$$\Rightarrow x^2 + 4x + 4 + y^2 + x^2 - 2x + 1 + y^2$$
$$+ 2\sqrt{[(x+2)^2 + y^2][(x-1)^2 + y^2]} = 25$$
$$2x^2 + 2y^2 + 2x + 5 - 25 = -2\sqrt{[(x+2)^2 + y^2][(x-1)^2 + y^2]}$$
$$(x^2 + y^2 + x - 10)^2 = [(x+2)^2 + y^2][(x-1)^2 + y^2]$$
$$x^4 + y^4 + x^2 + 100 + 2x^2y^2 + 2xy^2 + 2x^3 - 20x^2 - 20y^2 - 20x$$
$$= (x+2)^2(x-1)^2 + y^2\left[(x+2)^2 + (x-1)^2\right] + y^4$$
$$x^4 + y^4 + 2x^2y^2 + 2xy^2 + 2x^3 - 19x^2 - 20y^2 + 100 - 20x$$
$$= (x^2 + 4x + 4)(x^2 - 2x + 1) + y^2(2x^2 + 2x + 5) + y^4$$
$$x^4 + 2x^3 - 19x^2 - 20y^2 - 20x + 100 = x^4 + 2x^3 - 3x^2 - 4x + 4 + 5y^2$$
$$100 = 16x^2 + 16x + 4 + 25y^2$$
$$100 = 4(4x^2 + 4x + 1) + 25y^2$$
$$1 = \frac{(2x+1)^2}{25} + \frac{y^2}{4}$$

$$\frac{(x-\frac{1}{2})^2}{25/4} + \frac{y^2}{4} = 1$$

which represents an ellipse, centre $(-\frac{1}{2}, 0)$ with semi-axes $2\frac{1}{2}$ and 2.

Complex transformations

$z \to \bar{z}$ takes the point $x + jy$ to $x - jy$ and $\bar{z}$ is the image of z after reflection in the real (x) axis.

The effect of multiplying by the complex number j is to rotate any object point through $90°$; $j(x + jy) = -y + jx$ so (x, y) is mapped onto $(-y, x)$, a rotation of $+90°$ about the origin.

The effect of multiplying by z is to enlarge by a scale factor mod z and rotate by an angle of arg z.

Example 16 $z \to 1/z$.

$\dfrac{1}{z} = \dfrac{\bar{z}}{z\bar{z}} = \dfrac{\bar{z}}{|z|^2}$ so if $|z| = 1$, the image of z is $\bar{z}$.

This means that points on the circle $|z| = 1$ stay on the circle although they are reflected. Since $|\bar{z}| = |z|$ points outside the circle $|z| = 1$ are mapped onto points inside the circle and vice-versa.

$z \to \dfrac{1}{\bar{z}} = \dfrac{z}{\bar{z}z} = \dfrac{z}{|z|^2}$ maps points P outside the circle $|z| = 1$ to points P' inside the circle and vice versa in such a way that O, P, P' are in the same straight line. This transformation is called **inversion** with respect to the circle $|z| = 1$.

Example 17 $z \to z^2$.

The point P with polar coordinates $[r, \theta]$ is mapped onto $[r^2, 2\theta]$ and the argument is doubled.

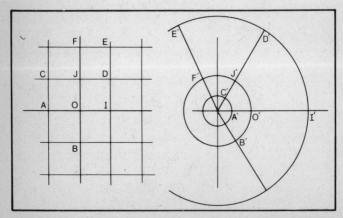

Figure 107

246

The quadrant $0 \leqslant \arg z \leqslant \pi/2$ is mapped onto the region. $0 \leqslant \arg z \leqslant \pi$, i.e. $y \geqslant 0$. The whole plane is mapped onto itself twice. Each image point has two object points corresponding to its two square roots.

Example 18 $z \rightarrow e^z$.

$x + jy \rightarrow e^{x+jy} = e^x . e^{jy} = e^x(\cos y + j \sin y) = [e^x, y]$ in polar coordinates. Lines $x = k$ are mapped onto points where $r = e^k$, i.e. circles of radius e^k. Lines $y = k$ are mapped onto points whose argument is equal to k, i.e. rays emanating from the origin. Figure 107a shows object points on the rectangular grid which is transformed into a radial circular grid shown in fig. 107b. For real x, e^x is always positive, but points on $y = 2\pi$ are mapped onto the positive real axis. The mapping is many to one and the region $0 \leqslant y \leqslant 2\pi$ is mapped onto the whole plane.

Key terms

A complex number is of the form $a + bj$ where a and b are real. If $z = a + bj$, then $z = a - bj$ is the complex conjugate of z.

Argand diagram Any complex number $x + yj$ can be represented by a point in a cartesian plane whose coordinates are (a, b). A complex number can be written in exponential form $e^{j\theta}$ or modulus-argument form $z = r(\cos \theta + j \sin \theta)$.
The modulus of $z = x + yj$ is $|z| = \sqrt{x^2 + y^2}$.

The argument of z is given by $\arg(z) = \tan^{-1}\left(\dfrac{y}{x}\right)$.

De Moivre's theorem
$(\cos \theta + j \sin \theta)^n = \cos n\theta + j \sin n\theta$ for n rational
$$e^{j\theta} = \cos \theta + j \sin \theta$$

Loci
$|z| = k$ represents a circle, centre the origin, radius k.

$|z - a| = k$ is a circle, centred on the point representing the complex number a, of radius k.

$|z - a| = |z - b|$ is the mediator of the line joining the points representing a and b.

$|z + a| + |z + b| = k$ is an ellipse.

Chapter 10
Differential Equations

Differential equations are equations in the usual sense but with the additional characteristic that at least one term contains a differential coefficient.

The following are all differential equations.

(i) $\dfrac{dy}{dx} = 2x^2 + 3$

(ii) $(1 + x)^2 \dfrac{dy}{dx} + y^2 = 1$

(iii) $\dfrac{d^2x}{dt^2} - 3\dfrac{dx}{dt} + 2x = 0$

The **order** of the differential equation is that of the highest derivative involved.
Hence equations (i) and (ii) are first order, whereas equation (iii) is second order.

The **degree** of the differential equation is the power of the highest derivative involved.
Hence the equations above are all of the first degree. The equation $\left(\dfrac{dx}{dt}\right)^2 = a^2(\omega^2 - x^2)$ is of second degree.

Now consider a very simple differential equation $\dfrac{dy}{dx} = 4$.

Clearly this is obtained from the equation $y = 4x + c$ which is the equation of all straight lines of gradient 4. We have obtained the result $y = 4x + c$ by direct integration which is possible in this trivial case. The result $y = 4x + c$ is called the **general solution of the differential equation** $\dfrac{dy}{dx} = 4$, i.e. the solution represents a family of lines all having a specific gradient. If some further data is given (usually called **boundary or initial conditions**) it is possible to determine the value of c.

For example, if $x = 1$ when $y = 7$ then substituting in the general solution we have $7 = 4(1) + c \Rightarrow c = 3$. Thus in this case the **particular solution** is $y = 4x + 3$. Compare the process shown on page 73 concerning basic integration. So we see that

(i) a **differential equation** defines some property common to a family of curves,

(ii) the **general solution** is the equation representing any member of the family and this must contain arbitrary constants equal in number to the order,

(iii) the **particular solution** gives the equation of one specific member of the family and does not contain any arbitrary constants.

We now consider the various types of differential equation and the methods for solving them.

First order equations

Type 1. Equations of the form $\dfrac{dy}{dx} = f(x)$

These can be solved by straightforward integration.

Example 1 Solve the differential equation $\dfrac{dy}{dx} = x^3 + \cos x + 2$.

The general solution is $y = \frac{1}{4}x^4 + \sin x + 2x + c$.

Type 2. Equations of the form $\dfrac{dy}{dx} = f(y)$

This can be written $\dfrac{1}{f(y)} \dfrac{dy}{dx} = 1$

Integrating with respect to x $\quad \displaystyle\int \dfrac{1}{f(y)}\, dy = \int 1\, dx$

i.e. $\quad \displaystyle\int \dfrac{1}{f(y)}\, dy = x + c$

Example 2 Solve the equation $\dfrac{dy}{dx} = 2 \sec y$.

Rearranging $\dfrac{1}{\sec y} \dfrac{dy}{dx} = 2 \Rightarrow \cos y \dfrac{dy}{dx} = 2$

Integrating with respect to x, $\quad \displaystyle\int \cos y\, dy = \int 2\, dx$

The general solution is $\sin y = 2x + c$.

Type 3. Variables separable

This type of equation contains variables in x and y but it is possible to separate them into two terms, one a function of x only and the other a function of y only, i.e. in the form

$$f(y)\frac{dy}{dx} + g(x) = 0$$

Integrating with respect to x yields

$$\int f(y)\,dy + \int g(x)\,dx = 0$$

Example 3 Solve the equation $\dfrac{dy}{dx} = \dfrac{y(y+1)}{2x}$.

Separating the variables $\dfrac{1}{y(y+1)}\dfrac{dy}{dx} = \dfrac{1}{2x}$

Integrating with respect to x $\quad \displaystyle\int \frac{1}{y(y+1)}\,dy = \int \frac{1}{2x}\,dx$

$$\Rightarrow \int \left(\frac{1}{y} - \frac{1}{y+1}\right) dy = \int \frac{1}{2x}\,dx$$

i.e. $\qquad \ln y - \ln(y+1) = \tfrac{1}{2}\ln x + c$

or $\qquad \ln\left(\dfrac{y}{y+1}\right) = \ln\sqrt{x} + c$

$$\ln\left(\frac{y}{y+1}\right) = \ln k\sqrt{x} \text{ writing } c \text{ as } \ln k$$

Hence the general solution is $\dfrac{y}{y+1} = k\sqrt{x}$.

Example 4 Find the general solution of the equation $(x+1)\dfrac{dy}{dx} - y = 0$ and sketch the family of solution curves. If $y = 3$ when $x = 0$, give the particular solution.

Separating the variables $\dfrac{1}{y}\dfrac{dy}{dx} = \dfrac{1}{x+1}$

Integrating with respect to x gives $\displaystyle\int \frac{1}{y}\,dy = \int \frac{1}{x+1}\,dx$

$\Rightarrow \ln y = \ln c(x+1)$ writing the constant $= \ln c$

The general solution is $y = c(x+1)$.

When $x = 0$, $y = 3$, hence $3 = c(0+1) \Rightarrow c = 3$.

The particular solution is $y = 3x + 3$.

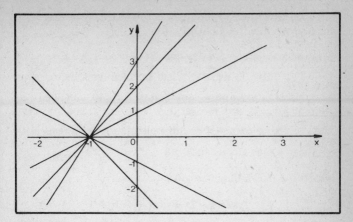

Figure 108

Figure 108 shows the family of solutions giving the cases when $c = -2, -1, 0, 1, 2, 3$. Notice the particular solution $y = 3x + 3$ passes through $(0, 3)$.

Type 4. Exact equations

Consider the differential equation $3xy^2 \dfrac{dy}{dx} + y^3 = e^x$.

This is an equation in which the variables cannot be separated but it will be noticed that the LHS is, in fact, $\dfrac{d}{dx}(xy^3)$ and hence this equation can be solved by direct integration,

i.e. $\dfrac{d}{dx}(xy^3) = e^x \Rightarrow xy^3 = e^x + c$

Thus a differential equation is said to be **exact** if it is formed from the primitive relating x and y by simple differentiation. Thus for the equation $x^4 + 3y^4 + 4xy = c$ we obtain, by differentiation with respect to x,

$$4x^3 + 12y^3 \frac{dy}{dx} + 4x \frac{dy}{dx} + 4y = 0$$

$$\Rightarrow (3y^3 + x) \frac{dy}{dx} + x^3 + y = 0$$

251

This can be written in the form $(x^3 + y)\,dx + (3y^3 + x)\,dy = 0$, and is an exact differential equation.

In general, a differential equation $M\,dx + N\,dy = 0$ is exact if $\dfrac{\partial M}{\partial y} = \dfrac{\partial N}{\partial x}$. This is a necessary and sufficient condition. $\dfrac{\partial M}{\partial y}$ is the **partial differential coefficient** of M obtained by differentiating M with respect to y, regarding x as a constant. Likewise $\dfrac{\partial N}{\partial x}$ is obtained by differentiating N with respect to x, regarding y as a constant.

Example 5 Solve the differential equation

$$y \cos x - 2x + \sin x \frac{dy}{dx} = 0.$$

Since the equation can be written as

$$(y \cos x - 2x)\,dx + \sin x\,dy = 0$$

we have

$$M = y \cos x - 2x \Rightarrow \frac{\partial M}{\partial y} = \cos x$$

$$N = \sin x \Rightarrow \frac{\partial N}{\partial x} = \cos x$$

The equation is exact and can be written as

$$\left(y \cos x + \sin x \frac{dy}{dx} \right) - 2x = 0$$

$$\frac{d}{dx}(y \sin x) - 2x = 0 \Rightarrow y \sin x - x^2 = c$$

Type 5. Equations requiring an integrating factor

Some differential equations are not exact as they stand but can be made exact by multiplying each side by a suitable factor, called the **integrating factor**.

Consider the equation $\dfrac{dy}{dx} + Py = Q$ where P and Q are constants or functions of x only. Such equations are called **linear differential equations**.

If the equation is not exact, assume that we can make it exact by multiplying by a function of x, say R.

Thus
$$R\frac{dy}{dx} + RPy = RQ \tag{1}$$

Now
$$\frac{d}{dx}(Ry) = R\frac{dy}{dx} + y\frac{dR}{dx}$$

Thus equation (1) can be written

$$\frac{d}{dx}(Ry) - y\frac{dR}{dx} + RPy = RQ$$

i.e.
$$\frac{d}{dx}(Ry) + y\left(PR - \frac{dR}{dx}\right) = RQ \tag{2}$$

Now choose R so that $\dfrac{dR}{dx} = PR$, i.e. the coefficient of $y = 0$.

$$\therefore \qquad \frac{1}{R}\frac{dR}{dx} = P \Rightarrow \ln R = \int P\,dx$$

Therefore R, the **integrating factor,** is given by $R = e^{\int P\,dx}$ and equation (2) becomes

$$\frac{d}{dx}(e^{\int P\,dx}y) = Qe^{\int P\,dx}$$

which can be integrated. Note that the integrating factor itself does not contain any arbitrary constants.

Example 6 Solve the equation $\dfrac{dy}{dx} + 4y = e^{3x}$. $\qquad$ (1)

The integrating factor is $e^{\int P\,dx} = e^{\int 4\,dx} = e^{4x}$. Multiplying equation (1) by this factor we have

$$e^{4x}\frac{dy}{dx} + 4e^{4x}y = e^{4x}.e^{3x} = e^{7x} \qquad \therefore \quad \frac{d}{dx}(e^{4x}y) = e^{7x}$$

Integrating with respect to x .

$$ye^{4x} = \frac{1}{7}e^{7x} + c \qquad \text{i.e.} \qquad y = \frac{1}{7}e^{3x} + ce^{-4x}$$

Example 7 Solve the equation $\dfrac{dy}{dx} + y\cot x = \sec x$.

The integrating factor is $e^{\int P\,dx} = e^{\int \cot x\,dx} = e^{\ln \sin x} = \sin x$
Hence, multiplying by $\sin x$ the equation becomes

$$\sin x\frac{dy}{dx} + y\cos x = \tan x$$

$$\frac{d}{dx}(y \sin x) = \tan x$$

Integrating with respect to x

$$y \sin x = \int \tan x \, dx$$

$$y \sin x = -\ln \cos x + k = \ln \sec x + k$$

The general solution is $y = k \operatorname{cosec} x + \operatorname{cosec} x \ln \sec x$.

Type 6. Equations reducible to linear equations

A more general case of the linear equation given in type 5 is

$$\frac{dy}{dx} + Py = Qy^n. \tag{1}$$

This is known as Bernoulli's equation and it can be reduced to a linear form by a substitution $z = y^{1-n}$.

$$z = y^{1-n} \Rightarrow \frac{dz}{dx} = (1-n)y^{-n} \frac{dy}{dx} \Rightarrow \frac{dy}{dx} = \frac{y^n}{1-n} \frac{dz}{dx}$$

Substituting in equation (1) gives

$$\frac{y^n}{1-n} \frac{dz}{dx} + Py = Qy^n$$

$$\therefore \qquad \frac{1}{1-n} \frac{dz}{dx} + Py^{1-n} = Q$$

$$\frac{1}{1-n} \frac{dz}{dx} + Pz = Q$$

This is clearly a linear differential equation since $1-n$ is a constant and hence it can be solved by use of the integrating factor.

Example 8 Solve the equation $\frac{dy}{dx} + \frac{y}{x} = 2x^3 y^4$.

Use the substitution $z = y^{-3}$ (i.e. $n = 4$).

Hence $\frac{dz}{dx} = -3y^{-4} \frac{dy}{dx} \Rightarrow \frac{dy}{dx} = \frac{-y^4}{3} \frac{dz}{dx}$

Thus the given equation becomes $\frac{-y^4}{3} \frac{dz}{dx} + \frac{y}{x} = 2x^3 y^4$

$$\therefore \qquad \frac{-1}{3} \frac{dz}{dx} + \frac{1}{x} \cdot y^{-3} = 2x^3 \quad \text{(dividing by } y^4\text{)}$$

$$\therefore \qquad \frac{dz}{dx} - \frac{3z}{x} = -6x^3$$

The integrating factor is $e^{\int -(3/x)\,dx} = e^{-3\ln x} = e^{\ln x^{-3}} = x^{-3}$.
Multiplying through by x^{-3} we have

$$x^{-3}\frac{dz}{dx} - 3x^{-4}z = -6 \qquad \text{i.e.} \qquad \frac{d}{dx}(x^{-3}z) = -6$$

Integrating with respect to x gives $x^{-3}z = -6x + c$

i.e. $$z = -6x^4 + cx^3$$

But $z = y^{-3}$ $\therefore$ $y^{-3} = -6x^4 + cx^3$

The general solution is $y^3 = 1/(cx^3 - 6x^4)$.

Type 7. Homogeneous equations

A homogeneous equation is of the form $P\dfrac{dy}{dx} + Q = 0$ where P
and Q are homogeneous expressions of the same degree in
variables x and y.

$x^2y + 7xy^2 + 3y^3$ is a homogeneous expression of degree 3, i.e.
the sum of the powers of x and y in each term is 3.

For this type of equation, it can be written as $\dfrac{dy}{dx} = \dfrac{-Q}{P} = f\left(\dfrac{y}{x}\right)$
by dividing by x^n where n is the degree of P and Q.

Now let $y = vx$, so that $\dfrac{dy}{dx} = v + x\dfrac{dv}{dx}$

Thus the equation becomes $v + x\dfrac{dv}{dx} = f(v)$

$$\therefore \qquad x\frac{dv}{dx} = \phi(v) \text{ where } \phi(v) = f(v) - v$$

We can now separate the variables and integrate to give

$$\int \frac{1}{\phi(v)}\,dv = \int \frac{1}{x}\,dx \qquad \text{i.e.} \qquad \int \frac{1}{\phi(v)}\,dv = \ln x + c$$

After integrating the function $1/\phi(v)$, v must be replaced by y/x
and the final solution is obtained. Remember that the expressions
for P and Q must have the same degree.

Example 9 Solve the differential equation

$$(x^2 - 2xy)\frac{dy}{dx} = y^2 - 2xy.$$

255

This is of the form $P \dfrac{dy}{dx} + Q = 0$ where P and Q are homogeneous expressions of degree 2. By dividing by x^2 and rearranging we obtain,

$$\frac{dy}{dx} = \frac{\left(\dfrac{y}{x}\right)^2 - 2\left(\dfrac{y}{x}\right)}{1 - 2\left(\dfrac{y}{x}\right)}$$

Use the substitution $y = vx \Rightarrow \dfrac{dy}{dx} = v + x\dfrac{dv}{dx}$

$$\therefore \qquad v + x\frac{dv}{dx} = \frac{v^2 - 2v}{1 - 2v}$$

$$x\frac{dv}{dx} = \frac{v^2 - 2v}{1 - 2v} - v = \frac{3v^2 - 3v}{1 - 2v}$$

Separating the variables and integrating with respect to x

$$\frac{1}{3} \int \frac{1 - 2v}{v(v-1)}\, dv = \int \frac{dx}{x}$$

Using partial fractions

$$\frac{1}{3} \int -\frac{1}{v}\, dv - \frac{1}{3} \int \frac{1}{v-1}\, dv = \int \frac{dx}{x}$$

$$\therefore \qquad -\frac{1}{3}\ln v - \frac{1}{3}\ln(v-1) = \ln kx$$

$$\therefore \qquad -\frac{1}{3}\ln v(v-1) = \ln kx$$

Hence $\qquad\qquad v(v-1) = (kx)^{-3}$

Since $v = \dfrac{y}{x}$ $\qquad \dfrac{y}{x}\left(\dfrac{y}{x} - 1\right) = (kx)^{-3}$

$$xy(y - x) = k^{-3}$$

Writing $c = -k^{-3}$ the general solution can be written

$$xy(x - y) = c$$

Type 8. Equations reducible to homogeneous equations

Equations of the form $\dfrac{dy}{dx} = \dfrac{a_1 x + b_1 y + c_1}{a_2 x + b_2 y + c_2}$ \hfill (1) can be made homogeneous by a process which effectively moves the origin of coordinates to the point of intersection of the lines $a_1x + b_1y + c_1 = 0$ and $a_2x + b_2y + c_2 = 0$.

Let $x = X + k$ and $y = Y + l$ and thus $\dfrac{dy}{dx} = \dfrac{dY}{dX}$

Thus equation (1) becomes

$$\frac{dY}{dX} = \frac{a_1(X + k) + b_1(Y + l) + c_1}{a_2(X + k) + b_2(Y + l) + c_2} \tag{2}$$

If k and l are chosen so that $a_1(X + k) + b_1(Y + l) + c_1 = 0$ and $a_2(X + k) + b_2(Y + l) + c_2 = 0$ pass through the origin we must have

$$a_1k + b_1l + c_1 = 0 \tag{3}$$

$$a_2k + b_2l + c_2 = 0 \tag{4}$$

which will reduce equation (2) to $\dfrac{dY}{dX} = \dfrac{a_1X + b_1Y}{a_2X + b_2Y}$

This is now homogeneous and can be solved by the method given under type 7.

Example 10 Solve the equation $\dfrac{dy}{dx} = \dfrac{2x + y - 1}{x + 2y + 1}$.

Let $x = X + k$ and $y = Y + l \Rightarrow \dfrac{dY}{dX} = \dfrac{dy}{dx}$

Choose k and l so that the lines $2(X + k) + (Y + l) - 1 = 0$ and $X + k + 2(Y + l) + 1 = 0$ meet at the origin. We have

$$2k + l - 1 = 0$$

$$k + 2l + 1 = 0$$

Solving gives $k = 1$ and $l = -1$.
Hence the substitutions required are $x = X + 1$ and $y = Y - 1$ which will transform the original equation into

$$\frac{dY}{dX} = \frac{2X + Y}{X + 2Y} = \frac{2 + \left(\dfrac{Y}{X}\right)}{1 + 2\left(\dfrac{Y}{X}\right)}$$

If $Y = vX$, $\dfrac{dY}{dX} = v + X\dfrac{dv}{dX}$

The equation becomes $v + X \dfrac{dv}{dX} = \dfrac{2+v}{1+2v}$

$\therefore \qquad X \dfrac{dv}{dX} = \dfrac{2+v}{1+2v} - v = \dfrac{2(1-v^2)}{1+2v}$

Separating the variables and integrating with respect to X we have

$$\int \frac{1+2v}{1-v^2}\, dv = \int \frac{2}{X}\, dX$$

$$\therefore \frac{1}{2} \int \left(\frac{3}{1-v} - \frac{1}{1+v} \right) dv = 2 \ln X + \ln k$$

$$-\tfrac{3}{2} \ln(1-v) - \tfrac{1}{2}\ln(1+v) = \ln kX^2$$

$$-\tfrac{1}{2} \ln[(1-v)^3(1+v)] = \ln kX^2$$

$$\ln[(1-v)^3(1+v)] = \ln(kX^2)^{-2}$$

i.e. $\qquad (1-v)^3(1+v) = cX^{-4}$ where $c = k^{-2}$

Now $v = \dfrac{Y}{X}$ $\quad \therefore \left(1 - \dfrac{Y}{X}\right)^3 \left(1 + \dfrac{Y}{X}\right) = cX^{-4}$

$$(X - Y)^3(X + Y) = c$$

But $X = x - 1$ and $Y = y + 1$ and the general solution is given by

$$(x - y - 2)^3(x + y) = c$$

A modification is required when the two lines are parallel since then there is no finite point of intersection. In this case we let

$$a_1 x + b_1 y = z \qquad (5)$$

Differentiating (5) with respect to x gives $a_1 + b_1 \dfrac{dy}{dx} = \dfrac{dz}{dx}$

and substituting into the equation (1) on page 256 we have

$$\frac{1}{b_1}\left(\frac{dz}{dx} - a_1 \right) = \frac{z + c_1}{\left(\dfrac{a_2}{a_1}\right) z + c_2}$$

i.e. $\qquad \dfrac{dz}{dx} = a_1 + \dfrac{b_1(z + c_1)}{\left(\dfrac{a_2}{a_1}\right) z + c_2}$

which can be solved directly giving z, and hence y, in terms of x only.

Example 11 Solve the equation $\dfrac{dy}{dx} = \dfrac{x+y+1}{x+y+3}$.

Make the substitution $x + y = z \Rightarrow 1 + \dfrac{dy}{dx} = \dfrac{dz}{dx}$

Hence the equation becomes

$$\frac{dz}{dx} - 1 = \frac{z+1}{z+3} \Rightarrow \frac{dz}{dx} = \frac{2z+4}{z+3}$$

Separating the variables and integrating

$$\int \frac{z+3}{z+2} \, dz = \int 2 \, dx$$

$$\int 1 + \frac{1}{z+2} \, dz = \int 2 \, dx$$

$$\therefore \qquad z + \ln(z+2) = 2x + c$$

But $z = x + y$, and so the general solution is

$$x + y + \ln(x + y + 2) = 2x + c$$

i.e. $\qquad y - x + \ln(x + y + 2) = c$

Second order equations
Type 9. Equations of the form $\dfrac{d^2y}{dx^2} = f(x)$

These can be solved by direct integration and will clearly give rise to two arbitrary constants.

Example 12 Solve the equation $\dfrac{d^2y}{dx^2} = \cos x$.

By integration with respect to x

$$\frac{dy}{dx} = \sin x + A$$

and $\qquad y = -\cos x + Ax + B$

Type 10. Equations of the form $\dfrac{d^2y}{dx^2} = f(y)$

To solve equations of this form we use a substitution. Let

$$p = \frac{dy}{dx} \Rightarrow \frac{d^2y}{dx^2} = \frac{dp}{dx} = \frac{dp}{dy} \cdot \frac{dy}{dx} = p \frac{dp}{dy}$$

Hence the equation can be written as $p\dfrac{dp}{dy} = f(y)$

Integrating with respect to y gives $\frac{1}{2}p^2 = \displaystyle\int f(y)\,dy + c$. This has reduced the equation to a first order equation which can be solved by one of the methods previously described.

Example 13 Solve the equation $\dfrac{d^2y}{dx^2} = 2y$ given that when

$x = 0$, $y = 2$ and $\dfrac{dy}{dx} = 0$.

Using the substitution given above the equation becomes

$$p\frac{dp}{dy} = 2y$$

Integrating with respect to y gives

$$\tfrac{1}{2}p^2 = y^2 + c \qquad (1)$$

Now when $y = 2$, $\dfrac{dy}{dx} = 0$, hence $c = -4$.

Since $p = \dfrac{dy}{dx}$ equation (1) may be written $\dfrac{1}{2}\left(\dfrac{dy}{dx}\right)^2 = y^2 - 4$

$$\frac{1}{\sqrt{2}}\frac{dy}{dx} = \sqrt{y^2 - 4}$$

Separating the variables and integrating

$$\int \frac{1}{\sqrt{y^2 - 4}}\,dy = \int \sqrt{2}\,dx$$

$$\cosh^{-1}\left(\frac{y}{2}\right) = \sqrt{2}x + c_1$$

$\therefore \qquad\qquad y = 2\cosh(\sqrt{2}x + c_1)$

when $y = 2$, $x = 0 \Rightarrow c_1 = 0$

$\therefore \qquad\qquad y = 2\cosh\sqrt{2}x = e^{\sqrt{2}x} + e^{-\sqrt{2}x}$

Type 11. Second order linear differential equations with constant coefficients

Consider the equation $\qquad a\dfrac{d^2y}{dx^2} + b\dfrac{dy}{dx} + cy = f(x) \qquad (1)$

where a, b and c are constants.

Before attempting to solve the complete differential equation we consider the case when $f(x) = 0$.

i.e. the equation can be simplified by dividing by a

$$\frac{d^2y}{dx^2} + m\frac{dy}{dx} + ny = 0 \qquad (2)$$

Clearly any solution to equation (2) must be such that the function itself, its first and second differentials, must satisfy the differential equation when substituted.

An obvious possibility is an exponential function, say $y = Ae^{kx}$.

Now $\frac{dy}{dx} = Ake^{kx}$ and $\frac{d^2y}{dx^2} = Ak^2e^{kx}$

Substituting into equation (2) gives, after dividing by Ae^{kx},

$$k^2 + mk + n = 0 \qquad (3)$$

This is known as the **auxiliary equation** of equation (2). If its roots are k_1 and k_2 then $y = Ae^{k_1x}$ and $y = Ae^{k_2x}$ will be solutions of equation (2).

The complete solution needs two arbitrary constants and hence a solution could be $y = Ae^{k_1x} + Be^{k_2x}$.

That this is the solution can be verified by substituting in equation (2).

The type of solution depends on the nature of the roots of the auxiliary equation. There are three cases:

(i) $m^2 > 4n$, i.e. the roots are real and distinct,

(ii) $m^2 = 4n$, i.e. the roots are real and equal,

(iii) $m^2 < 4n$, i.e. the roots are complex.

(i) **Real unequal roots** mean that the auxiliary equation gives two real unequal solutions and hence the solution of the differential can be written down.

Example 14 Solve the equation $\frac{d^2x}{dt^2} + \frac{dx}{dt} - 6x = 0$.

The auxiliary equation will be $k^2 + k - 6 = 0$ having assumed a solution $x = Ae^{kt}$.

Hence $(k+3)(k-2) = 0 \Rightarrow k = 2$ or -3.

The complete solution is $x = Ae^{2t} + Be^{-3t}$.

(ii) **Real equal roots,** mean that the auxiliary equation gives repeated roots.

Let the roots be k_1, twice. In this case it is not possible to form a general solution with two arbitrary constants since it would be

$$y = Ae^{k_1 x} + Be^{k_1 x} = (A + B)e^{k_1 x}$$

i.e. $y = Ce^{k_1 x}$ and thus there is only one arbitrary constant.
In this situation we have to use a modified solution in the form $y = (A + Bx)e^{k_1 x}$.
Again this can be verified by substituting into the original equation.

Example 15 Solve the equation $\dfrac{d^2y}{dx^2} + 6\dfrac{dy}{dx} + 9y = 0$.

The auxiliary equation is $k^2 + 6k + 9 = 0$.
Hence $(k + 3)^2 = 0 \Rightarrow k = -3$ twice.
The complete general solution is $y = (A + Bx)e^{-3x}$.

(iii) **Complex roots** mean that the auxiliary equation has no real solutions.

However, the roots must occur as a conjugate pair since their sum is minus the coefficient of k, which is real. Let the roots be $k + jn$ and $k - jn$ where $j^2 = -1$.
Hence the general solution will be

$$y = Ae^{(k+jn)x} + Be^{(k-jn)x}.$$

This appears to be a complex solution but it need not necessarily contain imaginary values.
Rewriting we have $y = (Ae^{jnx} + Be^{-jnx})e^{kx}$.
But $e^{jnx} = \cos nx + j \sin nx$ and $e^{-jnx} = \cos nx - j \sin nx$

$\therefore \qquad y = e^{kx}[(A + B)\cos nx + j(A - B)\sin nx]$

Writing $A + B = C$ and $j(A - B) = D$ we have

$$y = e^{kx}[C \cos nx + D \sin nx]$$

Example 16 Solve the equation $\dfrac{d^2y}{dx^2} + 2\dfrac{dy}{dx} + 2y = 0$.

The auxiliary equation is $k^2 + 2k + 2 = 0$

$$k = \frac{-2 \pm \sqrt{2^2 - 4(1)(2)}}{2} = -1 \pm j$$

Hence the general solution is $y = Ae^{(-1+j)x} + Be^{(-1-j)x}$

$$= e^{-x}(Ae^{jx} + Be^{-jx})$$

This can be rearranged as above ($k = -1, n = 1$) to give a solution in the form $y = e^{-x}(C \cos x + D \sin x)$.

We now return to consider the full differential equation in the form

$$\frac{d^2y}{dx^2} + m\frac{dy}{dx} + ny = f(x) \qquad (1)$$

Let $y = v$ be a **particular integral** of the equation. Consider a solution of the form $y = u + v$ where u contains two arbitrary constants.

Substituting in equation (1)

$$\frac{d^2u}{dx^2} + \frac{d^2v}{dx^2} + m\left(\frac{du}{dx} + \frac{dv}{dx}\right) + n(u + v) = f(x)$$

But since $y = v$ is a particular solution

$$\frac{d^2v}{dx^2} + m\frac{dv}{dx} + nv = f(x)$$

Hence $\qquad\qquad \dfrac{d^2u}{dx^2} + m\dfrac{du}{dx} + nu = 0 \qquad (2)$

i.e. u satisfies the **reduced equation** (2). But this equation is simply the original with the RHS = zero. The general solution of this equation can be found by one of the above methods. $y = u$ is a general solution and is called the **complementary function**.

Hence the complete solution is $y = u + v$ where u is the complementary function (i.e. the solution of the reduced equation) and v is a particular solution.

To find a **particular solution** it is possible to use the theory of operators but we can determine particular integrals of some specific functions by trial. The usual types of functions for $f(x)$ are given below.

(i) $f(x) = c$ (a constant). A particular integral would be $y = c/n$ since both derivatives are zero.

(ii) $f(x) = $ polynomial of degree n. A trial solution would also be a polynomial in x of degree n, the coefficients being found by substitution. Note that if $n = 0$ a polynomial of degree $(n + 1)$ should be tried.

(iii) $f(x) =$ exponential (Ae^{ax}). A trial solution would also be an exponential since $\frac{d}{dx}(e^{ax}) = ae^{ax}$. Note that if e^{ax} is a solution of the complementary function this solution fails and we try Bxe^{ax}.

(iv) $f(x) = A \sin x$ or $B \cos x$ or $A \sin x + B \cos x$. A trial solution would also be a function in this form, say $C \sin x + D \cos x$.

(v) $f(x)$ is the sum of any of these cases, then the particular integrals for each term can be found and the results summed.

Example 17 Solve the equation $\dfrac{d^2y}{dx^2} + \dfrac{dy}{dx} - 6y = 12x + 4$.

For the particular integral try $y_1 = ax + b$.
Substituting in given equation

$$a - 6(ax + b) = 12x + 4$$

Equating coefficients $-6a = 12 \Rightarrow a = -2$

$$a - 6b = 4 \Rightarrow b = -1$$

$\therefore$ the particular integral is $y_1 = -2x - 1$.
Now the reduced equation is

$$\frac{d^2y}{dx^2} + \frac{dy}{dx} - 6y = 0$$

Its auxiliary equation will be $k^2 + k - 6 = 0$.
Thus $(k + 3)(k - 2) = 0 \Rightarrow k = 2$ or -3.
Hence the complementary function is $y_2 = Ae^{2x} + Be^{-3x}$.
The complete solution is

$$y = Ae^{2x} + Be^{-3x} - 2x - 1$$

Example 18 Solve the equation $\dfrac{d^2y}{dt^2} + 2\dfrac{dy}{dt} + y = 6 \cos t$ given that $y = 0$, $\dfrac{dy}{dt} = 0$ when $t = 0$.

For the particular integral try $y_1 = A \cos t + B \sin t$.
Substituting we have

$$(-A \cos t - B \sin t) + 2(-A \sin t + B \cos t)$$
$$+ (A \cos t + B \sin t) = 6 \cos t$$

Equating coefficients of $\cos t$ and $\sin t$

$$-A + 2B + A = 6 \Rightarrow 2B = 6 \Rightarrow B = 3$$
$$-B - 2A + B = 0 \Rightarrow A = 0$$

$\therefore$ The particular integral is $y_1 = 3 \sin t$.

The reduced equation is

$$\frac{d^2y}{dt^2} + 2\frac{dy}{dt} + y = 0$$

Its auxiliary equation is $k^2 + 2k + 1 = 0 \Rightarrow (k+1)^2 = 0$.
The complementary function is thus $y_2 = (At + B)e^{-t}$.
The complete general solution is $y = (At + B)e^{-t} + 3 \sin t$.
Now $y = 0$ when $t = 0$ $\quad \therefore \quad B = 0$

$$\frac{dy}{dt} = -Ate^{-t} + Ae^{-t} - Be^{-t} + 3\cos t \Rightarrow A = -3 \text{ as } \frac{dy}{dt} = 0 \text{ when}$$
$t = 0$

The particular solution is $y = 3(\sin t - te^{-t})$

Numerical solution of differential equations

The equations so far have yielded exact solutions but there are many differential equations for which such solutions cannot be found. In these cases we use a numerical approximation process which produces the coordinates of points which *nearly* lie on the solution curve. This process is sometimes called a **step-by-step solution.**

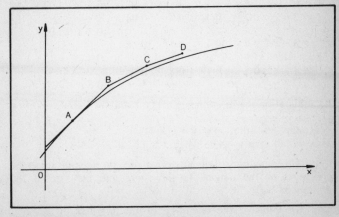

Figure 109

Consider fig. 109 which shows a solution curve to a given differential equation. We shall try to find points on this curve starting at a known point A. If, from A, we move along the tangent at A a displacement δx in the x direction then the increment in y, $\delta y \simeq \dfrac{dy}{dx} \cdot \delta x$.

This will define point B. We now move from B along a line whose direction is found by substituting the coordinates of B into the differential equation. Note that this is only approximately the same line as the tangent at B. Thus we shall determine the coordinates of C and by repetition the coordinates of D and so on. The following example will illustrate the technique.

Example 19 Solve the equation $\dfrac{dy}{dx} = 2x$ for
$x = 0 \cdot 0(0 \cdot 1)0 \cdot 5$ given that $y = 4$ when $x = 0$.

The notation $x = 0 \cdot 0(0 \cdot 1)0 \cdot 5$ means x takes the values from $0 \cdot 0$ to $0 \cdot 5$ in steps of $0 \cdot 1$.
The working is best done in tabular form

interval	δx	$\dfrac{dy}{dx}$	δy	x	y
				0	4·00
$0 \cdot 0 \leqslant x \leqslant 0 \cdot 1$	0·1	0	0		
				0·1	4·00
$0 \cdot 1 \leqslant x \leqslant 0 \cdot 2$	0·1	0·2	0·02		
				0·2	4·02
$0 \cdot 2 \leqslant x \leqslant 0 \cdot 3$	0·1	0·4	0·04		
				0·3	4·06
$0 \cdot 3 \leqslant x \leqslant 0 \cdot 4$	0·1	0·6	0·06		
				0·4	4·12
$0 \cdot 4 \leqslant x \leqslant 0 \cdot 5$	0·1	0·8	0·08		
				0·5	4·20

By integration the solution is $y = x^2 + c$ and the initial condition gives $c = 4$.
The analytical solution is $y = x^2 + 4$.
This would give $y = 4 \cdot 25$ when $x = 0 \cdot 5$.

Clearly the accuracy of the numerical process depends upon the closeness of the tangent to the curve and the size of the increment δx.

We can improve the accuracy of the method by using a quadratic approximation instead of a linear one. So instead of using $\delta y \simeq \dfrac{dy}{dx} \delta x = f'(x)\delta x$ we make use of the second term of the Taylor series, i.e. $\delta y \simeq f'(x)\delta x + \frac{1}{2}f''(x)(\delta x)^2$

Example 20 Solve the differential equation $\dfrac{dy}{dx} = x + y$ for $0{\cdot}0(0{\cdot}4)2{\cdot}0$ given that when $x = 0$, $y = 1$.

We shall need an initial value for $\dfrac{d^2y}{dx^2} = 1 + \dfrac{dy}{dx}$

$f'(x)$	$f''(x)$	δx	$f'(x)\delta x$	$\frac{1}{2}f''(x)(\delta x)^2$	δy	x	y
						0	1
1	2	0·4	0·4	0·16	0·56		
						0·4	1·56
1·96	2·96	0·4	0·7840	0·2368	1·0208		
						0·8	2·5808
3·3808	4·3808	0·4	1·3523	0·3505	1·7028		
						1·2	4·2836
5·4836	6·4836	0·4	2·1934	0·5187	2·7121		
						1·6	6·9957
8·5957	9·5957	0·4	3·4383	0·7677	4·2060		
						2·0	11·2017

This equation can be solved by using the integrating factor,

$$\text{i.e. } \frac{dy}{dx} - y = x$$

The integrating factor is $e^{\int -1\,dx} = e^{-x}$.

$$\therefore \quad e^{-x}\frac{dy}{dx} - e^{-x}y = e^{-x}x \Rightarrow \frac{d}{dx}(e^{-x}y) = xe^{-x}$$

Integrating $ye^{-x} = \displaystyle\int xe^{-x}\,dx$

$$= \int -xe^{-x} + \int e^{-x}\,dx \qquad = -xe^{-x} - e^{-x} + C$$

The general solution is $y = -x - 1 + Ce^x$

But $x = 0$ when $y = 1$ $\quad \therefore \quad C = 2 \quad\quad y = 2e^x - x - 1$

Comparing the values obtained from the numerical method with those obtained by the analytical method we can see the error involved. This can be seen from the table below where values are given to two decimal places.

x	0·4	0·8	1·2	1·6	2·0
Numerical y	1·56	2·58	4·28	7·00	11·20
Analytical y	1·58	2·65	4·44	7·31	11·78

Key terms

First order 1. $\dfrac{dy}{dx} = f(x)$ can be solved by direct integration.

2. $\dfrac{dy}{dx} = f(y) \Rightarrow \displaystyle\int \dfrac{1}{f(y)} \, dy = \int dx$ which can be integrated directly.

3. Variables separable These equations can be written in the form $f(y)\dfrac{dy}{dx} + g(x) = 0 \Rightarrow \displaystyle\int f(y) \, dy + \int g(x) \, dx = C.$

4. Exact equations A differential equation is exact if it is formed from the equation of the curve by direct differentiation. In general $M + N \dfrac{dy}{dx} = 0$ is exact if and only if $\dfrac{\partial M}{\partial y} = \dfrac{\partial N}{\partial x}.$

5. Linear equations $\dfrac{dy}{dx} + Py = Q$ is called linear if P and Q are constants or functions of x only. They are solved by multiplying by $e^{\int P \, dx}$ which is called the integrating factor.

6. Homogeneous equations $P \dfrac{dy}{dx} + Q = 0$ where P and Q are homogeneous expressions of the same degree is solved by substituting $y = vx$.

Second order

7. $\dfrac{d^2y}{dx^2} = f(x)$ is solved by direct integration.

8. $\dfrac{d^2y}{dx^2} = f(y)$ is solved by using the substitution $p = \dfrac{dy}{dx}.$

9. Linear equations $\dfrac{d^2y}{dx^2} + m \dfrac{dy}{dx} + ny = 0$. If m and n are constants the **auxiliary equation** is $k^2 + mk + n = 0$ and, depending on the roots, will give solutions of

$y = Ae^{kx} + Be^{kx}$, $y = (A + Bx)e^{kx}$ or
$y = e^{kx}(C \cos nx + D \sin nx)$.

The solution of $\dfrac{d^2y}{dx^2} + b \dfrac{dy}{dx} + cy = f(x)$ is $y = u + v$ where u is the **complementary function** (i.e. the solution of the **reduced equation** obtained by putting $f(x) = 0$) and v is a **particular integral** found by trial.

Chapter 11
Vectors

Many physical quantities that are studied in mechanics are vectors. For example, velocities, accelerations and forces are vector quantities, and in the physical sense a vector is a quantity which has direction as well as magnitude. It is not only necessary to know how large a force is but also to know in which direction the force acts and even through which point it acts. However, the properties of vectors are not confined to mechanics. Vector methods are so powerful that one finds many applications in geometry and complex numbers and they comprise a part of pure mathematics on their own.

Two dimensional vectors can be represented with the aid of cartesian or polar coordinates. In fig. 110a the displacement **OP** can be represented by the 2×1 matrix $\begin{pmatrix} x \\ y \end{pmatrix}$ which is known as a column vector. **OP** could also be represented by its polar coordinates $[r, \theta]$ but this representation is only convenient when we multiply complex numbers (Chapter 9).

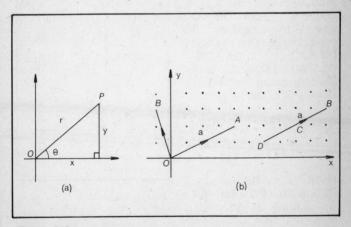

Figure 110

Two dimensional vectors can be represented by displacements which can be represented by column vectors. In fig. 110b

$$\mathbf{OA} = \begin{pmatrix} 4 \\ 2 \end{pmatrix}, \qquad \mathbf{OB} = \begin{pmatrix} -1 \\ 3 \end{pmatrix}, \qquad \mathbf{DB} = \begin{pmatrix} 4 \\ 2 \end{pmatrix}$$

There are many notations used for vectors and it is wise to be aware of most of them. Vectors are denoted by boldface type **OA** or as directed line segments $\overrightarrow{OA}$ or sometimes by a single letter **a**. When writing them they must be underlined to distinguish them from numbers (which only have magnitude).

OA and **OB** are the **position vectors** of the points A and B with respect to the origin O. In general the vector **OA** = **DB** but care must be taken if these vectors represent forces (as they act through different points).

Multiplication by a scalar (number)

In fig. 110b $\mathbf{DB} = \begin{pmatrix} 4 \\ 2 \end{pmatrix} = 2 \begin{pmatrix} 2 \\ 1 \end{pmatrix} = 2\mathbf{DC}$

which represents a vector in the same direction with double the magnitude. $k\mathbf{DB}$ represents a vector in the same direction as **DB** but k times larger.

Addition of vectors

Vectors are added by the parallelogram law. From fig. 111 **OA** + **OB** = **OD** where D completes the parallelogram $OADB$. Since **OB** = **AD**, **OA** + **OB** = **OA** + **AD** = **OD** and vectors can be added by drawing them nose to tail.

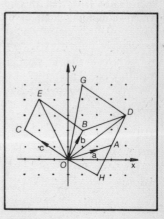

Figure 111

$\mathbf{OA} = \begin{pmatrix} 3 \\ 1 \end{pmatrix}$, $\mathbf{OB} = \begin{pmatrix} 1 \\ 2 \end{pmatrix}$, $\mathbf{OD} = \begin{pmatrix} 4 \\ 3 \end{pmatrix}$

$$\begin{pmatrix} 3 \\ 1 \end{pmatrix} + \begin{pmatrix} 1 \\ 2 \end{pmatrix} = \begin{pmatrix} 4 \\ 3 \end{pmatrix}$$

If **OA** = **a** and **OB** = **b**
OD = **a** + **b** = **OB** + **BD** = **b** + **a**,
so addition of vectors is **commutative**.

If **OC** = **c**, (**a** + **b**) + **c** = **OD** + **c** = **OD** + **DG** = **OG**.
a + (**b** + **c**) = **OA** + **OE** = **OA** + **AG** = **OG**
(**a** + **b**) + **c** = **a** + (**b** + **c**) and vector addition is **associative**.

Subtraction of vectors

$$\mathbf{OB} = \mathbf{b} = \begin{pmatrix} 1 \\ 2 \end{pmatrix}, \ \mathbf{BO} = -\mathbf{b} = -\begin{pmatrix} 1 \\ 2 \end{pmatrix} = \begin{pmatrix} -1 \\ -2 \end{pmatrix}$$

$$\mathbf{a} - \mathbf{b} = \mathbf{OA} - \mathbf{OB} = \mathbf{OA} + \mathbf{BO} = \mathbf{OA} + \mathbf{AH} = \mathbf{OH},$$

$$\mathbf{a} - \mathbf{b} = \begin{pmatrix} 3 \\ 1 \end{pmatrix} - \begin{pmatrix} 1 \\ 2 \end{pmatrix} = \begin{pmatrix} 2 \\ -1 \end{pmatrix}$$

$$\mathbf{a} - \mathbf{b} = -\mathbf{b} + \mathbf{a} = \mathbf{BO} + \mathbf{OA} = \mathbf{BA} \text{ and similarly } \mathbf{AB} = \mathbf{b} - \mathbf{a}.$$

These results have been derived for two dimensional vectors which are easy to represent with the aid of coordinates in the plane. Many of the results derived for the vectors in two dimensions are true in three dimensions although these results may be difficult to illustrate. The distinction between two and three dimensional vectors will be made when it is necessary. It is difficult to illustrate three dimensional vectors on the two dimensional plane of the paper and students find difficulty in visualizing three dimensional properties of figures, lines and planes. Consequently the study of vectors can appear theoretical when it should be practically illustrated.

Ratio theorem

The vectors $\mathbf{a}$ and $\mathbf{b}$ define a plane containing O, A and B (fig. 112). Any linear combination of $\mathbf{a}$ and $\mathbf{b}$ (e.g. $3\mathbf{a} + 2\mathbf{b}$) will define a vector lying in the same plane of OAB.

If P divides AB in the ratio $\lambda : \mu$ so that $\dfrac{AP}{PB} = \dfrac{\lambda}{\mu}$ then

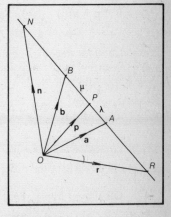

Figure 112

$$\mathbf{OP} = \mathbf{p} = \mathbf{a} + \mathbf{AP}$$

$$= \mathbf{a} + \frac{\lambda}{\lambda + \mu} \times \mathbf{AB}$$

$$= \mathbf{a} + \frac{\lambda}{\lambda + \mu} (\mathbf{b} - \mathbf{a})$$

$$= \left(1 - \frac{\lambda}{\lambda + \mu}\right) \mathbf{a} + \frac{\lambda}{\lambda + \mu} \mathbf{b}$$

So $\mathbf{p} = \dfrac{\mu}{\lambda + \mu} \mathbf{a} + \dfrac{\lambda}{\lambda + \mu} \mathbf{b}$

Ratio theorem P divides AB in the ratio $s:t \Leftrightarrow \mathbf{p} = t\mathbf{a} + s\mathbf{b}$ where $s + t = 1$.

In particular if M is the mid-point of AB, $\mathbf{OM} = \mathbf{m} = \frac{1}{2}\mathbf{a} + \frac{1}{2}\mathbf{b}$. If Q is the point of trisection nearer A, $\dfrac{AQ}{QB} = \dfrac{1}{2}$ so $\mathbf{OQ} = \mathbf{q} = \frac{2}{3}\mathbf{a} + \frac{1}{3}\mathbf{b}$.

If $BA = AR$ then $AR:RB = -1:2$ so $\mathbf{r} = 2\mathbf{a} - \mathbf{b}$. (fig. 112)
If $AB = BN$ then $AN:NB = 2:-1$ so $\mathbf{n} = -\mathbf{a} + 2\mathbf{b}$.
Points on the line AB but outside AB (like R and N) divide AB in a negative ratio. If this is not clearly understood it is easier to say A is the mid-point of BR

$$\Rightarrow \mathbf{a} = \tfrac{1}{2}\mathbf{r} + \tfrac{1}{2}\mathbf{b} \Rightarrow \mathbf{r} = 2\mathbf{a} - \mathbf{b}$$

B is the mid-point of AN

$$\Rightarrow \mathbf{b} = \tfrac{1}{2}\mathbf{a} + \tfrac{1}{2}\mathbf{n} \Rightarrow \mathbf{n} = 2\mathbf{b} - \mathbf{a} = -\mathbf{a} + 2\mathbf{b}$$

Alternative approach Any point P on the line AB satisfies $\mathbf{p} = \mathbf{a} + s \times \mathbf{AB} = \mathbf{a} + s(\mathbf{b} - \mathbf{a})$ where $\mathbf{AP} = s \times \mathbf{AB} \Rightarrow s = \dfrac{AP}{AB}$ (the ratio of $AP:AB$ is $s:1$)

$$\mathbf{p} = \mathbf{a} + s(\mathbf{b} - \mathbf{a}) = \mathbf{a} + s\mathbf{b} - s\mathbf{a} = (1-s)\mathbf{a} + s\mathbf{b}$$

$\mathbf{p} = (1 - s)\mathbf{a} + s\mathbf{b}$ is the vector equation of the line AB.

$$s = 0 \Rightarrow \mathbf{p} = \mathbf{a}, \quad s = 1 \Rightarrow \mathbf{p} = \mathbf{b}, \quad s = \tfrac{1}{2} \Rightarrow \mathbf{p} = \tfrac{1}{2}\mathbf{a} + \tfrac{1}{2}\mathbf{b}$$

$$s = 2 \Rightarrow \mathbf{p} = -\mathbf{a} + 2\mathbf{b}, \quad s = -1 \Rightarrow \mathbf{p} = 2\mathbf{a} - \mathbf{b}$$

Each value of s defines a point on the line AB and as s takes values from $-\infty$ to $+\infty$ all points on the line AB are described. With the notation of fig. 112

$$\mathbf{AP} = \frac{\lambda}{\lambda + \mu} \times \mathbf{AB} \Rightarrow \mathbf{p} = \left(1 - \frac{\lambda}{\lambda + \mu}\right)\mathbf{a} + \left(\frac{\lambda}{\lambda + \mu}\right)\mathbf{b}$$

$$= \frac{\mu}{\lambda + \mu}\mathbf{a} + \frac{\lambda}{\lambda + \mu}\mathbf{b}$$

and P divides AB in the ratio $\lambda : \mu$. From fig. 113.

$$\mathbf{OB} = \mathbf{b} = \binom{5}{2} = \binom{5}{0} + \binom{0}{2} = 5\binom{1}{0} + 2\binom{0}{1} = 5\mathbf{i} + 2\mathbf{j}$$

$$\mathbf{i} = \binom{1}{0}, \quad \mathbf{j} = \binom{0}{1}, \quad \mathbf{a} = \binom{1}{4}$$

$\mathbf{i}$ is the unit vector (length 1) in the direction of the x axis. $\mathbf{j}$ is the unit vector (length 1) in the direction of the y axis.

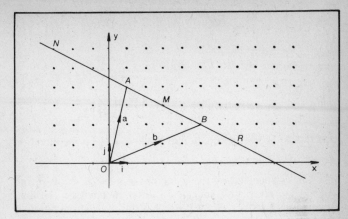

Figure 113

In three dimensions $\mathbf{i} = \begin{pmatrix} 1 \\ 0 \\ 0 \end{pmatrix}$, $\mathbf{j} = \begin{pmatrix} 0 \\ 1 \\ 0 \end{pmatrix}$, $\mathbf{k} = \begin{pmatrix} 0 \\ 0 \\ 1 \end{pmatrix}$

$\mathbf{k}$ is the unit vector of the z axis.

(Some examining boards use the notation $\mathbf{i}$, $\mathbf{j}$, $\mathbf{k}$, instead of column vectors.)

From fig. 113 M is the mid-point of AB

$$\mathbf{OM} = \tfrac{1}{2}\mathbf{a} + \tfrac{1}{2}\mathbf{b} = \tfrac{1}{2}(\mathbf{a} + \mathbf{b}) = \frac{1}{2}\left[\begin{pmatrix} 1 \\ 4 \end{pmatrix} + \begin{pmatrix} 5 \\ 2 \end{pmatrix}\right] = \frac{1}{2}\begin{pmatrix} 6 \\ 6 \end{pmatrix} = \begin{pmatrix} 3 \\ 3 \end{pmatrix} = 3\mathbf{i} + 3\mathbf{j}$$

M has coordinates $(3, 3)$.

The point Q dividing AB in the ratio $1:3$ is given by

$$\mathbf{OQ} = \tfrac{3}{4}\mathbf{a} + \tfrac{1}{4}\mathbf{b} = \tfrac{1}{4}(3\mathbf{a} + \mathbf{b}) = \frac{1}{4}\left[\begin{pmatrix} 3 \\ 12 \end{pmatrix} + \begin{pmatrix} 5 \\ 2 \end{pmatrix}\right] = \frac{1}{4}\begin{pmatrix} 8 \\ 14 \end{pmatrix} = \begin{pmatrix} 2 \\ 3\tfrac{1}{2} \end{pmatrix}$$

i.e. Q is $(2, 3\tfrac{1}{2})$.

To find N: A is the mid-point of BN so $\mathbf{a} = \tfrac{1}{2}\mathbf{ON} + \tfrac{1}{2}\mathbf{b}$

$$\mathbf{ON} = 2\mathbf{a} - \mathbf{b} = \begin{pmatrix} 2 \\ 8 \end{pmatrix} - \begin{pmatrix} 5 \\ 2 \end{pmatrix} = \begin{pmatrix} -3 \\ 6 \end{pmatrix} \quad \text{i.e.} \quad N \text{ is } (-3, 6)$$

To find R; B divides AR in the ratio $2:1$, so $\mathbf{b} = \tfrac{2}{3}\mathbf{r} + \tfrac{1}{3}\mathbf{a}$, $3\mathbf{b} = 2\mathbf{r} + \mathbf{a}$.

$$\mathbf{r} = \tfrac{3}{2}\mathbf{b} - \tfrac{1}{2}\mathbf{a} = \tfrac{1}{2}(3\mathbf{b} - \mathbf{a}) = \frac{1}{2}\left[\begin{pmatrix} 15 \\ 6 \end{pmatrix} - \begin{pmatrix} 1 \\ 4 \end{pmatrix}\right] = \frac{1}{2}\begin{pmatrix} 14 \\ 2 \end{pmatrix} = \begin{pmatrix} 7 \\ 1 \end{pmatrix}$$

$$\mathbf{OM} = \begin{pmatrix} 1 \\ 4 \end{pmatrix} + \begin{pmatrix} 2 \\ -1 \end{pmatrix}, \quad \mathbf{OB} = \begin{pmatrix} 1 \\ 4 \end{pmatrix} + 2\begin{pmatrix} 2 \\ -1 \end{pmatrix}, \quad \mathbf{OR} = \begin{pmatrix} 1 \\ 4 \end{pmatrix} + 3\begin{pmatrix} 2 \\ -1 \end{pmatrix},$$

$$\mathbf{ON} = \begin{pmatrix} 1 \\ 4 \end{pmatrix} + -2\begin{pmatrix} 2 \\ -1 \end{pmatrix}$$

All points on AB satisfy $\mathbf{r} = \begin{pmatrix} x \\ y \end{pmatrix} = \begin{pmatrix} 1 \\ 4 \end{pmatrix} + t\begin{pmatrix} 2 \\ -1 \end{pmatrix}$

and this is the vector equation of the line; $\begin{pmatrix} 2 \\ -1 \end{pmatrix}$ is the direction vector of the line.

The line is specified by the equations $x = 1 + 2t$, $y = 4 - t$ and each value given to t specifies a different point on the line. Eliminating t gives $x + 2y = 9$ which is the cartesian equation of the line.

The vector equation of a line is not unique. $\begin{pmatrix} -2 \\ 1 \end{pmatrix}$ can be used as a direction vector for the line AB (fig. 113).

$$\mathbf{OA} = \begin{pmatrix} 1 \\ 4 \end{pmatrix} + 0\begin{pmatrix} -2 \\ 1 \end{pmatrix}, \; \mathbf{OB} = \begin{pmatrix} 1 \\ 4 \end{pmatrix} + -2\begin{pmatrix} -2 \\ 1 \end{pmatrix}, \; \mathbf{ON} = \begin{pmatrix} 1 \\ 4 \end{pmatrix} + 2\begin{pmatrix} -2 \\ 1 \end{pmatrix}$$

This can be summarized as $r = \begin{pmatrix} x \\ y \end{pmatrix} = \begin{pmatrix} 1 \\ 4 \end{pmatrix} + s\begin{pmatrix} -2 \\ 1 \end{pmatrix}$

or $x = 1 - 2s$, $y = 4 + s$. Eliminating s gives $x + 2y = 9$, the same equation as before in its cartesian form.

Comparing $\begin{pmatrix} x \\ y \end{pmatrix} = \begin{pmatrix} 1 \\ 4 \end{pmatrix} + t\begin{pmatrix} 2 \\ -1 \end{pmatrix}$

with $\begin{pmatrix} x \\ y \end{pmatrix} = \begin{pmatrix} 1 \\ 4 \end{pmatrix} + s\begin{pmatrix} -2 \\ 1 \end{pmatrix}$

the point A is given by $t = 0$ and $s = 0$

$\qquad B$ is given by $t = 2$ and $s = -2$

$\qquad N$ is given by $t = -2$ and $s = 2$

In general $\mathbf{r} = \mathbf{a} + t\mathbf{d}$ where $\mathbf{a}$ is the position vector of any point on the line and $\mathbf{d}$ represents a direction vector for the line. $\mathbf{d}$ must be a multiple of $\begin{pmatrix} 2 \\ -1 \end{pmatrix}$ for the line AB in fig. 113.

Vectors in three dimensions

Vectors are specified relative to mutually perpendicular axes Ox, Oy and Oz.

If the point P has coordinates $(1, 2, 3)$ then the vector **OP**

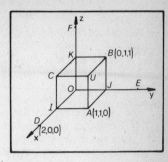

$$= \begin{pmatrix} 1 \\ 2 \\ 3 \end{pmatrix} = \begin{pmatrix} 1 \\ 0 \\ 0 \end{pmatrix} + \begin{pmatrix} 0 \\ 2 \\ 0 \end{pmatrix} + \begin{pmatrix} 0 \\ 0 \\ 3 \end{pmatrix}$$

$$= \mathbf{i} + 2\mathbf{j} + 3\mathbf{k}$$

In three dimensions

$$\mathbf{i} = \begin{pmatrix} 1 \\ 0 \\ 0 \end{pmatrix}, \quad \mathbf{j} = \begin{pmatrix} 0 \\ 1 \\ 0 \end{pmatrix}, \quad \mathbf{k} = \begin{pmatrix} 0 \\ 0 \\ 1 \end{pmatrix}$$

Figure 114

Vector equation of a line in three dimensions

Points on the line CJ (fig. 114) can be specified by adding multiples of **JC** to the vector **OJ**.

$$\mathbf{r} = \begin{pmatrix} x \\ y \\ z \end{pmatrix} = \mathbf{OJ} + t\,\mathbf{JC} = \begin{pmatrix} 0 \\ 1 \\ 0 \end{pmatrix} + t \begin{pmatrix} 1 \\ -1 \\ 1 \end{pmatrix} \quad \text{or} \quad \begin{aligned} y &= 0 + t \\ y &= 1 - t \\ z &= 0 + t \end{aligned}$$

where $J(0, 1, 0)$ is a point on the line and $\mathbf{i} - \mathbf{j} + \mathbf{k}$ is the direction vector.

The equation of the line through the point $P(1, 2, 3)$ in the direction $\begin{pmatrix} 4 \\ 5 \\ 6 \end{pmatrix}$ will be $\begin{pmatrix} x \\ y \\ z \end{pmatrix} = \begin{pmatrix} 1 \\ 2 \\ 3 \end{pmatrix} + t \begin{pmatrix} 4 \\ 5 \\ 6 \end{pmatrix}$ or $\begin{aligned} x &= 1 + 4t \\ y &= 2 + 5t \\ z &= 3 + 6t \end{aligned}$

These equations can be rearranged in the form

$t = \dfrac{x-1}{4} = \dfrac{y-2}{5} = \dfrac{z-3}{6}$ which are equivalent to two equations

$5x - 4y = -3$ and $6y - 5z = -3$.

Intersection of two lines in three dimensions

From fig. 114 we have seen that the line CJ has equation

$$\mathbf{r} = \begin{pmatrix} x \\ y \\ z \end{pmatrix} = \begin{pmatrix} 0 \\ 1 \\ 0 \end{pmatrix} + t \begin{pmatrix} 1 \\ -1 \\ 1 \end{pmatrix}$$

The line AK has direction vector $\mathbf{AK} = -\mathbf{i} - \mathbf{j} + \mathbf{k}$.

$$\mathbf{r} = \begin{pmatrix} x \\ y \\ z \end{pmatrix} = \begin{pmatrix} 1 \\ 1 \\ 0 \end{pmatrix} + s\begin{pmatrix} -1 \\ -1 \\ 1 \end{pmatrix} \quad \text{or} \quad \begin{aligned} x &= 1 - s \\ y &= 1 - s \\ z &= s \end{aligned}$$

To find the intersection of CJ with AK equate the x, y and z values

$$\begin{aligned} x &= 1 - s = 0 + t \\ y &= 1 - s = 1 - t \\ z &= t \quad\;\; = s \end{aligned}$$
Three equations with 2 unknowns! But the second and third equations give $t = s$ and using this in the first equation $s = \frac{1}{2} = t$.

CJ and AK intersect at the point where $s = \frac{1}{2}$, i.e. $(x, y, z) = (\frac{1}{2}, \frac{1}{2}, \frac{1}{2})$. Obviously $t = \frac{1}{2}$ must give the same point and it can be seen from the diagram that CJ and AK intersect at the centre of the unit cube, where the diagonals intersect. If two lines have the same direction vector they are parallel or they coincide for all points. The line CU could be written

$$\begin{pmatrix} x \\ y \\ z \end{pmatrix} = \begin{pmatrix} 1 \\ 0 \\ 1 \end{pmatrix} + t\begin{pmatrix} 0 \\ 1 \\ 0 \end{pmatrix} \quad \text{or} \quad \begin{pmatrix} x \\ y \\ z \end{pmatrix} = \begin{pmatrix} 1 \\ 1 \\ 1 \end{pmatrix} + s\begin{pmatrix} 0 \\ 1 \\ 0 \end{pmatrix}$$

by first using C as the point on the line where $t = 0$, then using U as the point on the line where $s = 0$. In fact $t = s + 1$ and the lines are coincident. The line CU has equation

$$\begin{pmatrix} x \\ y \\ z \end{pmatrix} = \begin{pmatrix} 1 \\ 0 \\ 1 \end{pmatrix} + t\begin{pmatrix} 0 \\ 1 \\ 0 \end{pmatrix} \quad \text{and } IA \text{ has equation } \begin{pmatrix} x \\ y \\ z \end{pmatrix} = \begin{pmatrix} 1 \\ 0 \\ 0 \end{pmatrix} + s\begin{pmatrix} 0 \\ 1 \\ 0 \end{pmatrix}$$

They intersect when
$$\begin{aligned} x &= 1 = 1 \\ y &= t = s \\ z &= 1 = 0 \end{aligned}$$
Since $1 \neq 0$ there is no solution to these equations and therefore they do not intersect.

CJ has equation
$$\begin{pmatrix} x \\ y \\ z \end{pmatrix} = \begin{pmatrix} 0 \\ 1 \\ 0 \end{pmatrix} + t\begin{pmatrix} 1 \\ -1 \\ 1 \end{pmatrix}$$

and IA has equation
$$\begin{pmatrix} x \\ y \\ z \end{pmatrix} = \begin{pmatrix} 1 \\ 0 \\ 0 \end{pmatrix} + s\begin{pmatrix} 0 \\ 1 \\ 0 \end{pmatrix}$$

They intersect when
$$\begin{aligned} x &= t = 1 \\ y &= 1 - t = s \\ z &= t = 0 \end{aligned}$$

The first and third equations are inconsistent and so have no solution.

In fact the lines do not intersect but are not parallel. They are **skew** lines.

Four cases arise for the intersection of lines in three dimensions:

(i) the lines meet in a single point, e.g. *CJ* and *AK*,
(ii) the lines are parallel, e.g. *JK* and *AC*,
(iii) the lines are coincident,
(iv) the lines are **skew**, e.g. *CK* and *IA*.

Direction cosines of a vector

$\mathbf{OU} = \begin{pmatrix} 1 \\ 1 \\ 1 \end{pmatrix}$ has length $\lambda = \sqrt{OI^2 + IA^2 + AU^2}$

$$= \sqrt{1^2 + 1^2 + 1^2} = \sqrt{3}$$

If *OU* makes an angle α with the *x* axis then in the right angled triangle *OIU*, $\cos \alpha = \dfrac{OI}{OU} = \dfrac{1}{\sqrt{3}}$; similarly if *OU* makes an angle β with the *y* axis and γ with the *z* axis then $\cos \beta = \dfrac{1}{\sqrt{3}} = \cos \gamma$ and *OU* is equally inclined to all three axes.

For the vector $\mathbf{OP} = \begin{pmatrix} 1 \\ 2 \\ 3 \end{pmatrix}$

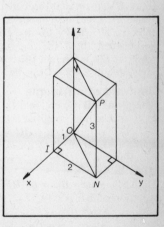

length $OP = \sqrt{ON^2 + NP^2}$
$= \sqrt{OI^2 + IN^2 + NP^2}$
$= \sqrt{1^2 + 2^2 + 3^2}$
$= \sqrt{14}$

If **OP** makes angles α, β, γ with the axes *Ox*, *Oy*, *Oz* then
$\alpha = $ angle *IOP*, $\beta = $ angle *POy*,
$\gamma = $ angle *POz*.
In $\triangle OIP$ (right-angled at *I*)
$\cos \alpha = \dfrac{OI}{OP} = \dfrac{1}{\sqrt{14}}$. Similarly

$\cos \beta = \dfrac{2}{\sqrt{14}}$, $\cos \gamma = \dfrac{3}{\sqrt{14}}$

Figure 115

The values $\dfrac{1}{\sqrt{14}}$, $\dfrac{2}{\sqrt{14}}$, $\dfrac{3}{\sqrt{14}}$ are the **direction cosines** of the vector **OP** which specify its direction with respect to the axes *Ox*, *Oy*, *Oz*. In general the length *OP* where *P* is (a, b, c) is

$l = \sqrt{a^2 + b^2 + c^2}$ and its direction cosines $\dfrac{a}{l}, \dfrac{b}{l}, \dfrac{c}{l}$.

Mid-point theorem

If D and E are the mid-points of AB and BC respectively, then DE is parallel to AC and $DE = \frac{1}{2}AC$.

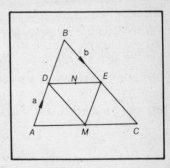

Figure 116

$\mathbf{AD} = \mathbf{a} \Rightarrow \mathbf{DB} = \mathbf{a} \Rightarrow \mathbf{AB} = 2\mathbf{a}$
$\mathbf{BE} = \mathbf{b} \Rightarrow \mathbf{EC} = \mathbf{b} \Rightarrow \mathbf{BC} = 2\mathbf{b}$
$\mathbf{DE} = \mathbf{a} + \mathbf{b}$,
$\mathbf{AC} = 2\mathbf{a} + 2\mathbf{b} = 2(\mathbf{a} + \mathbf{b})$
$\mathbf{AC} = 2\mathbf{DE}$ which implies that $\mathbf{AC}$ is parallel to $\mathbf{DE}$ and $AC = 2DE$. This 'proof' assumes the property that $2\mathbf{a} + 2\mathbf{b} = 2(\mathbf{a} + \mathbf{b})$ which is the essence of the mid-point theorem.

That $\mathbf{AC} = 2\mathbf{DE}$ is a consequence of the properties of enlargement. Alternatively, if M is the mid-point of AC, a half-turn about N (the mid-point of DE) will map $\triangle DEM$ onto $\triangle EDB$. Since EM is rotated through $180°$, $\mathbf{EM} = \mathbf{BD}$ and $\mathbf{DM} = \mathbf{BE}$.

Similarly, a half-turn about the mid-point of EM maps $\triangle DEM$ onto $\triangle CME$; $\mathbf{DM} = \mathbf{EC}$ and $\mathbf{DE} = \mathbf{MC}$.

A half-turn about the mid-point of DM maps $\triangle DEM$ onto $\triangle MAD$; $\mathbf{EM} = \mathbf{DA}$ and $\mathbf{DE} = \mathbf{AM}$.

Consequently $\mathbf{AM} = \mathbf{DE} = \mathbf{MC}$ and the theorem is proved.

Centroid of a triangle

If $\mathbf{OA} = \mathbf{a}$, $\mathbf{OB} = \mathbf{b}$ etc. and D, E, F are the mid-points of AB, AC, BC respectively, $\mathbf{OD} = \mathbf{d} = \frac{1}{2}\mathbf{a} + \frac{1}{2}\mathbf{b}$. If $CG = s \times CD$, equation of CD is $\mathbf{r} = \frac{1}{2}s(\mathbf{a} + \mathbf{b}) + (1 - s)\mathbf{c}$.

Similarly if $BG = t \times BE$ equation of BE is

$$\mathbf{r} = \frac{t}{2}(\mathbf{a} + \mathbf{c}) + (1 - t)\mathbf{b}.$$

BE and CD intersect at G where
$$\frac{1}{2}s(\mathbf{a} + \mathbf{b}) + (1 - s)\mathbf{c}$$
$$= \frac{1}{2}t(\mathbf{a} + \mathbf{c}) + (1 - t)\mathbf{b}$$ Figure 117

Since the vector representations must be equal $s = t$ (coefficients of **a**)

$s/2 = 1 - t$ (coefficients of **b**) $1 - s = t/2$ (coefficients of **c**)

Hence $s = 2/3 = t$ and $BG = 2/3\ BE$ or G divides BE in the ratio $2:1$ Similarly G divides CD in the ratio $2:1$ and AF in the ratio $2:1$. The medians of a triangle CD, BE, AF are concurrent at G and G is the centroid (centre of mass) of the triangle.

$$\mathbf{OG} = \mathbf{g} = \tfrac{1}{3}\mathbf{a} + \tfrac{1}{3}\mathbf{b} + \tfrac{1}{3}\mathbf{c} = \tfrac{1}{3}(\mathbf{a} + \mathbf{b} + \mathbf{c})$$

Example 1 Referring back to fig. 113 on page 273, the centroid of the triangle OAB is given by

$$\mathbf{OG} = \tfrac{1}{3}\left[\binom{0}{0} + \binom{1}{4} + \binom{5}{2}\right] = \tfrac{1}{3}\binom{6}{6} = \binom{2}{2}$$

It can be seen that $G(2, 2)$ divides OM in the ratio $2:1$ and similarly for the other medians from A and B.

Example 2 Referring back to fig. 114 on page 275, the centroid of triangle IJK is given by

$$\mathbf{OG} = \tfrac{1}{3}\left[\begin{pmatrix}1\\0\\0\end{pmatrix} + \begin{pmatrix}0\\1\\0\end{pmatrix} + \begin{pmatrix}0\\0\\1\end{pmatrix}\right] = \tfrac{1}{3}\begin{pmatrix}1\\1\\1\end{pmatrix} = \begin{pmatrix}\tfrac{1}{3}\\\tfrac{1}{3}\\\tfrac{1}{3}\end{pmatrix}$$

Similarly the centroid of $\triangle ABC$ has coordinates $(\tfrac{2}{3}, \tfrac{2}{3}, \tfrac{2}{3})$.

Centroid of a tetrahedron

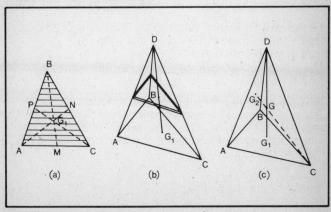

Figure 118

The centroid of a triangle is at the intersection of the medians. This is also the centre of mass (centre of gravity). Imagine the triangle divided into narrow strips parallel to AC (fig. 118a). The centre of mass of each strip is at its centre so the centre of mass of the triangle lies on the locus of the mid-points of the strips, which is the median BM. Similarly the centre of mass of the whole triangle lies on all three medians and is therefore at G_1 the centroid. Imagine the tetrahedron divided into narrow triangles parallel to the base ABC. The centre of mass of each triangle is the centroid of the triangle. The centre of mass of the tetrahedron lies on the locus of these centroids and by the properties of enlargement, or similar figures, these centroids lie on the line joining D to G_1 the centroid of $\triangle ABC$. Similarly the centre of mass G lies on the line joining each vertex to the centroid of the opposite face.

Equation of DG_1 is $\mathbf{r} = \mathbf{d} + s(\mathbf{d} - \frac{1}{3}(\mathbf{a} + \mathbf{b} + \mathbf{c}))$ using $\mathbf{OA} = \mathbf{a}$, $\mathbf{OB} = \mathbf{b}$ etc. with origin O. Equation of CG_2 is $\mathbf{r} = \mathbf{c} + t(\mathbf{c} - \frac{1}{3}(\mathbf{a} + \mathbf{b} + \mathbf{d}))$.
At G, the intersection of these lines, $\mathbf{d} + s(\mathbf{d} - \frac{1}{3}(\mathbf{a} + \mathbf{b} + \mathbf{c})) = \mathbf{c} + t(\mathbf{c} - \frac{1}{3}(\mathbf{a} + \mathbf{b} + \mathbf{d}))$ from which $1 + s = -t/3$, $-s/3 = -t/3$, $-\frac{1}{3}s = 1 + t$.
$s = t$ gives $s = t = -\frac{3}{4}$ so

$$\mathbf{r} = \mathbf{c} - \tfrac{3}{4}(\mathbf{c} - \tfrac{1}{3}(\mathbf{a} + \mathbf{b} + \mathbf{d})) = \tfrac{1}{4}\mathbf{c} + \tfrac{1}{4}\mathbf{a} + \tfrac{1}{4}\mathbf{b} + \tfrac{1}{4}\mathbf{d}.$$

Since

$$OG_1 = \tfrac{1}{3}(\mathbf{a} + \mathbf{b} + \mathbf{c}), \; \mathbf{OG} = \tfrac{1}{4}\mathbf{d} + \tfrac{3}{4} \times \tfrac{1}{3}(\mathbf{a} + \mathbf{b} + \mathbf{c}) = \tfrac{1}{4}\mathbf{d} + \tfrac{3}{4} \times \mathbf{OG_1}$$

By the ratio theorem G divides DG_1 in the ratio of $3:1$.
If M is the mid-point of AC then $\mathbf{OM} = \frac{1}{2}(\mathbf{a} + \mathbf{c})$
If N is the mid-point of BD then $\mathbf{ON} = \frac{1}{2}(\mathbf{b} + \mathbf{d})$
The mid-point of MN is given by $\frac{1}{2}[\frac{1}{2}(\mathbf{a} + \mathbf{c}) + \frac{1}{2}(\mathbf{b} + \mathbf{d})] = \frac{1}{4}(\mathbf{a} + \mathbf{b} + \mathbf{c} + \mathbf{d})$

So G is the mid-point of MN and will similarly lie at the mid-point of lines joining mid-points of opposite edges AB and CD and AD and BC. These properties of the tetrahedron illustrate the geometrical power of vector methods, especially the ratio theorem.

Referring to fig. 119
$BX : XC = 1:4 \Rightarrow \mathbf{OX}$ $AZ : ZB = 2:3 \Rightarrow \mathbf{OZ}$
$= \frac{4}{5}\mathbf{b} + \frac{1}{5}\mathbf{c}$ $= \frac{3}{5}\mathbf{a} + \frac{2}{5}\mathbf{b}$

Equation of AX is
$$\mathbf{r} = \mathbf{a} + t\left(\frac{4}{5}\mathbf{b} + \frac{1}{5}\mathbf{c} - \mathbf{a}\right)$$
Equation of CZ is
$$\mathbf{r} = \mathbf{c} + s\left(\frac{3}{5}\mathbf{a} + \frac{2}{5}\mathbf{b} - \mathbf{c}\right)$$

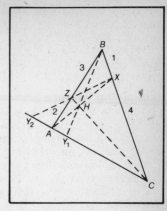

Figure 119

At H these are equal
$$(1-t)\mathbf{a} + \frac{4}{5}t\mathbf{b} + \frac{t}{5}\mathbf{c}$$
$$= \frac{3}{5}s\mathbf{a} + \frac{2}{5}s\mathbf{b} + (1-s)\mathbf{c}$$

$$1 - t = \frac{3}{5}s, \ \frac{4}{5}t = \frac{2}{5}s \Rightarrow s = 2t$$

$$5 - 5t = 6t \Rightarrow t = \frac{5}{11} \text{ and } s = \frac{10}{11}$$

H is $\mathbf{r} = \frac{6}{11}\mathbf{a} + \frac{4}{11}\mathbf{b} + \frac{1}{11}\mathbf{c}$

BY_1 is a multiple of BH so $BY_1 = k \times BH$

$$\Rightarrow \mathbf{OY_1} = \mathbf{b} + k \times \mathbf{BH} = \mathbf{b} + k\left(\frac{6}{11}\mathbf{a} - \frac{7}{11}\mathbf{b} + \frac{1}{11}\mathbf{c}\right)$$

Y_1 is on $AC \Rightarrow 1 - \frac{7k}{11} = 0 \Rightarrow k = \frac{11}{7} \Rightarrow \mathbf{OY_1} =$

$\frac{6}{7}\mathbf{a} + \frac{1}{7}\mathbf{c} \Rightarrow CY_1 : Y_1 A = 6 : 1$

$$\frac{AZ}{ZB} \times \frac{BX}{XC} \times \frac{CY_1}{Y_1 A} = \frac{2}{3} \times \frac{1}{4} \times \frac{6}{1} = 1$$

and this is the converse of **Ceva's theorem** which states that if points Z, X and Y divide the sides AB, BC and CA in ratios such that $\frac{AZ}{ZB} \times \frac{BX}{XC} \times \frac{CY}{YA} = +1$ then AX, BY and CZ are concurrent (meet in the same point H). If XZ is produced to meet CA at Y_2 then $XY = l \times XZ$.

$$\mathbf{OY_2} = \mathbf{OX} + l \times \mathbf{XZ} = \frac{4}{5}\mathbf{b} + \frac{1}{5}\mathbf{c} + l\left(\frac{3}{5}\mathbf{a} - \frac{2}{5}\mathbf{b} - \frac{1}{5}\mathbf{c}\right)$$

But Y_2 lies on CA so $\frac{4}{5} - \frac{2}{5}l = 0 \Rightarrow l = 2$

and $\mathbf{OY_2} = \frac{6}{5}\mathbf{a} - \frac{1}{5}\mathbf{c} \Rightarrow CY_2 : Y_2 A = 6 : -1$

$$\frac{AZ}{ZB} \times \frac{BX}{XC} \times \frac{CY_2}{Y_2 A} = \frac{2}{3} \times \frac{1}{4} \times \frac{6}{-1} = -1.$$

281

This is the converse of **Menelaus' theorem** which states that if points Z, X, and Y divide the sides AB, BC and CA in ratios such that $\dfrac{AZ}{ZB} \times \dfrac{BX}{XC} \times \dfrac{CY}{YA} = -1$ then X, Y, Z are collinear (in the same line).

Multiplying vectors

Vectors can be multiplied in two ways, the first producing a number (scalar) and is called the **scalar product** (or dot product) and the second produces a vector and is called the **vector product** (or cross product). Both products have important applications especially in mechanics. The study of their properties gives valuable insight into the algebraic structure of vectors and mathematical structures in general. Two dimensional vectors are easy to represent and understand while three dimensional vectors are difficult to draw on plane paper and three dimensional ideas difficult to convey. Algebraic derivations are included to help the understanding of three dimensional work. The many geometrical applications of vector methods are emphasized whenever possible. This also helps understanding and conveys an idea of the power and simplicity of working with vectors.

Scalar product

The scalar product (or dot product) of two vectors **a** and **b** is denoted by **a.b** where **a.b** $= ab \cos \theta$ where a, b are the lengths of the vectors **a**, **b** and θ is the angle between them, measured positively with an anticlockwise rotation from OA to OB (fig. 120). If the angle is measured from OB to OA as $-\theta$, since $\cos(-\theta) = \cos \theta$, the scalar product takes the same value.

Example 3 If $a = 2$, $b = 3$ and $\theta = 60°$ ($\pi/3$ radians).

a.b $= 2 \times 3 \times \cos 60° = 3$

Example 4 If $a = 2$, $b = 3$ and $\theta = 120°$.

a.b $= 2 \times 3 \times \cos 120° = -3$

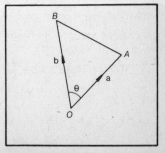

Figure 120

282

Example 5

$(2\mathbf{a}) \cdot (3\mathbf{b}) = 2a \times 3b \times \cos\theta = 6ab\cos\theta \Rightarrow (2\mathbf{a}) \cdot (3\mathbf{b}) = 6(\mathbf{a} \cdot \mathbf{b})$.

From fig. 120 $\mathbf{b} \cdot \mathbf{a} = ba\cos(360 - \theta)$ measuring the angle positively. Since $\cos(360 - \theta) = \cos\theta$, $\mathbf{b} \cdot \mathbf{a} = ba\cos\theta = \mathbf{a} \cdot \mathbf{b}$ so the scalar product of two vectors is commutative. An operation (denoted by $*$) combining two elements p and q is commutative if $p*q = q*p$ (see Chapter 12 on Groups).

For the vectors $\mathbf{i} = \begin{pmatrix} 1 \\ 0 \\ 0 \end{pmatrix}$, $\mathbf{j} = \begin{pmatrix} 0 \\ 1 \\ 0 \end{pmatrix}$, $\mathbf{k} = \begin{pmatrix} 0 \\ 0 \\ 1 \end{pmatrix}$

since the angle between any two is 90°

$$\mathbf{i} \cdot \mathbf{j} = \mathbf{j} \cdot \mathbf{k} = \mathbf{i} \cdot \mathbf{k} = 0 \tag{1}$$

and $\mathbf{i} \cdot \mathbf{i} = 1 \times 1 \times \cos 0° = 1$. So

$$\mathbf{i} \cdot \mathbf{i} = \mathbf{j} \cdot \mathbf{j} = \mathbf{k} \cdot \mathbf{k} = 1 \tag{2}$$

If the angle between two (non-zero) vectors $\mathbf{a}$ and $\mathbf{b}$ is 90° then $\mathbf{a} \cdot \mathbf{b} = ab\cos 90° = 0$ and this property provides an easy way of telling whether two vectors are at right angles.

Length of a vector $\mathbf{a} \cdot \mathbf{a} = a \times a \times \cos 0° = a^2$, the square of the length of $\mathbf{a}$.

If $\mathbf{a} = \begin{pmatrix} a_1 \\ a_2 \\ a_3 \end{pmatrix}$, $\mathbf{a} \cdot \mathbf{a} = \begin{pmatrix} a_1 \\ a_2 \\ a_3 \end{pmatrix} \cdot \begin{pmatrix} a_1 \\ a_2 \\ a_3 \end{pmatrix} \Rightarrow a^2 = a_1{}^2 + a_2{}^2 + a_3{}^2$

Scalar product in component form

Can we find an expression for $\mathbf{a} \cdot \mathbf{b}$ if

$a = \begin{pmatrix} a_1 \\ a_2 \\ a_3 \end{pmatrix}$ and $b = \begin{pmatrix} b_1 \\ b_2 \\ b_3 \end{pmatrix}$?

Writing $\mathbf{a} = a_1\mathbf{i} + a_2\mathbf{j} + a_3\mathbf{k}$ and $\mathbf{b} = b_1\mathbf{i} + b_2\mathbf{j} + v_3\mathbf{k}$
$$\mathbf{a} \cdot \mathbf{b} = (a_1\mathbf{i} + a_2\mathbf{j} + a_3\mathbf{k}) \cdot (b_1\mathbf{i} + b_2\mathbf{j} + b_3\mathbf{k})$$
To multiply out the brackets we need to establish that $\mathbf{a} \cdot (\mathbf{b} + \mathbf{c}) = \mathbf{a} \cdot \mathbf{b} + \mathbf{a} \cdot \mathbf{c}$, i.e. that the scalar product is distributive over vector addition. The next section will show that this is true and it enables the brackets to be multiplied out just as in the algebra of numbers.

$\mathbf{a} \cdot \mathbf{b} = a_1b_1\mathbf{i} \cdot \mathbf{i} + a_1b_2\mathbf{i} \cdot \mathbf{j} + a_1b_3\mathbf{i} \cdot \mathbf{k} + a_2b_1\mathbf{j} \cdot \mathbf{i} + a_2b_2\mathbf{j} \cdot \mathbf{j}$
$\qquad + a_2b_3\mathbf{j} \cdot \mathbf{k} + a_3b_1\mathbf{k} \cdot \mathbf{i} + a_3b_2\mathbf{k} \cdot \mathbf{j} + a_3b_3\mathbf{k} \cdot \mathbf{k}$
$\qquad = a_1b_1 + a_2b_2 + a_3b_3$ using $\mathbf{i} \cdot \mathbf{j} = \mathbf{j} \cdot \mathbf{k} = \mathbf{k} \cdot \mathbf{i} = 0$ from (1)
$\qquad\qquad\qquad$ and $\mathbf{i} \cdot \mathbf{i} = \mathbf{j} \cdot \mathbf{j} = \mathbf{k} \cdot \mathbf{k} = 1$ from (2)

Scalar product is distributive over vector addition

$\mathbf{a} \cdot (\mathbf{b} + \mathbf{c}) = \mathbf{a} \cdot \mathbf{b} + \mathbf{a} \cdot \mathbf{c}$

Let $\mathbf{OA} = \mathbf{a}$ be along the x axis.
Let $\mathbf{OB} = \mathbf{b}$ be in the xy plane.

$\mathbf{a} \cdot \mathbf{b} = ab \cos \theta = aOF$
where OF is the projection of
$\mathbf{OB}$ onto the x axis. Similarly
$\mathbf{a} \cdot (\mathbf{b} + \mathbf{c}) = a \times OD \cos$ angle
DOG where OG is the pro-
jection of $\mathbf{b} + \mathbf{c} = \mathbf{OD}$ onto the
x axis.
So $\mathbf{a} \cdot (\mathbf{b} + \mathbf{c}) = aOG$

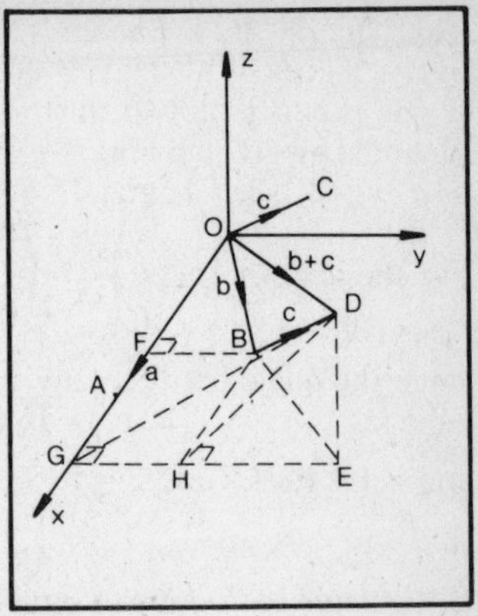

Let $\mathbf{c} = \mathbf{OC} = \mathbf{BD}$
D is not in the xy plane.

$\mathbf{a} \cdot \mathbf{c} = aOC \cos$ angle COA
$\quad = a \times BD \cos$ angle DBH
$\quad = a \times BH$

Since $\mathbf{OG} = \mathbf{OF} + \mathbf{BH}$
$\quad = aOG = a \times OF + a \times BH$
$\mathbf{a} \cdot (\mathbf{b} + \mathbf{c}) = \mathbf{a} \cdot \mathbf{b} + \mathbf{a} \cdot \mathbf{c}$

Figure 121

The diagram in fig. 121 is difficult to draw and visualize and students are not easily convinced by this proof. It is easier to understand an algebraic proof from an algebraic definition.

Alternative definition of scalar product

If $a = \begin{pmatrix} a_1 \\ a_2 \\ a_3 \end{pmatrix}$ and $b = \begin{pmatrix} b_1 \\ b_2 \\ b_3 \end{pmatrix}$ then $\mathbf{a} \cdot \mathbf{b} = a_1 b_1 + a_2 b_2 + a_3 b_3$

Distributive property i.e. $\mathbf{a} \cdot (\mathbf{b} + \mathbf{c}) = \mathbf{a} \cdot \mathbf{b} + \mathbf{a} \cdot \mathbf{c}$.

If $\mathbf{c} = \begin{pmatrix} c_1 \\ c_2 \\ c_3 \end{pmatrix}$ then $\mathbf{b} + \mathbf{c} = \begin{pmatrix} b_1 + c_1 \\ b_2 + c_2 \\ b_3 + c_3 \end{pmatrix}$

$\mathbf{a} \cdot (\mathbf{b} + \mathbf{c}) = \begin{pmatrix} a_1 \\ a_2 \\ a_3 \end{pmatrix} \cdot \begin{pmatrix} b_1 + c_1 \\ b_2 + c_2 \\ b_3 + c_3 \end{pmatrix}$

$\quad = a_1(b_1 + c_1) + a_2(b_2 + c_2) + a_3(b_3 + c_3)$

$\quad = a_1 b_1 + a_2 b_2 + a_3 b_3 + a_1 c_1 + a_2 c_2 + a_3 c_3$

$\quad = \mathbf{a} \cdot \mathbf{b} + \mathbf{a} \cdot \mathbf{c}$

which seems simple enough. However if we start with this definition we shall need to show that it is equivalent to our previous definition or rather that $\mathbf{a} \cdot \mathbf{b} = a_1b_1 + a_2b_2 + a_3b_3 = ab \cos \theta$. With the algebraic definition,
$$\mathbf{a} \cdot \mathbf{a} = a_1{}^2 + a_2{}^2 + a_3{}^2 = a^2$$
the square of the length. Referring to fig. 120 and using the cosine rule in triangle OAB
$$AB^2 = OA^2 + OB^2 - 2OA \times OB \cos \theta$$
$$(\mathbf{b} - \mathbf{a}) \cdot (\mathbf{b} - \mathbf{a}) = a^2 + b^2 - 2ab \cos \theta$$
$$\mathbf{b} \cdot \mathbf{b} - \mathbf{b} \cdot \mathbf{a} - \mathbf{a} \cdot \mathbf{b} + \mathbf{a} \cdot \mathbf{a} = a^2 + b^2 - 2ab \cos \theta$$
From the algebraic definition we know $\mathbf{a} \cdot \mathbf{a} = a^2$ and $\mathbf{b} \cdot \mathbf{b} = b^2$ and it is easy to show that $\mathbf{b} \cdot \mathbf{a} = \mathbf{a} \cdot \mathbf{b} = a_1b_1 + a_2b_2 + a_3b_3$. So
$$b^2 - 2 \times \mathbf{a} \cdot \mathbf{b} + a^2 = a^2 + b^2 - 2ab \cos \theta$$
$$\Rightarrow \mathbf{a} \cdot \mathbf{b} = ab \cos \theta$$
and the two definitions are equivalent.

Angle between two vectors
When two vectors are given in component form it is easy to find their scalar product.

Example 6 If $\mathbf{a} = \begin{pmatrix} 1 \\ 2 \\ 3 \end{pmatrix}$ and $\mathbf{b} = \begin{pmatrix} 2 \\ 3 \\ 4 \end{pmatrix}$

then $\mathbf{a} \cdot \mathbf{b} = 1 \times 2 + 2 \times 3 + 3 \times 4 = 20$.
Since the length of $\mathbf{a}$, $a = \sqrt{1^2 + 2^2 + 3^2} = \sqrt{14}$ and $b = \sqrt{2^2 + 3^2 + 4^2} = \sqrt{29}$, $ab \cos \theta = \sqrt{14}\sqrt{29} \cos \theta = 20$ (from the component definition)
$$\cos \theta = \frac{20}{\sqrt{14}\sqrt{29}} = 0 \cdot 99258 \Rightarrow \theta = 6 \cdot 98°$$

Example 7 $\mathbf{a} = \begin{pmatrix} 1 \\ 2 \\ 2 \end{pmatrix}$ $\mathbf{b} = \begin{pmatrix} 2 \\ 1 \\ -2 \end{pmatrix} \Rightarrow \mathbf{a} \cdot \mathbf{b} = 2 + 2 - 4 = 0$

and $\mathbf{a}$ and $\mathbf{b}$ are perpendicular.

Example 8 Referring to fig. 114 on page 275 find the angle between (a) a diagonal and a plane face of the cube and (b) the angle between two diagonals (a diagonal joins opposite vertices).

(a) The angle between the diagonal OU and the plane face $OIAJ$ is the angle between the vectors $\mathbf{OU} = \mathbf{i} + \mathbf{j} + \mathbf{k}$ and $\mathbf{OA} = \mathbf{i} + \mathbf{j}$
$OU = \sqrt{1^2 + 1^2 + 1^2} = \sqrt{3}$ $OA = \sqrt{1^2 + 1^2 + 0^2} = \sqrt{2}$;
$\mathbf{OU} \cdot \mathbf{OA} = 2$; $\mathbf{OU} \cdot \mathbf{OA} = \sqrt{3}\sqrt{2} \cos AOU = 2$

$$\cos AOU = \frac{2}{\sqrt{3}\sqrt{2}} = \frac{2}{\sqrt{6}} = \frac{2\sqrt{6}}{6} = \frac{\sqrt{6}}{3} = 0.8165$$

angle $AOU = 35.3°$.

(b) The angle α between two diagonals is equal to the angle between **OU** and **IB**.

OU $= \mathbf{i} + \mathbf{j} + \mathbf{k}$, **IB** $= -\mathbf{i} + \mathbf{j} + \mathbf{k}$, $OU = IB = \sqrt{3}$

$$\mathbf{OU} \cdot \mathbf{IB} = \sqrt{3}\sqrt{3} \cos \alpha = \begin{pmatrix} 1 \\ 1 \\ 1 \end{pmatrix}\begin{pmatrix} -1 \\ 1 \\ 1 \end{pmatrix}$$

$$\cos \alpha = \frac{1}{\sqrt{3}\sqrt{3}} = \frac{1}{3} \Rightarrow \alpha = 70.5°$$

In general the angle between two vectors **a** and **b** is given by $\cos \theta = \dfrac{\mathbf{a} \cdot \mathbf{b}}{ab} = \dfrac{a_1 b_1 + a_2 b_2 + a_3 b_3}{\sqrt{a_1^2 + a_2^2 + a_3^2}\sqrt{b_1^2 + b_2^2 + b_3^2}}$

if $\mathbf{a} = \begin{pmatrix} a_1 \\ a_2 \\ a_3 \end{pmatrix}$ and $\mathbf{b} = \begin{pmatrix} b_1 \\ b_2 \\ b_3 \end{pmatrix}$

Equation of a plane in three dimensions

The points IJK define a plane which includes the mid-points of JK, IJ, IK, i.e. $(0, \frac{1}{2}, \frac{1}{2})(\frac{1}{2}, \frac{1}{2}, 0)(\frac{1}{2}, 0, \frac{1}{2})$ respectively. The centroid of triangle IJK $(\frac{1}{3}, \frac{1}{3}, \frac{1}{3})$ must also lie in the plane. This suggests the equation of the plane is $x + y + z = 1$.

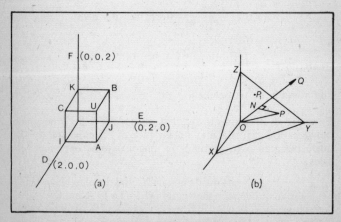

Figure 122

Similarly the plane through $A(1, 1, 0)$ $B(0, 1, 1)$ and $C(1, 0, 1)$ includes the points $D(2, 0, 0)$, $E(0, 2, 0)$ and $F(0, 0, 2)$. All the points satisfy the equation $x + y + z = 2$. A little knowledge of the geometry and symmetry of the cube will suffice to see that these two planes are parallel.

Consider the plane passing through XYZ. This plane is defined by specifying a vector $\mathbf{OQ}$ perpendicular to the plane and a point lying on the plane $P_1(x_1, y_1, z_1)$.

Let $\mathbf{OQ} = \begin{pmatrix} a \\ b \\ c \end{pmatrix}$ which cuts the plane at N (where $ON = h$).

For any point $P(x, y, z)$ on the plane, $\mathbf{NP}$ is perpendicular to $\mathbf{OQ}$

$$\Rightarrow \mathbf{NP} \cdot \mathbf{OQ} = 0$$

$$\Rightarrow (\mathbf{OP} - \mathbf{ON}) \cdot \mathbf{OQ} = 0$$

$$\mathbf{OP} \cdot \mathbf{OQ} = \mathbf{ON} \cdot \mathbf{OQ} \Rightarrow ax + by + cz = hOQ = d$$

which is the equation of the plane XYZ. d can be calculated if P_1 is known and $d = ax_1 + by_1 + cz_1$.

$d = h \times OQ = h\sqrt{a^2 + b^2 + c^2}$ so the distance of the plane from the origin

$$h = \frac{d}{\sqrt{a^2 + b^2 + c^2}}$$

$d = 0 \Rightarrow$ plane passes through the origin.

The vector $\begin{pmatrix} a \\ b \\ c \end{pmatrix}$ is perpendicular to the plane and is called the **normal vector** of the plane.

Example 9 Find the equation of the plane IJK (fig. 114).

The normal vector is $\mathbf{OU} = \mathbf{i} + \mathbf{j} + \mathbf{k}$. This can be checked by seeing that

$$\mathbf{OU} \cdot \mathbf{IJ} = \begin{pmatrix} 1 \\ 1 \\ 1 \end{pmatrix} \cdot \begin{pmatrix} -1 \\ 1 \\ 0 \end{pmatrix} = 0 \text{ and } \mathbf{OU} \cdot \mathbf{IK} = \begin{pmatrix} 1 \\ 1 \\ 1 \end{pmatrix} \cdot \begin{pmatrix} -1 \\ 0 \\ 1 \end{pmatrix} = 0$$

so $\mathbf{OU}$ is perpendicular to $\mathbf{IJ}$ and $\mathbf{IK}$ and so is perpendicular to all points in the plane IJK.

The equation of plane IJK is $x + y + z = d$.
$I(1, 0, 0)$ lies on the plane so $d = 1 + 0 + 0 = 1$ and $x + y + z = 1$.

Similarly **OU** is the normal vector of plane ABC so the equation of plane ABC is $x + y + z = k = 2 + 0 + 0 = 2$ since $A(2, 0, 0)$ lies on plane ABC.

The two planes are parallel as they have the same normal vector.

Example 10 Find the equation of the plane cutting the x axis at $(1, 0, 0)$ the y axis at $(0, 2, 0)$ and the z axis at $(0, 0, 3)$.

Equation of plane is $ax + by + cz = d$. Point $(1, 0, 0)$ lies on the plane $\Rightarrow a = d$ and similarly $b = \frac{1}{2}d$ and $c = \frac{1}{3}d$, so $dx + \frac{1}{2}dy + \frac{1}{3}dx = d$ or $6x + 3y + 2z = 6$.

Example 11 Find the equation of the plane passing through the three points $P(1, 1, 1)$, $Q(1, 2, 3)$ and $R(3, 2, 1)$.

The normal vector of the plane must be perpendicular to **PQ** and to **PR**.

$$\begin{pmatrix} a \\ b \\ c \end{pmatrix} \cdot \begin{pmatrix} 0 \\ 1 \\ 2 \end{pmatrix} = 0 \Rightarrow b + 2c = 0 \qquad \begin{pmatrix} a \\ b \\ c \end{pmatrix} \cdot \begin{pmatrix} 2 \\ 1 \\ 0 \end{pmatrix} = 0 \Rightarrow 2a + b = 0$$

$$\Rightarrow b \quad = -2c \qquad\qquad\qquad \Rightarrow b \quad = -2a$$

Since the normal vector can have any magnitude choose $a = 1$ then $b = -2$, $c = 1$. The normal vector is $\mathbf{i} - 2\mathbf{j} + \mathbf{k}$ and the equation of the plane is $x - 2y + z = d$. Since P lies on the plane $d = 1 - 2 + 1 = 0$. The equation of plane PQR is $x - 2y + z = 0$.

Distances from the origin

All points on **OU** have coordinates which are multiples of $(1, 1, 1)$.

OU meets the plane IJK $(x + y + z = 1)$ at $G = (\frac{1}{3}, \frac{1}{3}, \frac{1}{3})$ and **OU** meets the plane ABC at $H = (\frac{2}{3}, \frac{2}{3}, \frac{2}{3})$.

$$OG = \sqrt{(\tfrac{1}{3})^2 + (\tfrac{1}{3})^2 + (\tfrac{1}{3})^2} = \sqrt{\tfrac{1}{3}} = \frac{1}{\sqrt{3}}$$

and $OH = \sqrt{(\tfrac{2}{3})^2 + (\tfrac{2}{3})^2 + (\tfrac{2}{3})^2} = \dfrac{2}{\sqrt{3}}$.

$OU = \sqrt{3} = \dfrac{3}{\sqrt{3}}$ so G and H are the points of trisection of OU.

This fact is reasonably obvious and can be proved using the ratio theorem.

Distance of a point $P_1(x_1, y_1, z_1)$ from the plane $ax + by + cz = d$

With the notation in fig. 123.

$OP_1 . OQ = (OM + MP_1) . OQ$
$\qquad = OM . OQ$
$\qquad\qquad + MP_1 . OQ$

OM . OQ $= d$ since M is on the plane and all points (x, y, z) satisfy $\begin{pmatrix} a \\ b \\ c \end{pmatrix} . \begin{pmatrix} x \\ y \\ z \end{pmatrix} = d$

$MP_1 . OQ = MP_1 \times OQ \cos 0°$ since MP_1 is parallel to OQ.

$OP_1 . OQ = \begin{pmatrix} x_1 \\ y_1 \\ z_1 \end{pmatrix} . \begin{pmatrix} a \\ b \\ c \end{pmatrix}$
$\qquad = d + MP_1 \times OQ$

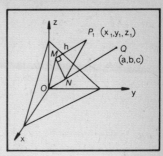

Figure 123

$$ax_1 + by_1 + cz_1 = d + MP_1 \times \sqrt{a^2 + b^2 + c^2}$$

$$h = MP_1 = \frac{ax_1 + by_1 + cz_1 - d}{\sqrt{a^2 + b^2 + c^2}}$$

Example 12 Find the distance of the point $(1, 1, 1)$ from the plane $x + 2y + 3z = 6$.

$h = \dfrac{ax_1 + by_1 + cz_1 - d}{\sqrt{a^2 + b^2 + c^2}}$ gives $h = \dfrac{1 + 2 + 3 - 6}{\sqrt{1^2 + 2^2 + 3^2}} = 0$ which means that $(1, 1, 1)$ lies on the plane $x + 2y + 3z = 6$ which can easily be seen since it satisfies the equation of the plane.

Intersection of a line with a plane

Three cases arise: (i) the line is parallel to the plane,
                   (ii) the line lies in the plane,
                   (iii) the line intersects the plane in a single point.

Example 13 Find the intersection of the line $x = 1 + 3t$, $y = 2 + 4t$, $z = 3 + 5t$ with the plane $x + 2y + 3z = 6$.

Substitute the x, y, z values of the line into the equation of the plane. This will give the value of t which specifies the point on the line and the plane

$$1 + 3t + 2(2 + 4t) + 3(3 + 5t) = 6 \Rightarrow 26t + 14 = 6 \Rightarrow t = -4/13$$

289

The point of intersection is $x = 1 - \frac{12}{13} = \frac{1}{13}$, $y = 2 - \frac{16}{13} = \frac{10}{13}$, $z = 3 - \frac{20}{13} = \frac{19}{13}$.

The line meets the plane at the point $\left(\frac{1}{13}, \frac{10}{13}, \frac{19}{13}\right)$, i.e. case (iii).

If the line is parallel to the plane its direction vector will be perpendicular to the normal vector of the plane. Consider the line $x = 1 + 3t$, $y = 2 + 4t$, $z = 3 + 5t$ and the plane $3x + 4y - 5z = 10$.

Line's direction vector is $\begin{pmatrix} 3 \\ 4 \\ 5 \end{pmatrix}$. Plane's normal vector is $\begin{pmatrix} 3 \\ 4 \\ -5 \end{pmatrix}$.

To find t value, $3(1 + 3t) + 4(2 + 4t) - 5(3 + 5t) = 10$

$$\Rightarrow 3 + 9t + 8 + 16t - 15 - 25t = 10$$

$$\Rightarrow 11 - 15 = 10 \; (not \; true)$$

Our assumption that the line meets the plane is false, so the line must be parallel to the plane.

Consider the same line with the plane $3x + 4y - 5z = -4$. Finding t leads to $11 - 15 = -4$ which is always true. In this case all values of t satisfy the equation of the plane and the line lies in the plane.

Intersection of two lines

Four cases arise:
 (i) the lines are parallel $\Rightarrow$ they have the same direction vector,
 (ii) the lines are coincident $\Rightarrow$ they have the same direction vector,
(iii) the lines intersect at a single point,
(iv) the lines do not intersect but are skew.

Example 14 Consider the two lines $x = 1 + 3t$, $y = 2 + 4t$, $z = 3 + 5t$ and $x = 5 - s$, $y = -2 + 8s$, $z = 1 + 7s$.

At their intersection
$$1 + 3t = 5 - s \quad \Rightarrow 3t + s = 4 \quad \Rightarrow 6t + 2s = 8 \qquad (1)$$
$$2 + 4t = -2 + 8s \Rightarrow 4t - 8s = -4 \Rightarrow t - 2s = -1 \qquad (2)$$
$$3 + 5t = 1 + 7s \qquad\qquad\qquad\qquad\qquad\qquad (3)$$
Adding (1) and (2) gives $7t = 7 \Rightarrow t = 1 \Rightarrow s = 1$.

Check that $t = 1$, $s = 1$ satisfies (3), so the two lines meet in a single point given by $t = 1$, i.e. the point $(4, 6, 8)$ (case (iii)).

If the second line had third equation $z = 1 - 7s$, equation (3) would be $3 + 5t = 1 - 7s$ which is not satisfied by $t = 1$, $s = 1$. In this case the lines do not intersect and are skew lines.

Example 15 Consider the two lines $x = 1 + 3t$, $y = 2 + 4t$, $z = 3 + 5t$ and $x = 7 - 6s$, $y = 10 - 8s$, $z = 13 - 10s$.

These lines intersect when
$$1 + 3t = 7 - 6s \Rightarrow t + 2s = 2 \quad (1)$$
$$2 + 4t = 10 - 8s \Rightarrow t + 2s = 2 \quad (2)$$
$$3 + 5t = 13 - 10s \Rightarrow t + 2s = 2 \quad (3)$$

There is a linear relation $t + 2s = 2$ connecting t with s and in this case since the direction vectors are multiples of each other the lines are coincident (case (ii)).

If the third equation of the second line were $z = 18 - 10s$ then equation (3) would be $3 + 5t = 18 - 10s \Rightarrow t + 2s = 3$ which is not consistent with (1) and (2) and the equations have no solution. In this case the lines are parallel (case (i)).

Intersection of two planes

Case (i) The planes are parallel, is easily recognized since the planes would have the same normal vector. For example the two planes $x + 2y + 3z = 6$ and $3x + 6y + 9z = 12$ are parallel since the second can be rewritten as $x + 2y + 3z = 4$ and the normal vector of each plane is $\mathbf{i} + 2\mathbf{j} + 3\mathbf{k}$. The two planes $x + 2y + 3z = 6$ and $3x + 6y + 9 = 18$ are coincident as the second equation is three times the first.

Case (ii) The planes are not parallel and intersect in a line. Consider the planes $x + 2y + 3z = 6$ (1)

and $x + y + z = 1$ (2)

$(1) - (2)$ gives $y + 2z = 5$

$2 \times (2) - (1)$ gives $x - z = -4$

Arrange these two equations in the form of a straight line

$$x + 4 = z = -\frac{y + 5}{2} \quad \Rightarrow \quad \frac{x + 4}{1} = \frac{z - 0}{1} = \frac{5 - y}{2} = t$$

which can be rewritten as $x = -4 + t$, $y = 5 - 2t$, $z = t$ which represents a straight line through the point $(-4, 5, 0)$ in the direction $\mathbf{i} - 2\mathbf{j} + \mathbf{k}$.

Intersection of three planes

The three equations (1), (2) and (3) represent three planes. To find points common to the three

$$x + y + z = 6 \quad (1)$$
$$x + 2y + 3z = 14 \quad (2)$$
$$x + 3y + 7z = 28 \quad (3)$$

planes the equations must be solved simultaneously.

$(2) - (1)$ gives $y + 2z = 8$ (4)

$(3) - (1)$ gives $2y + 6z = 22 \Rightarrow y + 3z = 11$ (5)

$(5) - (4)$ gives $z = 3$

Substituting $z = 3$ in (4) gives $y + 6 = 8 \Rightarrow y = 2$

Substituting $z = 3$, $y = 2$ in (1) gives $x + 2 + 3 = 6 \Rightarrow x = 1$

The three planes intersect in the single point $(1, 2, 3)$.

Special case A $x + y + z = 6$ (6)

                              $x + 2y + 3z = 14$ (7)

                              $x + 3y + 5z = 28$ (8)

The same analysis gives

$$y + 2z = 8 \text{ and } 2y + 4z = 22 \Rightarrow y + 2z = 11$$

and these two equations have no solution.

Planes (6) and (7) intersect in a line given by $y + 2z = 8$ and $x - z = -2$.

$$x + 2 = z = \frac{8 - y}{2} \Rightarrow \frac{x + 2}{1} = \frac{y - 8}{-2} = \frac{z - 0}{1}$$

(6) and (7) intersect in the line through $(-2, 8, 0)$ in the direction $\mathbf{i} - 2\mathbf{j} + \mathbf{k}$. (7) and (8) intersect in a line given by $y + 2z = 14$ and $x - z = -14$.

$$x + 14 = z = \frac{14 - y}{2} \Rightarrow \frac{x + 14}{1} = \frac{y - 14}{-2} = \frac{z}{1}$$

which represents a line through $P_2(-14, 14, 0)$ in the direction $\mathbf{i} - 2\mathbf{j} + \mathbf{k}$.

(6) and (8) intersect in a line given by $y + 2z = 11$ and $x - z = -5$.

$$x + 5 = z = \frac{11 - y}{2} \Rightarrow \frac{x + 5}{1} = \frac{y - 11}{-2} = \frac{z}{1}$$

which is a line through $P_3(-5, 11, 0)$ in the direction $\mathbf{i} - 2\mathbf{j} + \mathbf{k}$. The three lines of intersection of the planes taken in pairs all have the same direction vector $\mathbf{i} - 2\mathbf{j} + \mathbf{k}$ and are therefore parallel. None of these three lines are coincident since $\mathbf{P_1P_2} = -12\mathbf{i} + 6\mathbf{j}$ is not a multiple of the direction vector $\mathbf{i} - 2\mathbf{j} + \mathbf{k}$; neither is $\mathbf{P_1P_3}$ nor $\mathbf{P_2P_3}$. The planes form a **prism**, the three lines of intersection forming the three edges of a triangular prism.

The normal vector of plane (8) is $\mathbf{i} + 3\mathbf{j} + 5\mathbf{k}$ which is perpendicular to the direction vector of the line of intersection of planes (6) and (7) $\mathbf{i} - 2\mathbf{j} + \mathbf{k}$ (scalar product zero). The line of intersection of planes (6) and (7) is parallel to plane (8). It follows that the line of intersection of each pair of planes is parallel to the third plane.

Special case B
$$x + y + z = 6 \tag{9}$$
$$x + 2y + 3z = 14 \tag{10}$$
$$x + 3y + 5z = 22 \tag{11}$$

$(10) - (9)$ gives $y + 2z = 8$ and $(11) - (9)$ gives $2y + 4z = 16$ which is the same equation. The three equations are not independent. (9) and (10) intersect in the line through $P_1(-2, 8, 0)$ in the direction $\mathbf{i} - 2\mathbf{j} + \mathbf{k}$. (10) and (11) give $y + 2z = 8$ and

$$x - z = -2 \Rightarrow x + 2 = z = \tfrac{1}{2}(8 - y) \Rightarrow \frac{x + 2}{1} = \frac{y - 8}{-2} = \frac{z}{1}$$

representing a line through $P_1(-2, 8, 0)$ in the direction $\mathbf{i} - 2\mathbf{j} + \mathbf{k}$ which is the same line. All three planes meet in the same line and this configuration is called a sheaf of planes (fig. 124b).

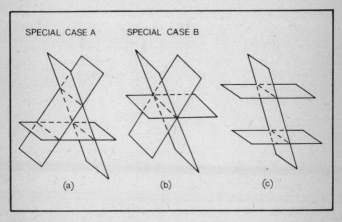

SPECIAL CASE A   SPECIAL CASE B

(a)   (b)   (c)

Figure 124

Figure 124c shows the situation when two of the three intersecting planes are parallel. The lines of intersection are then parallel.

Matrix solution The matrix solution of the three simultaneous equations summarize different situations neatly (method in Chapter 7).

General case

$$\begin{pmatrix} 1 & 1 & 1 \\ 1 & 2 & 3 \\ 1 & 3 & 7 \end{pmatrix} \begin{pmatrix} x \\ y \\ z \end{pmatrix} = \begin{pmatrix} 6 \\ 14 \\ 28 \end{pmatrix}$$

$$\begin{pmatrix} 1 & 1 & 1 \\ 0 & 1 & 2 \\ 0 & 2 & 6 \end{pmatrix} \begin{pmatrix} x \\ y \\ z \end{pmatrix} = \begin{pmatrix} 6 \\ 8 \\ 22 \end{pmatrix}$$

$\Rightarrow 2z = 6$
$\Rightarrow z = 3$
$y + 2z = 8 \Rightarrow y = 2$
$x + y + z = 6 \Rightarrow x = 1$

$$\begin{pmatrix} 1 & 1 & 1 \\ 0 & 1 & 2 \\ 0 & 0 & 2 \end{pmatrix} \begin{pmatrix} x \\ y \\ z \end{pmatrix} = \begin{pmatrix} 6 \\ 8 \\ 6 \end{pmatrix}$$

The planes meet in a single point $(1, 2, 3)$

Case A	*Case B*

Case A

$$\begin{pmatrix} 1 & 1 & 1 \\ 1 & 2 & 3 \\ 1 & 3 & 5 \end{pmatrix} \begin{pmatrix} x \\ y \\ z \end{pmatrix} = \begin{pmatrix} 6 \\ 14 \\ 28 \end{pmatrix}$$

$$\begin{pmatrix} 1 & 1 & 1 \\ 0 & 1 & 2 \\ 0 & 2 & 4 \end{pmatrix} \begin{pmatrix} x \\ y \\ z \end{pmatrix} = \begin{pmatrix} 6 \\ 8 \\ 22 \end{pmatrix}$$

$$\begin{pmatrix} 1 & 1 & 1 \\ 0 & 1 & 2 \\ 0 & 0 & 0 \end{pmatrix} \begin{pmatrix} x \\ y \\ z \end{pmatrix} = \begin{pmatrix} 6 \\ 8 \\ 6 \end{pmatrix}$$

$\Rightarrow 0z = 6$

No solutions
Planes do not
intersect, i.e.
prism formation

Case B

$$\begin{pmatrix} 1 & 1 & 1 \\ 1 & 2 & 3 \\ 1 & 3 & 5 \end{pmatrix} \begin{pmatrix} x \\ y \\ z \end{pmatrix} = \begin{pmatrix} 6 \\ 14 \\ 22 \end{pmatrix}$$

$$\begin{pmatrix} 1 & 1 & 1 \\ 0 & 1 & 2 \\ 0 & 2 & 4 \end{pmatrix} \begin{pmatrix} x \\ y \\ z \end{pmatrix} = \begin{pmatrix} 6 \\ 8 \\ 16 \end{pmatrix}$$

$$\begin{pmatrix} 1 & 1 & 1 \\ 0 & 1 & 2 \\ 0 & 0 & 0 \end{pmatrix} \begin{pmatrix} x \\ y \\ z \end{pmatrix} = \begin{pmatrix} 6 \\ 8 \\ 0 \end{pmatrix}$$

$\Rightarrow 0z = 0$ many solutions
Let $z = t$
$y + 2z = 8 \Rightarrow y = 8 - 2t$ Solution
$x + y + z = 6 \Rightarrow$ $x = -2 + t$
     $x = 6 + 2t - 8 - t$ $y = 8 - 2t$
        $= t - 2$ $z = 0 + t$

the line of intersection of the planes.

If any of the planes are parallel this should be recognized at once, by looking at the coefficients of x, y and z. The distance of a point from a line is the distance from the point P to the nearest point D of the line (fig. 125). This occurs when PD is perpendicular to the line.

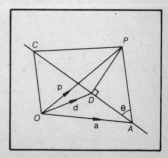

Figure 125

Example 16 Let P be $(11, 7, 3)$ and the line AC

294

be $\begin{pmatrix} x \\ y \\ z \end{pmatrix} = \begin{pmatrix} 1 \\ 2 \\ 3 \end{pmatrix} + t \begin{pmatrix} 3 \\ 4 \\ 5 \end{pmatrix}$

If D is given by $t = t_1$, $PD \perp AC \Rightarrow (\mathbf{OD} - \mathbf{OP}) . \mathbf{b} = 0$

$\begin{pmatrix} 1 + 3t - 11 \\ 2 + 4t - 7 \\ 3 + 5t - 3 \end{pmatrix} . \begin{pmatrix} 3 \\ 4 \\ 5 \end{pmatrix} = 0$

$3(3t_1 - 10) + 4(4t_1 - 5) + 5(5t_1) = 0 \Rightarrow 50t_1 = 50 \Rightarrow t_1 = 1$

D is the point $(4, 6, 8)$

$\mathbf{PD} = \mathbf{OD} - \mathbf{OP} = 4\mathbf{i} + 6\mathbf{j} + 8\mathbf{k} - (11\mathbf{i} + 7\mathbf{j} + 3\mathbf{k}) = -7\mathbf{i} - \mathbf{j} + 5\mathbf{k}$

Distance $PD = \sqrt{(-7)^2 + (-1)^2 + 5^2} = \sqrt{75} = 5\sqrt{3}$

In general, to find the distance of $P_1(\mathbf{OP_1} = \mathbf{p_1})$ from the line $\mathbf{r} = \mathbf{a} + t\mathbf{b}$, D is given by $\mathbf{PD} . \mathbf{b} = 0 \Rightarrow (\mathbf{d} - \mathbf{p_1}) . \mathbf{b} = 0$ $\Rightarrow (\mathbf{a} + t_1\mathbf{b} - \mathbf{p_1}) . \mathbf{b} = 0$

$\Rightarrow (\mathbf{a} - \mathbf{p_1}) . \mathbf{b} + t_1\mathbf{b} . \mathbf{b} = 0 \Rightarrow t_1 = \dfrac{(\mathbf{p_1} - \mathbf{a})}{b^2} . \mathbf{b}$ since $\mathbf{b} . \mathbf{b} = b^2$

$PD = |\mathbf{PD}| = |\mathbf{d} - \mathbf{p_1}| = \left| \mathbf{a} + \left[\dfrac{(\mathbf{p_1} - \mathbf{a}) . \mathbf{b}}{b^2} \right] \mathbf{b} - \mathbf{b_1} \right|$

but this is an unwieldy formula. A better one derives from the vector product.

Vector Product

The vector product of $\mathbf{a}$ and $\mathbf{b}$ (fig. 126) is denoted by $\mathbf{a} \wedge \mathbf{b}$ (a vector b) or $\mathbf{a} \times \mathbf{b}$ (a cross b) and is also called the cross product. $\mathbf{a} \wedge \mathbf{b}$ has magnitude $ab \sin \theta$ where a, b are the lengths of $\mathbf{a}$, $\mathbf{b}$ and θ the angle between them, measured from $\mathbf{a}$ to $\mathbf{b}$. The direction of $\mathbf{a} \wedge \mathbf{b}$ is in the direction of $\mathbf{n}$ where $\mathbf{n}$ is perpendicular to both $\mathbf{a}$ and $\mathbf{b}$ such that $\mathbf{a}$, $\mathbf{b}$ and $\mathbf{n}$ form a right hand set.

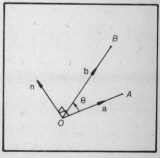

Figure 126

This means that a rotation from $\mathbf{a}$ to $\mathbf{b}$ would take a right-handed screw along the direction of $\mathbf{n}$ increasing. (If $\mathbf{a}$ is along the x axis and $\mathbf{b}$ in the positive x, y quadrant then $\mathbf{n}$ is in the direction of the positive z axis).

$\mathbf{a} \wedge \mathbf{b} = ab \sin \theta \mathbf{n}$ if $\mathbf{n}$ is a unit vector (length 1).

$\mathbf{b} \wedge \mathbf{a} = ab \sin \theta(-\mathbf{n})$ since a rotation from $\mathbf{b}$ to $\mathbf{a}$ takes a right hand screw in the direction of $-\mathbf{n}$.

$\mathbf{a} \wedge \mathbf{b} = -\mathbf{b} \wedge \mathbf{a}$ and the vector product is not commutative.

$\mathbf{i} \wedge \mathbf{j} = 1 \times 1 \times \sin 90°\mathbf{k} = \mathbf{k}$ (since $\mathbf{i}$, $\mathbf{j}$ and $\mathbf{k}$ form a right hand set)

$\mathbf{j} \wedge \mathbf{i} = 1 \times 1 \times \sin 90°(-\mathbf{k}) = -\mathbf{k}$ (since $\mathbf{j}$, $\mathbf{i}$ and $-\mathbf{k}$ form a right hand set).

Similarly $\mathbf{j} \wedge \mathbf{k} = \mathbf{i}$, $\mathbf{k} \wedge \mathbf{j} = -\mathbf{i}$, $\mathbf{k} \wedge \mathbf{i} = \mathbf{j}$, $\mathbf{i} \wedge \mathbf{k} = -\mathbf{j}$

$|\mathbf{i} \wedge \mathbf{i}| = 1 \times 1 \times \sin 0° \Rightarrow \mathbf{i} \wedge \mathbf{i} = \mathbf{0}$ and $\mathbf{j} \wedge \mathbf{j} = \mathbf{k} \wedge \mathbf{k} = \mathbf{0}$.

Vector product in component form

If $\mathbf{a} = a_1\mathbf{i} + a_2\mathbf{j} + a_3\mathbf{k}$ and $\mathbf{b} = b_1\mathbf{i} + b_2\mathbf{j} + b_3\mathbf{k}$

then $\mathbf{a} \wedge \mathbf{b} = (a_1\mathbf{i} + a_2\mathbf{j} + a_3\mathbf{k}) \wedge (b_1\mathbf{i} + b_2\mathbf{j} + b_3\mathbf{k})$

$\qquad = a_1b_1\mathbf{i} \wedge \mathbf{i} + a_1b_2\mathbf{i} \wedge \mathbf{j} + a_1b_3\mathbf{i} \wedge \mathbf{k} + a_2b_1\mathbf{j} \wedge \mathbf{i} + a_2b_2\mathbf{j} \wedge \mathbf{j}$

$\qquad \quad + a_2b_3\mathbf{j} \wedge \mathbf{k} + a_3b_1\mathbf{k} \wedge \mathbf{i} + a_3b_2\mathbf{k} \wedge \mathbf{j} + a_3b_3\mathbf{k} \wedge \mathbf{k}$

since vector product is distributive over addition

$\mathbf{a} \wedge \mathbf{b} = a_1b_2\mathbf{k} - a_1b_3\mathbf{j} - a_2b_1\mathbf{k} + a_2b_3\mathbf{i} + a_3b_1\mathbf{j} - a_3b_2\mathbf{i}$

$\qquad = (a_2b_3 - a_3b_2)\mathbf{i} + (a_3b_1 - a_1b_3)\mathbf{j} + (a_1b_2 - a_2b_1)\mathbf{k}$

$\qquad = \begin{vmatrix} \mathbf{i} & \mathbf{j} & \mathbf{k} \\ a_1 & a_2 & a_3 \\ b_1 & b_2 & b_3 \end{vmatrix}$

which is in a much easier form to remember.

Starting with this definition, i.e. $\mathbf{a} \wedge \mathbf{b} = \begin{pmatrix} a_2b_3 - a_3b_2 \\ a_3b_1 - a_1b_3 \\ a_1b_2 - a_2b_1 \end{pmatrix}$

$(\mathbf{a} \wedge \mathbf{b}) . (\mathbf{a} \wedge \mathbf{b}) = (a_2b_3 - a_3b_2)^2 + (a_3b_1 - a_1b_3)^2 + (a_1b_2 - a_2b_1)^2$

$\qquad = a_2{}^2b_3{}^2 + a_3{}^2b_2{}^2 + a_3{}^2b_1{}^2 + a_1{}^2b_3{}^2 + a_1{}^2b_2{}^2$

$\qquad \quad + a_2{}^2b_1{}^2 - 2a_2a_3b_2b_3 - 2a_1a_3b_1b_3 - 2a_1a_2b_1b_2$

$\qquad = (a_1{}^2 + a_2{}^2 + a_3{}^2)(b_1{}^2 + b_2{}^2 + b_3{}^2) - a_1{}^2b_1{}^2 - a_2{}^2b_2{}^2$

$\qquad \quad - a_3{}^2b_3{}^2 - 2a_2a_3b_2b_3 - 2a_1a_3b_1b_3 - 2a_1a_2b_1b_2$

$\qquad = (a_1{}^2 + a_2{}^2 + a_3{}^2)(b_1{}^2 + b_2{}^2b_3{}^2)$

$\qquad \quad - (a_1b_1 + a_2b_2 + a_3b_3)^2$

$\qquad = a^2b^2 - (\mathbf{a} . \mathbf{b})^2$

$\qquad = a^2b^2 \left[1 - \dfrac{(\mathbf{a} . \mathbf{b})^2}{a^2b^2} \right]$

$\qquad = a^2b^2[1 - \cos^2 \theta]$ since

$\qquad \qquad \cos \theta = \dfrac{\mathbf{a} . \mathbf{b}}{ab}$ from scalar product

$\qquad = a^2b^2 \sin^2 \theta$

$|\mathbf{a} \wedge \mathbf{b}| = ab \sin \theta$

$$\mathbf{a} \cdot (\mathbf{a} \wedge \mathbf{b}) = \begin{pmatrix} a_1 \\ a_2 \\ a_3 \end{pmatrix} \cdot \begin{pmatrix} a_2 b_3 - a_3 b_2 \\ a_3 b_1 - a_1 b_3 \\ a_1 b_2 - a_2 b_1 \end{pmatrix}$$

$$= a_1 a_2 b_3 - a_1 a_3 b_2 + a_2 a_3 b_1 - a_2 a_1 b_3 + a_3 a_1 b_2 - a_3 a_2 b_1$$

$= 0$ so $\mathbf{a} \wedge \mathbf{b}$ is perpendicular to $\mathbf{a}$.

Similarly it can be shown that $\mathbf{a} \wedge \mathbf{b}$ is perpendicular to $\mathbf{b}$ so $\mathbf{a} \wedge \mathbf{b}$ is perpendicular to $\mathbf{a}$ and $\mathbf{b}$ and we have $\mathbf{a} \wedge \mathbf{b} = ab \sin \theta \mathbf{n}$ where $\mathbf{n}$ is perpendicular to $\mathbf{a}$ and $\mathbf{b}$.

Since $\mathbf{a} \wedge \mathbf{b} = \begin{vmatrix} \mathbf{i} & \mathbf{j} & \mathbf{k} \\ a_1 & a_2 & a_3 \\ b_1 & b_2 & b_3 \end{vmatrix} = - \begin{vmatrix} \mathbf{i} & \mathbf{j} & \mathbf{k} \\ b_1 & b_2 & b_3 \\ a_1 & a_2 & a_3 \end{vmatrix} = -\mathbf{b} \wedge \mathbf{a}$

from the properties of determinants in Chapter 7 (interchanging two rows changes the sign of the determinant)

$$\mathbf{a} \wedge (\mathbf{b} + \mathbf{c}) = \begin{vmatrix} \mathbf{i} & \mathbf{j} & \mathbf{k} \\ a_1 & a_2 & a_3 \\ b_1 + c_1 & b_2 + c_2 & b_3 + c_3 \end{vmatrix}$$

$$= \begin{vmatrix} \mathbf{i} & \mathbf{j} & \mathbf{k} \\ a_1 & a_2 & a_3 \\ b_1 & b_2 & b_3 \end{vmatrix} - \begin{vmatrix} \mathbf{i} & \mathbf{j} & \mathbf{k} \\ a_1 & a_2 & a_3 \\ c_1 & c_2 & c_3 \end{vmatrix} \text{ if } c = \begin{pmatrix} c_1 \\ c_2 \\ c_3 \end{pmatrix}$$

$$= \mathbf{a} \wedge \mathbf{b} + \mathbf{a} \wedge \mathbf{c}$$

So the fact that the vector product is distributive over vector addition can be proved easily by using the properties of determinants. From the definition of the vector product in component form the following facts emerge;

(a) $\mathbf{a} \wedge \mathbf{b}$ is perpendicular to both $\mathbf{a}$ and $\mathbf{b}$.

(b) $\mathbf{a} \wedge (\mathbf{b} + \mathbf{c}) = \mathbf{a} \wedge \mathbf{b} + \mathbf{a} \wedge \mathbf{c}$,

(c) $\mathbf{a} \wedge \mathbf{b} = -\mathbf{b} \wedge \mathbf{a} = ab \sin \theta \mathbf{n}$ where $\mathbf{n}$ is a unit vector perpendicular to $\mathbf{a}$ and $\mathbf{b}$ and $\mathbf{a}$, $\mathbf{b}$ and $\mathbf{n}$ form a right-handed set.

By convention when θ is acute if $\mathbf{a}$, $\mathbf{b}$ and $\mathbf{n}$ are taken to form a right hand set then $\mathbf{b} \wedge \mathbf{a} = -ab \sin \theta \mathbf{n}$ and we have all the properties of the definition $\mathbf{a} \wedge \mathbf{b} = ab \sin \theta \mathbf{n}$.

Example 17 Find the vector product of $\mathbf{a} = \mathbf{i} + 2\mathbf{j} + 3\mathbf{k}$ and $\mathbf{b} = 2\mathbf{i} + 3\mathbf{j} + 4\mathbf{k}$.

$\mathbf{a} \wedge \mathbf{b} = \begin{vmatrix} \mathbf{i} & \mathbf{j} & \mathbf{k} \\ 1 & 2 & 3 \\ 2 & 3 & 4 \end{vmatrix} = -\mathbf{i} + 2\mathbf{j} - \mathbf{k}$ which is also perpendicular to both $\mathbf{a}$ and $\mathbf{b}$.

Example 11 repeated Find the equation of the plane through the three points $P(1, 1, 1)$, $Q(1, 2, 3)$ and $R(3, 2, 1)$.

The normal vector is perpendicular to both **PQ** and **PR** and so

it can be taken as $\mathbf{PQ} \wedge \mathbf{PR} = \begin{vmatrix} \mathbf{i} & \mathbf{j} & \mathbf{k} \\ 0 & 1 & 2 \\ 2 & 1 & 0 \end{vmatrix} = -2\mathbf{i} + 4\mathbf{j} - 2\mathbf{k}$.

Plane PQR has equation $-2x + 4y - 2z = d = 0$ since P lies on plane. Equation of PQR is $x - 2y + z = 0$.

Example 16 repeated Find the distance from the point $P(11, 7, 3)$ to the line through $(1, 2, 3)$ with direction vector $3\mathbf{i} + 4\mathbf{j} + 5\mathbf{k}$. With the notation of fig. 125

$$|\mathbf{AP} \wedge \mathbf{AC}| = AC \times AP \sin \theta = AC \times PD$$

which is true for any vector **AC**. Take $\mathbf{AC} = 3\mathbf{i} + 4\mathbf{j} + 5\mathbf{k}$, the direction vector of the line and

$$PD = \frac{|\mathbf{AP} \wedge (3\mathbf{i} + 4\mathbf{j} + 5\mathbf{k})|}{|3\mathbf{i} + 4\mathbf{j} + 5\mathbf{k}|} \qquad \begin{aligned} \mathbf{AP} &= 11\mathbf{i} + 7\mathbf{j} + 3\mathbf{k} - (\mathbf{i} + 2\mathbf{j} + 3\mathbf{k}) \\ &= 10\mathbf{i} + 5\mathbf{j} \end{aligned}$$

$$\mathbf{AP} \wedge (3\mathbf{i} + 4\mathbf{j} + 5\mathbf{k}) = \begin{vmatrix} \mathbf{i} & \mathbf{j} & \mathbf{k} \\ 10 & 5 & 0 \\ 3 & 4 & 5 \end{vmatrix} = 25\mathbf{i} - 50\mathbf{j} + 25\mathbf{k}$$

$$= 25(\mathbf{i} - 2\mathbf{j} + \mathbf{k})$$

$PD = \dfrac{25\sqrt{6}}{\sqrt{50}} = 5\sqrt{3}$ as before.

In general the distance of the point $P(\mathbf{OP} = \mathbf{b})$ from the line $\mathbf{r} = \mathbf{a} + t\mathbf{b}$ is $\dfrac{|(\mathbf{p} - \mathbf{a}) \wedge \mathbf{b}|}{|\mathbf{b}|}$.

Triple products

The fact that $\mathbf{a} \wedge \mathbf{b}$ is a vector enables triple products to be defined. The scalar product of $(\mathbf{a} \wedge \mathbf{b}) \cdot \mathbf{c}$ is called the scalar triple product. $(\mathbf{a} \wedge \mathbf{b}) \cdot \mathbf{c} = \mathbf{c} \cdot (\mathbf{a} \wedge \mathbf{b})$ since scalar product is commutative. Since $\mathbf{a} \wedge (\mathbf{b} \cdot \mathbf{c})$ is meaningless as $\mathbf{b} \cdot \mathbf{c}$ is a scalar we can safely write $\mathbf{a} \wedge \mathbf{b} \cdot \mathbf{c}$ with no brackets.

$$\mathbf{a} \wedge \mathbf{b} \cdot \mathbf{c} = \begin{pmatrix} a_2 b_3 - a_3 b_2 \\ a_3 b_1 - a_1 b_3 \\ a_1 b_2 - a_2 b_1 \end{pmatrix} \cdot \begin{pmatrix} c_1 \\ c_2 \\ c_3 \end{pmatrix}$$

$$= c_1(a_2 b_3 - a_3 b_2) - c_2(a_1 b_3 - a_3 b_1) + c_3(a_1 b_2 - a_2 b_1)$$

$$= \begin{vmatrix} c_1 & c_2 & c_3 \\ a_1 & a_2 & a_3 \\ b_1 & b_2 & b_3 \end{vmatrix} = \begin{vmatrix} a_1 & a_2 & a_3 \\ b_1 & b_2 & b_3 \\ c_1 & c_2 & c_3 \end{vmatrix} \quad \text{by interchanging two rows.}$$

$$\mathbf{a} \wedge \mathbf{b} \cdot \mathbf{c} = \begin{vmatrix} a_1 & a_2 & a_3 \\ b_1 & b_2 & b_3 \\ c_1 & c_2 & c_3 \end{vmatrix} = \mathbf{a} \cdot \mathbf{b} \wedge \mathbf{c}$$

and the vector and scalar products are interchanged. (Note that the cyclic order of a, b and c is unchanged.)

This means $\mathbf{a} \wedge \mathbf{b} \cdot \mathbf{c} = \mathbf{a} \cdot \mathbf{b} \wedge \mathbf{c} = \mathbf{c} \cdot \mathbf{a} \wedge \mathbf{b} = \mathbf{c} \wedge \mathbf{a} \cdot \mathbf{b}$ etc.

Interchanging two rows of the determinant changes its sign so

$$\mathbf{b} \wedge \mathbf{a} \cdot \mathbf{c} = -\mathbf{a} \wedge \mathbf{b} \cdot \mathbf{c}$$

Volume of a parallelepiped

Consider the parallelepiped defined by the vectors $\mathbf{a}$, $\mathbf{b}$ and $\mathbf{c}$ (fig. 127) where $\mathbf{a}$ and $\mathbf{b}$ are in the xy plane and $\mathbf{c}$ makes an angle θ with OZ.

$\mathbf{a} \wedge \mathbf{b} = $ (area of parallelogram $OADB$) $\mathbf{k}$.

Volume of parallelepiped = (area of base) × height.

(This can be proved by shearing the parallelepiped into a prism of height CN with base $OADB$).

$CN = OC \cos \theta$, so the volume of the parallelepiped
= (area of $OADB$) × $OC \cos \theta$
= (area of $OADB$) $\mathbf{k} \cdot \mathbf{c}$
= $\mathbf{a} \wedge \mathbf{b} \cdot \mathbf{c}$

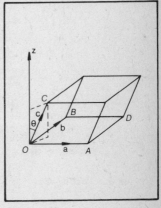

Figure 127

Shortest distance between two skew lines

Referring to fig. 128 let the two lines be AP with equation $\mathbf{r} = \mathbf{a} + t\mathbf{b}$ ($\mathbf{b}$ is in direction AP), and CQ with equation $\mathbf{r} = \mathbf{c} + s\mathbf{d}$ ($\mathbf{d}$ is in direction CQ). The shortest distance from C to the line AP is given by CP where angle $CPA = 90°$. The shortest distance from P to the line CQ is given by PQ where angle $PQC = 90°$. The shortest distance between AP and CQ is given by PQ which is perpendicular to AP

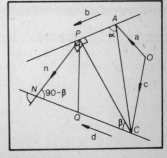

Figure 128

and CQ. $\mathbf{AC} \wedge \mathbf{b} = AC \times b \sin \alpha \mathbf{n}$ where $\mathbf{n}$ is a unit vector, perpendicular to plane PAC, i.e. angle $CPN = 90°$.

$$AC \sin \alpha = CP \Rightarrow \mathbf{AC} \wedge \mathbf{b} = b \times CP\mathbf{n}$$
$$\Rightarrow \mathbf{AC} \wedge \mathbf{b} \cdot \mathbf{d} = bCP \times \mathbf{n} \cdot \mathbf{d}$$
$$= b \times CP \times d \cos(90-\beta) = bd \times CP \sin \beta$$
$$CP \sin \beta = PQ \Rightarrow \mathbf{AC} \wedge \mathbf{b} \cdot \mathbf{d} = bd \times PQ$$
$$\Rightarrow PQ = \left| \frac{(\mathbf{c}-\mathbf{a}) \wedge \mathbf{b} \cdot \mathbf{d}}{bd} \right|$$

the modulus sign taken to give positive distance.

Example 18 Find the distance between the lines

$$\mathbf{r}_1 = \begin{pmatrix} 1 \\ 2 \\ 3 \end{pmatrix} + t\begin{pmatrix} 3 \\ 4 \\ 5 \end{pmatrix} \text{ and } \mathbf{r}_2 = \begin{pmatrix} 2 \\ 1 \\ 3 \end{pmatrix} + s\begin{pmatrix} 4 \\ 5 \\ 3 \end{pmatrix}.$$

Using the notation from fig. 134 $\mathbf{c}-\mathbf{a} = \begin{pmatrix} 2 \\ 1 \\ 3 \end{pmatrix} - \begin{pmatrix} 1 \\ 2 \\ 3 \end{pmatrix} = \begin{pmatrix} 1 \\ -1 \\ 0 \end{pmatrix}$

$$(\mathbf{c}-\mathbf{a}) \wedge \mathbf{b} \cdot \mathbf{d} = \begin{vmatrix} 1 & -1 & 0 \\ 3 & 4 & 5 \\ 4 & 5 & 3 \end{vmatrix} = \begin{vmatrix} 1 & 0 & 0 \\ 3 & 7 & 5 \\ 4 & 9 & 3 \end{vmatrix} = 21-45 = -24$$

$$b = \sqrt{3^2+4^2+5^2} = \sqrt{50} = d.$$

Distance between lines $= \left| \frac{-24}{\sqrt{50}\sqrt{50}} \right| = \left| \frac{-24}{50} \right| = 0.48$

Vector triple product

$\mathbf{a} \wedge (\mathbf{b} \wedge \mathbf{c})$ is a vector triple product of the vectors $\mathbf{a}$, $\mathbf{b}$ and $\mathbf{c}$. $\mathbf{b} \wedge \mathbf{c}$ is perpendicular to the plane of $\mathbf{b}$ and $\mathbf{c}$ which means that $\mathbf{a} \wedge (\mathbf{b} \wedge \mathbf{c})$ lies in the plane of $\mathbf{b}$ and $\mathbf{c}$ and can be expressed in terms of $\mathbf{b}$ and $\mathbf{c}$.

$$\mathbf{a} \wedge (\mathbf{b} \wedge \mathbf{c}) = \lambda\mathbf{b} + \mu\mathbf{c}$$
$$\mathbf{a} \cdot \mathbf{a} \wedge (\mathbf{b} \wedge \mathbf{c}) = \lambda\mathbf{a} \cdot \mathbf{b} + \mu\mathbf{a} \cdot \mathbf{c} = 0$$

since $\mathbf{a} \cdot \mathbf{a} \wedge (\mathbf{b} \wedge \mathbf{c}) = 0$ because its determinant would have two rows the same.

Hence $\lambda\mathbf{a} \cdot \mathbf{b} = -\mu\mathbf{a} \cdot \mathbf{c}$ and $\dfrac{\lambda}{\mathbf{a} \cdot \mathbf{c}} = -\dfrac{\mu}{\mathbf{a} \cdot \mathbf{b}} = \alpha$, say

$\mathbf{a} \wedge (\mathbf{b} \wedge \mathbf{c}) = \alpha(\mathbf{a} \cdot \mathbf{c})\mathbf{b} - \alpha(\mathbf{a} \cdot \mathbf{b})\mathbf{c} = \alpha[(\mathbf{a} \cdot \mathbf{c})\mathbf{b} - (\mathbf{a} \cdot \mathbf{b})\mathbf{c}]$
where α is the same for all vectors $\mathbf{a}$, $\mathbf{b}$ and $\mathbf{c}$.
Let $\mathbf{a} = \mathbf{k}$, $\mathbf{b} = \mathbf{j}$, $\mathbf{c} = \mathbf{k}$. Then

$$\mathbf{b} \wedge \mathbf{c} = \mathbf{j} \wedge \mathbf{k} = \mathbf{i} \text{ and } \mathbf{a} \wedge (\mathbf{b} \wedge \mathbf{c}) = \mathbf{k} \wedge \mathbf{i} = \mathbf{j}$$
$$\mathbf{j} = \alpha(1) \times \mathbf{j} - \alpha \times 0 \times \mathbf{k} = \alpha\mathbf{j} \Rightarrow \alpha = 1$$

So $\mathbf{a} \wedge (\mathbf{b} \wedge \mathbf{c}) = (\mathbf{a} . \mathbf{c})\mathbf{b} - (\mathbf{a} . \mathbf{b})\mathbf{c}$

$(\mathbf{a} \wedge \mathbf{b}) \wedge \mathbf{c} = -\mathbf{c} \wedge (\mathbf{a} \wedge \mathbf{b}) = -(\mathbf{b} . \mathbf{c})\mathbf{a} + (\mathbf{c} . \mathbf{a})\mathbf{b}$ using (12)

$$= (\mathbf{a} . \mathbf{c})\mathbf{b} - (\mathbf{b} . \mathbf{c})\mathbf{a} \qquad (13)$$

In general $\mathbf{a} \wedge (\mathbf{b} \wedge \mathbf{c}) \neq (\mathbf{a} \wedge \mathbf{b}) \wedge \mathbf{c}$.

Key terms

Notation If the coordinates of A are $(1, 2, 3)$ then

$$\mathbf{OA} = OA = \mathbf{i} + 2\mathbf{j} + \mathbf{k} = \begin{pmatrix} 1 \\ 2 \\ 3 \end{pmatrix}.$$

Addition and Subtraction

$$\mathbf{OA} = \begin{pmatrix} a_1 \\ a_2 \\ a_3 \end{pmatrix}, \ \mathbf{OB} = \begin{pmatrix} b_1 \\ b_2 \\ b_3 \end{pmatrix}, \ \mathbf{OA} + \mathbf{OB} = \begin{pmatrix} a_1 + b_1 \\ a_2 + b_2 \\ a_3 + b_3 \end{pmatrix},$$

$$\mathbf{AB} = \mathbf{OB} - \mathbf{OA} = \begin{pmatrix} b_1 - a_1 \\ b_2 - a_2 \\ b_3 + a_3 \end{pmatrix}$$

Ratio theorem P divides AB in ratio

$$p : q \Rightarrow \mathbf{OP} = \frac{q}{p + q} \cdot \mathbf{OA} + \frac{p}{p + q} \cdot \mathbf{OB}$$

$\mathbf{OA} = \mathbf{a}$ and $\mathbf{OB} = \mathbf{b}$. Mid-point of AB is $\frac{1}{2}\mathbf{a} + \frac{1}{2}\mathbf{b}$ and the points of trisection of $\mathbf{AB}$ are $\frac{2}{3}\mathbf{a} + \frac{1}{3}\mathbf{b}$ (nearer A), $\frac{1}{3}\mathbf{a} + \frac{2}{3}\mathbf{b}$ (nearer B).

Vector equation of a line $\begin{pmatrix} x \\ y \\ z \end{pmatrix} = \begin{pmatrix} 1 \\ 2 \\ 3 \end{pmatrix} + t \begin{pmatrix} 3 \\ 4 \\ 5 \end{pmatrix}$

represents a straight line through the point $(1, 2, 3)$ in the direction of the vector $3\mathbf{i} + 4\mathbf{j} + 5\mathbf{k}$.

Scalar product of $\mathbf{a} = \begin{pmatrix} a_1 \\ a_2 \\ a_3 \end{pmatrix}$ and $\mathbf{b} = \begin{pmatrix} b_1 \\ b_2 \\ b_3 \end{pmatrix}$ is

$\mathbf{a} . \mathbf{b} = ab \cos \theta = a_1 b_1 + a_2 b_2 + a_3 b_3$.

Equation of a plane is $ax + by + cz = d$

where $a\mathbf{i} + b\mathbf{j} + c\mathbf{k}$ is the normal vector of the plane.

Distance of the point (x_1, y_1, z_1) to the plane $ax + by + cz = d$ is $\dfrac{ax_1 + by_1 + cz_1 - d}{\sqrt{a^2 + b^2 + c^2}}$.

Vector product of $\mathbf{a} = \begin{pmatrix} a_1 \\ a_2 \\ a_3 \end{pmatrix}$ and $\mathbf{b} = \begin{pmatrix} b_1 \\ b_2 \\ b_3 \end{pmatrix}$ is $\mathbf{a} \times \mathbf{b} = \mathbf{a} \wedge \mathbf{b} = $

$ab \sin \theta \mathbf{n}$ where $\mathbf{n}$ is a unit vector perpendicular to $\mathbf{a}$ and $\mathbf{b}$.

Chapter 12
Groups

Algebra is usually thought of as long complicated formulae involving letters which stand for numbers. This is the algebra of numbers and the numbers obey certain rules. The study of mathematics involves much more than numbers and the mathematician works with sets, functions, transformations, matrices, vectors and other mathematical tools and ideas. These other 'elements' obey rules similar to those for numbers and the rules need to be established before mathematical work can be done.

Algebra of numbers

The most elementary numbers are the natural or counting numbers $N = \{1, 2, 3, 4, \ldots\}$. These form an infinite set and if we add two members of the set together we always obtain another member of the set.

N is said to be **closed** under the operation of addition. N is not closed under subtraction since $3 - 4 = -1$ which is not a member of the set N. N is closed under multiplication, but not closed under division.

The operation of subtraction on the members of N produces new elements like $0, -1, -2$, etc. and these together with N form a new set whose members are called the **integers**. Integers are denoted by $Z = \{\ldots -3, -2, -1, 0, 1, 2, \ldots\}$. The operation of division on Z produces numbers like $\frac{1}{2}, \frac{2}{3}$ which are fractions.

The fractions belong to a set of numbers called the **rational** numbers. The set of rational numbers, Q, consists of numbers of the form a/b where $a, b \in Z$ and $b \neq 0$. So the rational numbers include positive and negative whole numbers and fractions. Even recurring decimals belong to the set of rational numbers since they can be changed into a rational number by the following process. If $x = 0 \cdot 123123 \ldots$ then $1000x = 123 \cdot 123123 \ldots$, $1000x - x = 123$ since the recurring decimal parts are equal and $999x = 123 \Rightarrow x = \frac{123}{999} = \frac{41}{333}$ and x is rational.

The rational numbers form the first set of numbers which are closed under the four rules of arithmetic.

Addition $\dfrac{p}{q} + \dfrac{r}{s} = \dfrac{ps + qr}{qs}$ which is rational.

$q \neq 0,\ s \neq 0 \Rightarrow qs \neq 0.$

Subtraction $\dfrac{p}{q} - \dfrac{r}{s} = \dfrac{ps - qr}{qs}$ which is rational.

Multiplication $\dfrac{p}{q} \times \dfrac{r}{s} = \dfrac{pr}{qs}$ which is rational.

Division $\dfrac{p}{q} \div \dfrac{r}{s} = \dfrac{ps}{qr}$ which is rational, provided $r \neq 0.$

The number system has now been extended to include all rational numbers both positive and negative and might be thought to be complete. All the rational numbers appear on the number line (fig. 129) but there are others as well.

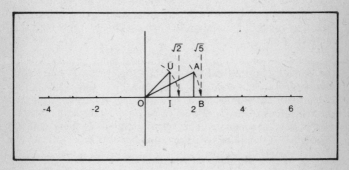

Figure 129

For instance the number $0 \cdot 1010010001 \ldots$ is not a recurring decimal. It has a pattern but cannot be converted into the ratio of two natural numbers. Similarly $\sqrt{2}$ and $\sqrt{5}$ are not rational numbers. We can calculate the value of $\sqrt{5}$ to any prescribed accuracy ($\sqrt{5} = 2 \cdot 23606798$ to 8 decimal places) but its decimal equivalent does not recur.

Proof that $\sqrt{5}$ is irrational
Assume $\sqrt{5}$ is rational and produce a contradiction. (The student may not be familiar with this kind of proof, but if a logically sound argument produces a false conclusion then the original assumption must have been wrong.)

$\sqrt{5}$ rational $\Rightarrow \sqrt{5}$ can be written as $\frac{p}{q}$ where p and q belong to N (whole numbers), $q \neq 0$ and p and q are coprime. (The fraction $\frac{p}{q}$ is in its lowest form; p and q have no common factors.)

$\sqrt{5} = \frac{p}{q} \Rightarrow 5 = \frac{p^2}{q^2} \Leftrightarrow 5q^2 = p^2 \Rightarrow p^2$ is divisible by 5

$p \times p$ has a factor of $5 \Rightarrow p$ has a factor of 5

$p = 5r$ where r is a whole number

$5q^2 = p^2 \Leftrightarrow 5q^2 = 25r^2 \Leftrightarrow q^2 = 5r^2 \Rightarrow q^2$ is divisible by 5
$$\Rightarrow q \text{ is divisible by } 5$$

p and q have both been shown to contain a factor 5 and this contradicts the fact that they were coprime. This is the contradiction proving the original assumption was false, i.e. $\sqrt{5}$ must be **irrational**. Other irrational numbers occur in mathematics like $\sqrt[3]{2}$, π and e. These numbers are on the number line (fig. 129) since $OA = \sqrt{5}$ (Pythagoras) and rotating OA down to meet the number line gives OB to represent the number $\sqrt{5}$.

The set of rational numbers and irrational numbers make up all the numbers on the real number line (fig. 129) since any real number which is not rational is irrational and this set is called the set of **real numbers**. The set of real numbers is closed under the four rules of arithmetic and for most purposes this set of numbers is adequate. The set of rational numbers will be denoted by Q and the set of real numbers by R.

Rules in the algebra of numbers

When numbers are combined under the operation of the four rules certain properties are obvious and are taken for granted. For example, $3 + 4 = 4 + 3 = 7$
$$3 \times 4 = 4 \times 3 = 12$$
The operations of addition and multiplication on the set of real numbers are **commutative**.

Subtraction is not commutative since $3 - 4 \neq 4 - 3$

Division is not commutative since $3 \div 4 \neq 4 \div 3$

When three numbers are combined why do we talk of "adding three numbers together" but not "subtracting three numbers together"?

$3 + 4 + 5$ either means $(3 + 4) + 5$ or $3 + (4 + 5)$ the brackets indicating which numbers are added first.

Since $(3 + 4) + 5 = 3 + (4 + 5) = 12$ the way of adding three numbers does not matter and addition of numbers is said to be **associative**.

Similarly the operation of multiplication on the set of real numbers is also associative since $(3 \times 4) \times 5 = 3 \times (4 \times 5) = 60$. However, subtraction is not associative since $5 - (4 - 3) \neq (5 - 4) - 3$ and neither is division since $12 \div (6 \div 2) \neq (12 \div 6) \div 2$.

The distributive laws

When two or more operations are being performed certain laws and conventions need to be established.

For example, what does $3 \times 4 + 5$ mean?

$3 \times 4 + 5$ should be punctuated by using brackets. The language of mathematics and its conventions must be clearly understood before progress can be made. Many students fail to appreciate basic rules which may only be conventions.

$3 \times 4 + 5$ either stands for $(3 \times 4) + 5 = 17$ or $3 \times (4 + 5) = 27$ and realistically the brackets must be left in. However, it has been agreed that the brackets are left out in the first case so that $3 \times 4 + 5$ means $(3 \times 4) + 5 = 17$ and $3 + 4 \times 5$ means $3 + (4 \times 5) = 23$. In other words multiplication is done first.

Also $3 - 4 \times 5 = 3 - (4 \times 5) = -17$ and $6 \div 2 - 3 = (6 \div 2) - 3 = 0$. Multiplication and division are done before addition and subtraction. Consider $3 - 4 + 5$ which could mean $(3 - 4) + 5 = 4$ or $3 - (4 + 5) = -6$. In the first case the brackets are omitted and they are inserted for the second.

Similarly $12 \div 6 \times 2$ could mean $(12 \div 6) \times 2 = 4$ or $12 \div (6 \times 2) = 1$. The brackets are retained only in the second case so that $24 \div 8 \times 2$ means $(24 \div 8) \times 2 = 6$.

All these conventions may seem puzzling but they form the language of mathematics, and language without punctuation can be meaningless.

$3 \times (4 + 5)$ can be written without brackets although the brackets are essential to convey the proper meaning.

$3 \times (4 + 5) = 3 \times 4 + 3 \times 5 = (3 \times 4) + (3 \times 5)$ the brackets being inserted to emphasise this **distributive** law.

Multiplication is distributive over addition for the set of real numbers.

Multiplication is also distributive over subtraction which can be summarized formally as

$$a \times (b - c) = (a \times b) - (a \times c)$$

Addition is *not* distributive over multiplication. If it were, $3 + (4 \times 5) = 23$ would equal $(3 + 4) \times (3 + 5) = 56$.

Multiplication is distributive over addition from the right hand side $(4 + 5) \times 3 = (4 \times 3) + (5 \times 3) = 27$.

Care must be taken with division.

$(6 + 9) \div 3 = (6 \div 3) + (9 \div 3) = 5$ is true

$3 \div (6 + 9) \neq (3 \div 6) + (3 \div 9)$

Many of these rules are assimilated carefully over a long period in one's learning and are taken for granted. Each time a new algebra is presented these rules of combination and procedure should be tested, otherwise wrong assumptions and mistakes are made.

Algebra of sets

Sets are made up of elements which may be numbers, letters, functions, cars or anything.

If set $V = \{a, e, i, o, u\}$ and $F = \{a, b, c, d, e\}$ the union of V and F, written $V \cup F = \{a, b, c, d, e, i, o, u\}$ i.e. the set containing elements which are in V, in F or both.

The intersection of V and F written $V \cap F = \{a, e\}$. The picture in fig. 130 is a Venn Diagram illustrating the situation clearly. $V \cap F$ is the overlapping region containing a and e. The universal set for this situation could be the letters of the alphabet. The complement of V denoted by V' is the set of elements which are not in V. $V' = \{$consonants$\}$

$F' = \{f, g, h, i, \ldots, x, y, z\}$

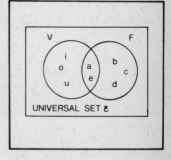

Figure 130

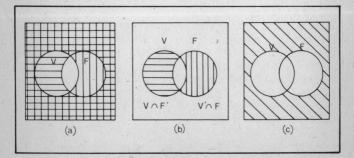

Figure 131

306

In fig. 131a $V \cap F$ is completely unshaded, V' is vertically shaded and F' is horizontally shaded.

In fig. 131b the region shaded horizontally is in V and outside F, so that it is in V and in F', hence it is in $V \cap F'$.

In fig. 131b the region shaded vertically is in F and outside V, so that it is in F and in V' and is therefore in $F \cap V'$.

Commutativity $V \cap F = \{a, c\} = F \cap V$ since they both describe those elements in both V and F.

Similarly $V \cup F = F \cup V$ and both the union and the intersection of sets are commutative.

The region unshaded in fig. 131c represents $V \cup F$. The region shaded is therefore the complement of this, $(V \cup F)'$. Referring to fig. 131a this same region is double shaded and is therefore in both V' and F'. $(V \cup F)' = V' \cap F'$. This is one of De Morgan's laws.

It can easily be verified by writing out the individual elements, but is quite easily illustrated by the diagrams. In fig. 133a the region unshaded is $V \cap F$ and the region outside it is $(V \cap F)'$. This region is shaded either vertically or horizontally and therefore is $V' \cup F'$. $(V \cap F)' = V' \cup F'$. This is the second of De Morgan's laws. Again this can be verified by listing the elements of each set.

Associativity

Figure 132 shows the Venn diagram for the possible situation with three sets. An element could be in A or in A', B or B', C or C' which gives eight possible combinations which correspond with the eight regions in fig. 132. These eight regions are listed in fig. 133.

The region $A \cap B$ is vertically shaded. Its intersection with C is the middle region (doubly shaded). $B \cap C$ is shaded

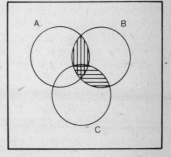

Figure 132

horizontally and its intersection with A is the central region.

$$(A \cap B) \cap C = A \cap (B \cap C)$$

since both represents the middle region; intersection is associative.

Similarly $(A \cup B) \cup C = A \cup (B \cup C)$ since both represent the seven regions inside the three circles. Union is associative and we can now talk about the intersection or union of three sets since there is no ambiguity.

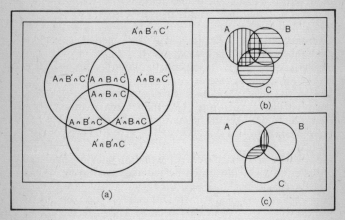

Figure 133

Figure 133a shows the eight possible regions that can result with three sets (seven within the circles and the outside region). Since the union of two sets will involve at least two regions these eight individual regions can only be described with the aid of intersections. The central region is in all three sets (as we have seen) and so is $A \cap B \cap C$ (no brackets since associative). The region directly above this is in A and B (and so $A \cap B$) but not in C (so in C'). This region is $A \cap B \cap C'$. The elements in C only are in A' and B' and C and this region is $A' \cap B' \cap C$. The complete list is in the diagram.

Distributivity
$A \cap (B \cup C) = (A \cap B) \cup (A \cap C)$.

In fig. 133b $B \cup C$ is shaded horizontally and A vertically, so $A \cap (B \cup C)$ is doubly shaded.

In fig. 133c $A \cap B$ is shaded vertically and $A \cap C$ horizontally, so the region with any shading at all represents $(A \cap B) \cup (A \cap C)$ and this is the same as $A \cap (B \cup C)$. So $A \cap (B \cup C) = (A \cap B) \cup (A \cap C)$.
$$A \cup (B \cap C) = (A \cup B) \cap (A \cup C).$$

308

By shading the appropriate diagrams the above rule can be shown to be true. For the operations of union and intersection of sets we have (a) union is distributive over intersection, (b) intersection is distributive over union. The algebraic laws of closure, commutativity, associativity and distributivity must be established before the usual operations are performed.

Finite or Modular arithmetic

The system of modular arithmetic may seem a little contrived but its extension has important applications in number theory and its simpler examples provide valuable insight into algebraic laws and structure. The set of integers can be partitioned into five sets (modulo 5) if they leave the same remainder after division by 5. The integers $1, 6, 11, \ldots$ leave a remainder 1 after dividing by 5. This set is denoted by **1**. Similarly the set denoted by **2** contains $2, 7, 12, 17, \ldots$ and also $-3, -8, \ldots$

$$\mathbf{0} \equiv \{0, \pm 5, \pm 10, \ldots\} \qquad \mathbf{1} \equiv \{\ldots, -9, -4, 1, 6, 11, 16, \ldots\}$$
$$\mathbf{2} \equiv \{\ldots -8, -3, 2, 7, 12, \ldots\} \qquad \mathbf{3} \equiv \{\ldots, -7, -2, 3, 8, 13, \ldots\}$$
$$\mathbf{4} \equiv \{\ldots, -6, -1, 4, 9, 14, \ldots\}$$

Alternatively the sets can be defined as sequences.

The members of the set denoted by **0** can be expressed as $5n$ where n is an integer.

The members of the set denoted by **1** can be expressed as $5n + 1$ where n is an integer.

The members of the set denoted by **2** can be expressed as $5n + 2$ where n is an integer, and similarly for **3** and **4**.

All the integers are now partitioned into the 5 sets **0, 1, 2, 3, 4** and each integer belongs to only one set.

Addition is defined in the usual way with the sum reduced modulo 5 to one of the five sets **0, 1, 2, 3, 4**.

$3 + 4 = 7 = 2$ (modulo 5) so $\mathbf{3} + \mathbf{4} = \mathbf{2}$. Any members of **3** and **4** would give the same result, since $5n + 3 + 5k + 4 = 5n + 5k + 5 + 2 = 2$ (mod 5). This system is also known as clock arithmetic and illustrated with a clock face with only 5 numbers (fig. 134a). The operation of addition (mod 5) can be summarized with the aid of a combination table (fig. 134b). $\mathbf{1} + \mathbf{2} = \mathbf{3}$ is in the 2nd row, 3rd column. $\mathbf{2} + \mathbf{1} = \mathbf{3}$ is in the 3rd row, 2nd column. $\mathbf{3} + \mathbf{4} = \mathbf{2}$ (4th row, 5th column).

The set $\{\mathbf{0}, \mathbf{1}, \mathbf{2}, \mathbf{3}, \mathbf{4}\}$ under addition (modulo 5) forms a closed set, the combination of any two giving another member in the set. When **0** is added to any element **a** the result is **a**, so **0** is the identity element for addition.

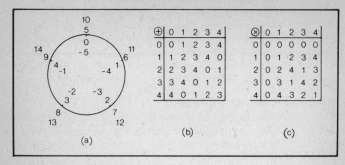

Figure 134

$3 + 4 = 4 + 3 = 2$ and this is true for every pair of elements and so addition is commutative. This is easily recognized from the table since $3 + 4$ and $4 + 3$ are symmetrically placed on either side of the main diagonal. The table is symmetrical about the leading diagonal.

$x + 4 = 2$ is solved from the table by looking in the last column for the number **2**. Since $3 + 4 = 2$, $x = 3$ is the solution. A more formal solution is as follows

$x + 4 = 2 \Rightarrow (x + 4) + 1 = 2 + 1$

$\qquad \Rightarrow x + (4 + 1) = 3$ if the operation is associative

$\qquad \Rightarrow x + 0 \qquad = 3$ since $4 + 1 = 0$

$\qquad \Rightarrow x \qquad\qquad = 3$ since $x + 0 = x$ and **0** is the identity

Any equation of the form $x + a = b$ (mod 5) can be solved provided that the operation of addition modulo 5 is associative and that there exists an element which combines with **4** to give **0**. **1** is called the inverse element of **4** since $4 + 1 = 0$. **4** is also the inverse of **1**, so **1** and **4** form an inverse pair. Similarly **2** and **3** form an inverse pair and **0** is its own inverse (self-inverse). To show that the operation is associative we must show that $(a + b) + c = a + (b + c)$ for all values of **a**, **b** and **c**.

Consider $(1 + 2) + 3 \equiv 1 + (2 + 3)$
1 can be written as $5m + 1$, **2** by $5k + 2$ and **3** by $5l + 3$.

$1 + (2 + 3) \equiv 5m + 1 + (5k + 2 + 5l + 3)$
$\qquad\qquad \equiv 5(m + k + l) + 1 + 2 + 3 = 1$ (mod 5)

$(1 + 2) + 3 \equiv (5m + 1 + 5k + 2) + 5l + 3$
$\qquad\qquad \equiv 5(m + k + l) + 1 + 2 + 3 = 1$ (mod 5)

310

It can be seen that the same result is obtained whatever values we take for **a**, **b** and **c**.

A set which is (a) closed for a certain operation, (b) associative, (c) contains an identity element and (d) possesses inverse elements for all its members is called a *Group* under that operation. The group properties enable equations with that operation to be solved uniquely with members from the set. The set Z of integers forms a group under addition with **0** as the identity and the inverse of **a** is $-$**a**. There are an infinite number of elements in the group so the group has infinite order. The **order** of a group is the number of elements in the group.

Figure 135b shows the set of integers (mod 5) under the operation of multiplication (mod 5). **0** multiplied with any elements gives **0**, and is called the zero member of the set.

If 0 is omitted the set, $Z_5^* = \{1, 2, 3, 4\}$ forms a group under multiplication mod 5. The suffix 5 indicates modulo 5 and * means that 0 is omitted.

$3 \times 3 = 9 \equiv 4$ (mod 5) so **3** $\times$ **3** = **4** (mod 5) can similarly for the other elements. The set is closed and this can be seen by the fact that only **1**, **2**, **3** and **4** occur in the multiplication table.

To verify the associative rule
$(\mathbf{3} \times \mathbf{4}) + \mathbf{2} = (5m + 3) \times (5k + 4) \times (5l + 2) = 3 \times 4 \times 2$
since all other terms will contain a multiple of 5 which is **0** (mod 5).

Thus $(\mathbf{3} \times \mathbf{4}) \times \mathbf{2} \equiv (3 \times 4) \times 2 = \mathbf{4}$ (mod 5)
and $\mathbf{3} \times (\mathbf{4} \times \mathbf{2}) \equiv 3 \times (4 \times 2) = \mathbf{4}$ (mod 5)
The associative rule holds for all triples.

The identity element is **1**; $a \times 1 = 1 \times a = a$ for all elements **a**.

The elements **2** and **3** form an inverse pair, and **1** and **4** are each self-inverse.

The set forms a group, under multiplication modulo 5, of order 4(4 elements). The operation is commutative.

A commutative group is called **Abelian**.

The equation $\mathbf{3} \times \mathbf{x} = \mathbf{2}$ has the solution $\mathbf{x} = \mathbf{4}$ (from the table).
More formally $\mathbf{3} \times \mathbf{x} = \mathbf{2}$
$\Rightarrow \mathbf{2} \times (\mathbf{3} \times \mathbf{x}) = \mathbf{2} \times \mathbf{2}$
$\Rightarrow (\mathbf{2} \times \mathbf{3}) \times \mathbf{x} = \mathbf{4}$ (associative rule)
$\Rightarrow \mathbf{1} \times \mathbf{x} = \mathbf{4}$
$\Rightarrow \quad \mathbf{x} = \mathbf{4}$ (identity is **1**)
An equation of this form would have a unique solution from the set $\{1, 2, 3, 4, 5\}$.

⊕	0	1	2	3	4	5
0	0	1	2	3	4	5
1	1	2	3	4	5	0
2	2	3	4	5	0	1
3	3	4	5	0	1	2
4	4	5	0	1	2	3
5	5	0	1	2	3	4

(a)

⊗	1	2	3	4	5
1	1	2	3	4	5
2	2	4	0	2	4
3	3	0	3	0	3
4	4	2	0	4	2
5	5	4	3	2	1

(b)

Figure 135

Fig. 135a shows the combination table for the set {**0, 1, 2, 3, 4, 5**} under addition modulo 6. The table is similar to the addition table modulo 5 with **0** as the identity element, but with 6 elements instead of 5. **1** and **5** form an inverse pair as do **2** and **4**. **0** and **3** are self-inverse. The equations $\mathbf{x}+\mathbf{a}=\mathbf{b}$ have unique solutions for all **a**, **b** in the set $Z_6=$ {**0, 1, 2, 3, 4, 5**} and the set forms an Abelian group under the operation of addition.

Fig. 135b shows the combination table for $Z_6^*=$ {**1, 2, 3, 4, 5**} under the operation of multiplication modulo 6.
The set is *not* closed under this operation since $\mathbf{2}\times\mathbf{3}=\mathbf{0}$ which is not a member of the set. The operation is still associative and the identity element is still **1**. The elements **1** and **5** are self-inverse but **2**, **3** and **4** do *not* have inverses.
Z_6^* does *not* form a group under multiplication.

The equation $\mathbf{3}\times\mathbf{x}=\mathbf{2}$ has no solutions in this set, while the equation $\mathbf{3}\times\mathbf{x}=\mathbf{1}$ has three solutions $\mathbf{x}=\mathbf{1}$, **3** or **5**.
Equations of the form $\mathbf{a}\times\mathbf{x}=\mathbf{b}$ can have no solution, or 1, 2 or 3 solutions. This type of equation does not have a unique solution. Even if $Z_6=$ {**0, 1, 2, 3, 4, 5**} under multiplication were considered, while this set under multiplication is closed, associative and has **1** as the identity, not all elements have an inverse and so Z_6 does not form a group under multiplication.
All sets Z_n ($n=2, 3, 4, \ldots$) form groups under addition modulo n and the combination tables are similar to those in fig. 135a and fig. 134b.

It is interesting to consider the sets Z_n^* under multiplication modulo n. Figure 136a shows the combination table for Z_7^* under multiplication modulo 7. Z_7^* is closed and associative with identity **1**. Inverse pairs are **2** and **4**, **3** and **5**; **1** and **6** are self-inverse. Z_7^* forms a group of order 6.

$\otimes$	1	2	3	4	5	6
1	1	2	3	4	5	6
2	2	4	6	1	3	5
3	3	6	2	5	1	4
4	4	1	5	2	6	3
5	5	3	1	6	4	2
6	6	5	4	3	2	1

(a)

$\otimes$	1	2	3	4	5	6	7	8
1	1	2	3	4	5	6	7	8
2	2	4	6	8	1	3	5	7
3	3	6	0	3	6	0	3	6
4	4	8	3	7	2	6	1	5
5	5	1	6	2	7	3	8	4
6	6	3	0	6	3	0	6	3
7	7	5	3	1	8	6	4	2
8	8	7	6	5	4	3	2	1

(b)

Figure 136

Figure 136b shows the combination table for Z_9^* under multiplication modulo 9. It is closed and **3** and **6** have no inverses. At first sight it might be thought that Z_n^* forms a group under multiplication modulo n if n is an odd number, but fig. 136b disproves this theory. In fact Z_n^* forms a group if p is prime since in this case there will be no zeros in the combination table. The table for Z_p^* (p prime) would all be Latin Squares, i.e. each row or column contains each element once only. The four group conditions are satisfied. In Z_9^*, if **3** and **6** are omitted the remaining 6 elements form a group. The sets Z_n of integers (modulo n) under addition all form cyclic groups. In fig. 135, in each row the elements stay in the same cyclic order with **0, 1, ...** appearing at the end. The multiplication tables provide more interesting structures which will be considered later.

Symmetry operations

Figure 137a shows an object, F, with three images F_1, F_2, F_3. F_1 is the image of F after reflection in the x axis. Denoting this transformation by X we have $X(F) = F_1$. If Y denotes reflection in the y axis then $Y(F) = F_2$ and if H denotes a half-term

313

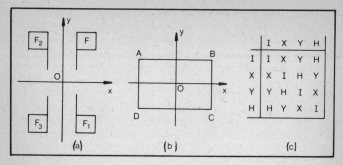

Figure 137

(rotation of 180°) about O, the origin $(0, 0)$ then $H(F) = F_3$.

$$Y(F_1) = F_3 \Rightarrow Y[X(F)] = F_3 = H(F) \Rightarrow YX = H$$

X followed by Y is equivalent to H. Similarly $XY = H$ since

$$XY(F) = X[Y(F)] = X(F_2) = F_3 = H(F)$$

The effect of XY (Y followed by X) is the same as H.

$$XH(F) = X[H(F)] = X(F_3) = F_2 = Y(F) \Rightarrow XH = Y$$
$$HH(F) = H[H(F)] = H(F_3) = F$$

$HH = I$ where I denotes the identity transformation i.e. $I(F) = F$.

These results for the combination of I, X, Y and H can be summarized in the form of a table (fig. 137c). The result of XY is placed in the second row, third column and the result of YX in the third row, second column. So the heading column represents the transformation which is written first in each product, while the heading row represents the transformation which is performed first, since XY means Y followed by X. This convention applies to all combination tables, but unfortunately in many of our examples like transformations, functions and matrices, in the product AB, B refers to the operation which is performed first.

Figure 137b shows a rectangle $ABCD$ which has two lines of symmetry Ox and Oy and rotational symmetry of order 2 (180°) about O. Denoting reflection in Ox by X, reflection in Oy by Y and a half-turn about O by H, the application of these transformations to the rectangle will leave it occupying its same outline with A, B, C and D occupying different positions

314

$X(ABCD) = DCBA$ (labelling the vertices clockwise from the top left-hand vertex).

$YX(ABCD) = Y(DCBA) = CDAB = H(ABCD)$. The combination of these transformations gives the same table of results, fig. 137c. These transformations are called the symmetry operations of the rectangle and they form a group of order 4 under the operation of successive application. (This is sometimes called 'followed by' since XY means 'Y followed by X'.) The set $\{I, X, Y, H\}$ is closed. This can be seen from the table (fig. 137b). To demonstrate the associative rule, consider $(AB)C$ and $A(BC)$ where A, B and C are isometries. $(AB)C$ means C fb (B fb A) where fb stands for 'followed by.' $A(BC)$ means (C fb B) fb A. In both cases C is performed first then B and A last, and will give the same final result. This is true of all transformations in space which are one to one (i.e. one object point gives one image point) since these then have an inverse transformation defined. (If the transformation is represented by a 3×3 matrix, the one to one property ensures that its determinant is not zero which implies that an inverse matrix exists.)

The set $\{I, X, Y, H\}$ has an identity element I, and all four elements are self-inverse, so the set forms an Abelian (commutative) group of order 4. (This particular group with all elements self-inverse is called the Klein group of order 4).

Symmetry group of the equilateral triangle

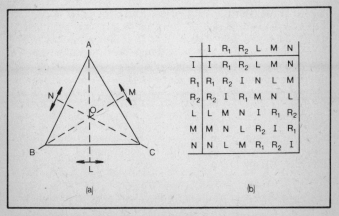

	I	R₁	R₂	L	M	N
I	I	R₁	R₂	L	M	N
R₁	R₁	R₂	I	N	L	M
R₂	R₂	I	R₁	M	N	L
L	L	M	N	I	R₁	R₂
M	M	N	L	R₂	I	R₁
N	N	L	M	R₁	R₂	I

(a)　　　　　(b)

Figure 138

The equilateral triangle has three lines of symmetry and order of symmetry 3(120°). The symmetry operations are R_1 (rotation of 120° about O), R_2 (rotation of +240° about O = rotation of −120° about O) and I

L, M, N represent these reflections in the lines of symmetry AO, BO and CO. The successive transformations are best evaluated by considering their effect on ABC, the vertices of the triangle. (Remember ML means L first, then M.)

$$ML = R_2;$$

$$R_1L = N$$

$$LR_1 = M;$$

$$R_1M = L$$

$$MR_1 = N;$$

$$M^2 = I$$

The full group table is in fig. 138b and four elements I, M, N, L are self-inverse, while R_1 and R_2 form an inverse pair. The group is not commutative since $MR_1 \neq R_1M$. (It is only necessary to find one pair of elements which do not commute to prove the group is not commutative, but when the group is commutative, all pairs must be commutative.) The set of $\{I, R_1, R_2\}$ form a group and this is called a subgroup of the main group $\{I, R_1, R_2, L, M, N\}$. Similarly $\{I, M\}$, $\{I, L\}$ and $\{I, N\}$ form subgroups of order 2.

A subgroup of a group is a sub-set of elements which form a group. The group $\{I, X, Y, H\}$ has 3 subgroups of order 2, i.e. $\{I, H\}$, $\{I, Y\}$ and $\{I, X\}$.

Example Groups of functions Show that the functions $f(x) = \dfrac{1}{x}$, $g(x) = -x$ and $h(x) = -\dfrac{1}{x}$ together with the identity function $e(x) = x$ form a group, the functions being applied successively.

$$fg(x) = f(g(x)) = f(-x) = -\frac{1}{x} = h(x)$$

$$gf(x) = g(f(x)) = g\left(\frac{1}{x}\right) = -\frac{1}{x} = h(x)$$

$$gh(x) = g(h(x)) = g\left(-\frac{1}{x}\right) = \frac{1}{x} = f(x)$$

$$hg(x) = h(g(x)) = h(-x) = \frac{1}{x} = f(x)$$

$$hf(x) = h\left(\frac{1}{x}\right) = -x = g(x)$$

$$fh(x) = f\left(-\frac{1}{x}\right) = -x = g(x)$$

$$ff(x) = f\left(\frac{1}{x}\right) = \frac{1}{1/x} = x = e(x)$$

$$gg(x) = g(-x) = -(-x) = x = e(x)$$

$$hh(x) = h\left(-\frac{1}{x}\right) = -\frac{1}{(-1/x)} = x = e(x)$$

	e	f	g	h
e	e	f	g	h
f	f	e	h	g
g	g	h	e	f
h	h	g	f	e

Figure 139

The results are summarized in the group table in fig. 139. All the four functions are self-inverse, and form the Klein group of order 4.

Key terms

Natural numbers $N = 1, 2, 3, 4, \ldots$
Integers $Z = \cdots \pm 3, \pm 2, \pm 1, 0 \cdots$

Rationals Q = numbers of the form $\frac{p}{q}$ where p, q, Z and $q \neq 0$

Real R rationals and irrationals.

Complex numbers C: $a + bj$ where a, b are real and $j^2 = 1$.
A set of elements $a, b, c, \ldots$ forms a group G under the operation $*$ if

(1) the set is closed $a * b \, G$, for all a, b
(2) $*$ is associative $a * (b * c) = (a * b) * c$, for all a, b, c
(3) there is an identity e such that $a * e = e * a = a$, for all a,
(4) every element a has an inverse a^{-1} such that

$$a * a^{-1} = a^{-1} * a = e$$

An Abelian group is commutative, $a * b = b * a$, for all a, b
$*$ is distributive over $\circ$ if $a * (b \circ c) = (a * b) \circ (a * c)$.

317

Examination Papers

Booklets of past papers can be ordered from bookshops or enquiries made at the addresses given below:

The School Mathematics Project (SMP)
SMP Office, Westfield College, Kidderpore Avenue,
  London, NW3 7ST

Joint Matriculation Board (JMB)
The Secretary, JMB, Manchester, M16 6EU

University of London
Publications Office, 52 Gordon Square, London, WC1 HOP5

University of Cambridge Local Examinations Syndicate
The Secretary, Syndicate Buildings, 17 Harvey Road,
  Cambridge, CB1 2EU

Mathematics in Education and Industry (MEI)
The Director, 57 High Street, Harrow, Middlesex.

Associated Examining Board (AEB)
The Secretary, Wellington House, Station Road, Aldershot,
  Hants, GU11 1BQ

Oxford and Cambridge Schools' Examination Board
The Secretary, 10 Trumpington Street, Cambridge, CB2 1QE
  and Elsfield Way, Oxford, OX2 8EP

Oxford Delegacy of Local Examinations
The Secretary, Oxford Local Examinations, Ewert Place,
  Summertown, Oxford, OX2 8EP

Index